The Beginnings of Agriculture

edited by
Annie Milles, Diane Williams
and Neville Gardner

Symposia of the
Association for Environmental Archaeology No. 8

BAR International Series 496
1989

B.A.R.

5, Centremead, Osney Mead, Oxford OX2 0DQ, England.

GENERAL EDITORS

A.R. Hands, B.Sc., M.A., D.Phil.
D.R. Walker, M.A.

BAR International Series 496, 1989 : The Beginnings of Agriculture'.
© The Individual Authors, 1989

The authors' moral rights under the 1988 UK Copyright,
Designs and Patents Act are hereby expressly asserted.

All rights reserved. No part of this work may be copied, reproduced, stored, sold, distributed, scanned, saved in any form of digital format or transmitted in any form digitally, without the written permission of the Publisher.

ISBN 9780860546368 paperback
ISBN 9781407347851 e-book
DOI https://doi.org/10.30861/9780860546368
A catalogue record for this book is available from the British Library
This book is available at www.barpublishing.com

Frontispiece Participants at the 'Beginnings of Agriculture' conference, Cardiff 1987.

CONTENTS

Foreward v
Annie Milles, Diane Williams and Neville Gardner

Conference report vii
Terry O'Connor

Addresses of contributors xi

I THEORETICAL APPROACHES TO THE BEGINNING, SPREAD, AND ORGANISATION OF AGRICULTURE

1   Towards the integration of social and ecological approaches to the study of early agriculture 3
Royston Clark
  1.1   Hunter-gatherer perspectives: an introduction 3
  1.2   Population growth processes 4
  1.3   Risk management 5
  1.4   Sequential and simultaneous hierarchies 7
  1.5   Archaeological correlates 8
  1.6   A case study 9
  1.7   Summary 18

2   Like rising damp? An ecological approach to the spread of farming in south east and central Europe 23
Paul Halstead
  2.1   Rising damp 23
  2.2   An ecological approach 24
  2.3   The spread of farming in south east and central Europe 26
  2.4   The annual scale: Greece 29
  2.5   The annual scale: the Balkans 32
  2.6   The annual scale: central Europe 33
  2.7   The annual scale: the Alpine Foreland 35
  2.8   The annual scale: summary 39
  2.9   The interannual scale: Greece 39
  2.10 The interannual scale: the Balkans, central Europe and the Alpine Foreland 40
  2.11 Conclusions 41

3   Hierarchical approaches to the evolution of complex agricultural systems 55
K. D. Thomas
  3.1   Introduction 55
  3.2   Complex agricultural systems 56
  3.3   Systems and ecosystems 58
  3.4   Agro-ecosystems 61
  3.5   Hierarchical models of ecosystems and agro-ecosystems 62

    3.6   Early agricultural systems as
process-functional hierarchies 64
    3.7   Integrated agricultural systems: possible
consequences for population and settlement 66
    3.8   Back down to earth 67
    3.9   Concluding comments 70

II   DOMESTICATION

**4   Size and sex - evidence for the domestication of
cattle in the Near East** 77
Caroline Grigson
    4.1   Introduction 77
    4.2   Material 79
    4.3   Method 82
    4.4   Results 82
    4.5   Conclusions 98
           Appendix 100

III   THE EARLIEST AGRICULTURE IN CONTINENTAL EUROPE: THEORY, METHODOLOGY AND AGRICULTURAL PRACTICE

5   The cereal pollen record and early agriculture 113
Kevin J. Edwards
    5.1   Introduction 113
    5.2   Identification 114
    5.3   Statistical and spatial considerations 118
    5.4   Pattern and the Cerealia pollen curve 120
    5.5   Species exclusivity and supporting
palaeo-ecological evidence 122
    5.6   Stratigraphic significance 125
    5.7   Conclusions 125

**6   Pollen analytical evidence for the beginning
of agriculture in south central Europe** 137
Hansjörg Küster
    6.1   Introduction 137
    6.2   Identification of cereal pollen grains 137
    6.3   The evidence in north central Europe 138
    6.4   The evidence in south central Europe 140
    6.5   Archaeological evidence 141

**7   Botanical investigations at the neolithic
lake village at Weier, N. E. Switzerland:
leaf hay and cereals as animal fodder** 149
David Robinson and Peter Rasmussen
    7.1   Introduction 149
    7.2   Recent work 152
    7.3   Conclusions 161

8	The evidence for early rye cultivation in north west Europe		165
	F. M. Chambers		
	8.1	Introduction	165
	8.2	Lines of evidence	166
	8.3	Uses for rye crops	170
	8.4	Crop husbandry practices	171
	8.5	Conclusions	172

IV THE ADOPTION OF AGRICULTURE IN THE BRITISH ISLES: THEORY AND EVIDENCE

9	Cattle and sheep in Britain and northern Europe up to the Atlantic period: a personal viewpoint		179
	Barbara Noddle		
	9.1	Introduction	179
	9.2	Wild cattle	180
	9.3	Domestication	185
	9.4	Transport of livestock across water	186
	9.5	Domestic cattle	187
	9.6	Wild and domestic sheep	193
	9.7	Concluding remarks	194
10	The evidence for cereal cultivation and animal husbandry in the southern British Neolithic and Bronze Age		203
	Roy Entwistle and Annie Grant		
	10.1	Introduction	203
	10.2	Cereal cultivation	203
	10.3	Animal husbandry	205
	10.4	Conclusions	207
		Appendix: interpreting mortality patterns; a preliminary note (Annie Grant)	208
11	Milking the evidence: a reply to Entwistle and Grant		217
	A.J. Legge		
	11.1	Introduction	217
	11.2	The scale of cultivation and the importance of cereals in the diet	217
	11.3	The significance of neolithic pits	220
	11.4	The importance of wild plant species in the diet	223
	11.5	The manner of cultivation	224
	11.6	Other environmental considerations	225
	11.7	Economy and society	225
	11.8	Neolithic-Bronze Age cattle as milking animals?	226
	11.9	Summary	236

12	Cereals, fruit and nuts: charred plant remains from neolithic sites in England and Wales and the neolithic economy	243
	L. Moffett, M.A. Robinson and V. Straker	
	12.1 Introduction	243
	12.2 The results	245
	12.3 Discussion of the results	245
	12.4 Neolithic landscape and agriculture	252
	12.5 Conclusions	254
	Appendix: The sites	256

V CONCLUDING REMARKS

Concluding remarks 265
W. Groenman-van Waateringe

FOREWORD

Annie Milles, Diane Williams and Neville Gardner
School of History and Archaeology, University of Wales College of Cardiff, PO Box 909, Cardiff CF1 3XU.

Between the 18th and 21st September 1987, the annual Association for Environmental Archaeology conference was held at University College Cardiff (now University of Wales College of Cardiff).

The title, and indeed the theme, of the conference 'The Beginnings of Agriculture' was decided upon at the 1986 annual general meeting held at Norwich. Most people felt that it was time to hold a series of chronologically based conferences, reviewing and presenting new evidence for specific archaeological periods or events. As a result the theme of early agriculture emerged, which was in fact devoid of both spatial or temporal limitations.

'The Beginnings of Agriculture' initially attracted a wide variety of papers, broadly divided into theoretical explanations and the archaeological evidence for the introduction and adoption of agriculture. Readers of this volume who were present in Cardiff will notice that the papers have been re-ordered and the themes structured slightly differently from those at the conference itself. The balance of papers eventually submitted for publication did not reflect the full spectrum of discussion at the conference. This was perhaps in part due to the demands on some of the speakers of editing various sections on domestication and agriculture from the World Archaeological Congress, these being the areas from which papers were not submitted for publication. Four sections were eventually distilled from the papers received: Theoretical approaches, Domestication, The earliest agriculture in continental Europe and The adoption of agriculture in the British Isles.

The papers were submitted, as requested on discs: however, they were written using a number of different word processing programs, and several operating systems. We were fortunate in the help given us by Dave Atkins and Simon Elliott in the Microcomputer Unit at Cardiff, who managed to read and transfer all of the contributions into Microsoft Word, which we ran on an IBM-compatible personal computer using MS-DOS. All editing, page layouts and so on were carried out using Word and Microsoft Pageview, printed on a laser printer, and submitted to BAR as camera ready copy. We did, however, have difficulty obtaining the full range of non-standard characters and apologise to authors for any such omissions.

Finally, in place of an introduction, we have included one person's impression of the conference, which appeared in *Circaea*, the Bulletin of the Association for Environmental Archaeology. We hope that this provides an insight into the spirit of the conferences and will encourage more people to participate in the future.

Acknowledgements:

We were very fortunate that Neville was willing to be co-opted as an editor when Diane started a new job in September, and was no longer able to devote as much time as she would have liked to the editing.

Particular thanks should go to Dr Willy Groenman-van Waateringe for summing up both the conference and its proceedings. We should also like to thank all those people whose efforts contributed to everyone's enjoyment of the conference, in particular the following: Rachel Hutton, Rosina Mount, Vanda Ringwood, Amanda Rouse; the Department of Archaeology for its hospitality; Dr Eurwyn Wiliam for shepherding an excitable and unruly gaggle of environmental archaeologists around the Welsh Folk Museum at St. Fagans; Barbara Noddle for herding a party around the National Museum of Wales 'Dinosaurs from China', exhibition; Bob Trett, Derek Upton and Peter Wardle for showing off the muddy delights of the intertidal areas on the Gwent Levels in the Severn Estuary, and the Glamorgan-Gwent Archaeological Trust for the use of their minibus. Special thanks should go to Dr J.G. Evans for contributing a paper at two days notice and for leading a fascinating and enjoyable field trip.

We would like to thank the editors of *Circaea* and Dr Terry O'Connor for allowing us to use his impression of the conference in this volume, and Nick Balaam for the photograph.

We should also like to thank Dave Atkins and Simon Elliott for all their help in converting the various word processing packages used by the authors into Microsoft Word.

AEA CONFERENCE: 'THE BEGINNINGS OF AGRICULTURE' AT UNIVERSITY COLLEGE, CARDIFF, 18th-21st SEPTEMBER 1987

T.P. O'Connor

First printed in *Circaea* **5** 57-59 (1988)

These conferences simply get better and better. Those AEA members who opted out of the 1987 Conference because it was "another early domestication talking shop and not my subject anyway" missed eighteen diverse papers of a high academic standard overall, some lively discussion of new data and ideas, and field excursions which met the essential criteria of being informative and fun. On the other hand, they also missed the hall of residence catering, which had served further to bond the conference in a spirit of shared adversity.

David Harris launched the meeting on Friday evening by describing his own initial misgivings about the conference theme, then surveyed the various theories and models for the emergence of agriculture on which the archaeological data have, often tenuously, been hung. Two topics were introduced which recurred throughout the meeting; the relative merits of gradualism as against the concept of a 'neolithic revolution', and the project which Harris and various colleagues are basing on material from Tel Abu Hureyrah.

Alert wakefulness was at a premium on Saturday morning. Ken Thomas assured us that he didn't "...want to get side-tracked into discussing what reality is at nine o'clock on a Saturday morning", but clearly felt no compunction about discussing various complex theoretical ecological models as bases for explaining why and at what rate agriculture was adopted. In a stimulating half-hour, we were introduced to a hierarchical model of interacting sub-systems. One participant subsequently summed up the feelings of many with the remark "I always thought a holon was a subatomic particle before I came here." From models to data, and accounts of current thoughts on the domestication of caprines and cereals were given by Tony Legge and Gordon Hillman, both of them using data from Abu Hureyrah. The next three papers gradually coaxed the conference nearer home, by way of Sebastian Payne's lucid account of his work at Franchthi Cave, Greece, Paul Halstead's creative use of ethnography to provide a model for the spread of farming onto the Central European loess, and John Evans' description of his recent work locating early neolithic land surfaces and valley fills in north Wiltshire. Evans' paper was a late addition to the programme, having been compiled at a mere two days notice, following the non-arrival of a promised contribution. The haste didn't show, and it was good to see snails intruding amongst the goats and grains.

A homicidally-spiced lasagne for lunch failed to dampen either the enthusiasm or the quality of the Saturday afternoon papers. Vertebrates were variously discussed by Caroline Grigson and Simon Davis (whose use of the educational cartoon is definitely to be encouraged), and two German colleagues then shed more light on some recent palaeobotanical work in Central Europe. Hansjörg Küster has been investigating pollen sequences from the northern fringe of the Alps, where the onset of agriculture is clearly marked in the pollen data, though virtually absent from the archaeological record.

Angela Kreuz drew murmurs of sympathy when she admitted to having identified 10,000 pieces of charcoal from a *Linearbandkeramic* settlement. Analysis of the spatial distribution of the taxa, and comparison with associated pollen sequences, have shown a remarkably sophisticated selection and use of different timbers. Keeping to the resource exploitation theme, David Robinson rounded off a very full day of lectures by describing his and Peter Rasmussen's work on botanical remains from the lake-village at Weier, Switzerland.

The second day of lectures started gently, with Barbara Noddle casting a vet's eye over the material remains of early domestic and wild ruminants from Britain, and introducing her pet mouflon along the way. Royston Clark chose a more theoretical framework for arguing that the whole process from palaeolithic hunting through to sedentary cultivation can be viewed as a series of risk-management exercises, stages within one continuous process rather than discrete cultural events. It was during Clark's lecture that the only projector jam of the conference occurred. Within seconds, a back-up machine was in position and normal service was resumed: well done Neville! Keeping a spare projector to hand is risk-management of the most sensible kind.

Risks were further to the fore when Annie Grant presented her own and Roy Entwistle's somewhat iconoclastic thoughts on the development of agriculture through the British Neolithic and Bronze Age. It was a challenging thesis, replacing the sedentary, cereal-based, plough-using Neolithic which we have grown to accept with a system of hoe-based horticulture and stock-rearing, in which cattle may have had a largely symbolic, even ritual, role. It was a brave thesis to present to an audience which included John Evans and Tony Legge, and in the heated discussion which ensued, an onslaught of data left it looking rather battered and bruised. None the less, it was good to see a sceptical eye being turned on received wisdom.

As blood pressures around the hall restabilised, Kevin Edwards reviewed the potential and difficulties of studying the sparse pre-*Ulmus* decline records of cereal pollen. Frank Chambers was more positive ("Good morning ladies and gentlemen, and anyone else that's crawled in...") in his examination of the evidence for the early exploitation of rye in north west Europe. Chambers stressed a couple of important points: the versatility of rye as a crop on poor soils, and the problems it would present to a community accustomed to non-free-threshing cereals. The final lecture ought perhaps to have been the first, as Susan Limbrey explored possible connections between soil types and the options which they presented to early farming communities, drawing particular attention to the potential value of self-mulching vertisols. Finally, Willy Groenman gave a very positive summing-up, detailing areas where studies can go forward, and anticipating another such conference in ten years time.

The Sunday afternoon outings were by way of gentle relief. A charabanc outing set off for the excellent Welsh Folk Museum, just outside Cardiff, where reconstructed buildings from all over Wales are set in a park landscape, and a variety of crafts and industries are carried on. There can't be many water-powered spinning frames still operating in Britain. Neither are there many Chinese dinosaurs, so a second party formed a neat crocodile and were taken to the National Museum of Wales to see this internationally-important exhibition. Sunday evening rounded off the conference in great style. Over thirty people engaged in a hugely enjoyable skittles match in a local pub, in which the Animals narrowly but convincingly defeated the Plants (naturally!). Given that conferences are at least in part social occasions, booking the skittles alley was a brilliant stroke of inspiration.

Monday's field trips were blessed with mild weather, which was just as well. One party travelled down the road to Newport to inspect the land surfaces and trackways which are being exposed and eroded along the Welsh side of the Severn Estuary. What effect might the proposed barrier have, one wonders? Meanwhile, a somewhat larger group toured Mid Glamorgan and neighbouring bits of Powys in search of glacial landforms and deposits. Traeth Mawr, near Brecon, provided a hands-on and feet-in mire experience which will not quickly be forgotten.

The 1987 AEA Conference thus managed to be stimulating, informative and enjoyable. The necessary planning had clearly been done well in advance, and in sufficient detail to ensure the smooth running of a packed schedule. When the proceedings are published, the papers will provide a valuable digest of current thoughts and progress in this fascinating area of research.

ADDRESSES OF CONTRIBUTORS

DR. F. M. CHAMBERS
Environmental Research Unit,
University of Keele,
Keele,
Staffordshire
ST5 5BG

ROYSTON CLARK
Department of Archaeology,
University of Southampton,
Southampton
SO9 5NH

DR. KEVIN J. EDWARDS
School of Geography,
University of Birmingham,
PO Box 363,
Birmingham
B15 2TT

ROY ENTWISTLE
Department of Archaeology,
University of Reading,
Whiteknights,
PO Box 218,
Reading
RG6 2AA

ANNIE GRANT
Department of Archaeology,
University of Reading,
Whiteknights,
PO Box 218,
Reading
RG6 2AA

DR. CAROLINE GRIGSON
Odontological Museum,
Royal College of Surgeons of England,
35-43 Lincoln's Inn Fields,
London
WC2 3PN

DR. W. GROENMAN-VAN
WAATERINGE
A.E. van Giffen Instituut voor Prae- en
Protohistorie,
Universiteit van Amsterdam,
Singel 453,
1012 WP Amsterdam
Netherlands

DR. PAUL HALSTEAD
Department of Archaeology and
Prehistory,
The University of Sheffield,
Sheffield
S10 2TN

DR. HANSJÖRG KÜSTER
Institut für Vor- und Frühgeschichte,
Universität München,
Schellingstrasse 5,
8000 München 40
W. Germany

A.J. LEGGE
Centre for Extra-Mural Studies,
Birkbeck College,
University of London,
26 Russell Square,
London
WC1B 5DQ

LISA MOFFETT
Biological Sciences,
University of Birmingham,
PO Box 363,
Birmingham
B15 2TT

BARBARA NODDLE
Department of Anatomy,
University of Wales College of Cardiff,
PO Box 900,
Cardiff
CF1 3YF

PETER RASMUSSEN
Naturvidenskabelig Afdeling,
Nationalmuseet,
Ny Vestergade 11,
1471 Copenhagen K,
Denmark

DR. DAVID ROBINSON
Naturvidenskabelig Afdeling,
Nationalmuseet,
Ny Vestergade 11,
1471 Copenhagen K
Denmark.

DR. MARK ROBINSON
University Museum,
Parks Road,
Oxford
OX1 3PW

VANESSA STRAKER
Department of Geography,
University of Bristol,
University Road,
Bristol
BS8 1SS

DR. K. THOMAS
Institute of Archaeology,
University College London,
31-34 Gordon Square,
London,
WC1H 0PY

I

THEORETICAL APPROACHES TO THE BEGINNING, SPREAD AND ORGANISATION OF AGRICULTURE

1

TOWARDS THE INTEGRATION OF SOCIAL AND ECOLOGICAL APPROACHES TO THE STUDY OF EARLY AGRICULTURE

Royston Clark

ABSTRACT

The aim of this paper is the formation of a theoretical perspective that integrates both social and ecological concepts in the understanding of the adoption of early farming techniques. In order to achieve this, hunter-gatherer subsistence and social organisation need to be examined in some detail. A case study using Italian data will examine some archaeological evidence with regard to the perspective on early farming.

1.1 HUNTER-GATHERER PERSPECTIVES: AN INTRODUCTION

Early agriculture in Europe has traditionally been viewed in diffusionist terms, spreading westwards from the Near East (e.g. Childe 1947; Ammerman and Cavalli-Sforza 1971, 1973). This is mainly because the earliest dates for domestic plants and animals, as well as other aspects of the farming package, such as ceramics, have much earlier radiocarbon dates in the Near East. Also, the general pattern of radiocarbon dates for the spread of the Neolithic into Europe tends to show a westward movement with progressively later dates in the western and northern regions. This is, however, a somewhat simplified perspective that requires critical analysis.

In contrast, what little evidence there is in Europe for independent or early domestication is ambiguous. For example, the mouflon in Sardinia are now thought to be descended from feral sheep, with the implication that there is no evidence for sheep domestication in the western Mediterranean (Geddes 1985). A further factor worthy of consideration is the possibility that wild species of cattle in central Europe bred with domesticated cattle (brought about intentionally by human groups, and occurring naturally in the wild through feral animals) and in so doing altered the relative size frequency of cattle bones. This can make the study of the adoption of early domesticates particularly difficult, because relative bone size is often the criteria used to determine whether the animal was wild or domesticated (Bökönyi 1969).

In addition, there is the question of non-utilitarian use of early domesticated animals, since surely their value would have been seen in terms of status and ownership, and only ultimately did they become a food source (e.g. Ducos 1978).

These points suggest that when we consider the adoption of farming in Europe, we need to consider in greater detail the role of indigenous hunter-gatherers. Earlier models, such as the diffusionist theories advocated by Ammerman and Cavalli-Sforza (1971, 1973) gave a minimal view of hunter-gatherer adaptations, since they were seen as impoverished groups unable to resist the 'wheels of progress' with regard to subsistence change. Such a view was largely based on the concepts of colonial movements by European settlers during the last century, when indigenous groups were overrun by technological superiority. More recent papers have given indigenous hunter-gatherers a less passive role in the adoption of agriculture (e.g. Zvelebil and Rowley-Conwy 1984; Dennell 1985; Lewthwaite 1986; Zvelebil 1986). This can partly be explained by changing anthropological perspectives of hunter-gatherer subsistence, with these societies now being seen as affluent (e.g. Sahlins 1972) or complex (e.g. Price and Brown 1985). This paper continues this approach by considering how long-term subsistence practices, together with emerging social institutions, could have facilitated the rate that farming was adopted in different parts of Europe.

If we adopt a broad diachronic perspective in order to look at development and variation as opposed to abrupt change, we can see a more continuous pattern beginning about 35,000 years ago with the emergence of anatomically modern humans. This period represented the transition between the Middle and Upper Palaeolithic, and it was perhaps at this time that settlement and exploitation patterns began to change (cf. Gamble 1983, 1986a). It also represents the first appearance of recognisable art forms and an associated social structure capable of supporting population groups in regions such as Europe, where the environment was relatively harsh and the resources difficult to exploit. Gamble (1983) has argued that mobile art forms, which first appear in the colder regions of northern Europe, could represent symbols of alliance networks used to combat the risks of exploiting unpredictable environments. Regional settlement histories and the use of regional resources suggest that upper palaeolithic groups had the ability to cope with the extreme cold. In the earlier period of the Middle Palaeolithic, many parts of Europe contain no evidence of occupation during the last interglacial (128,000-118,000 bp), yet in the broadly comparable climatic conditions of the early Postglacial, the archaeological record shows abundant evidence for human occupation in Europe. The resources may have been similar in both periods, but the adaptive responses were not.

Once the Pleistocene megafauna was replaced by smaller sized animals and a richer plant biomass in the Holocene, new adaptive strategies capable of exploiting more local resources were required by upper palaeolithic groups. We can look at a range of subsistence patterns practised throughout the world in the post-Pleistocene as adaptations to different environments containing specialised animal and plant communities. Examples include the patterns of exploitation that resulted in animal and plant domestication in regions such as the Near East and Mesoamerica, the intensive exploitation of coastal resources such as in Denmark, and the herding of animals such as reindeer in more northerly latitudes. This variation can be seen as the behavioural adaptations being applied to specific environments. It is argued here that the origins of these adaptive responses emerged with the advent of *Homo sapiens sapiens* about 35,000 years ago.

1.2 POPULATION GROWTH PROCESSES

The ability to exploit these 'novel resources' (Dennell 1983) had additional effects. Apart from the seasonality of resources, to overcome which human organisation was needed, there would have been a reduction in long distance group mobility in areas such as Europe. Instead of following herds of animals that migrated over great distances (for example, reindeer), these 'novel resources' would have occupied smaller annual territories, thus reducing the distances that hunting groups would have needed to travel.

I suggest that human territories could therefore have become more identifiable as discrete areas within a region. This would have implications for population growth processes that occurred throughout the Upper Palaeolithic. A possible effect could have been an increase in the number of population groups inhabiting a region. Such a rise in population numbers must be seen as distinct from the more general processes cited by writers like Cohen (1977). Archaeologically, there is very little evidence for large scale population growth immediately outside areas of early domestication such as the Near East. Such regions would be precisely the areas where we might expect general population growth to occur first. These early neolithic groups, however, would almost certainly have had cultural mechanisms for maintaining a certain population level (for example, contraception, abortion and infanticide). An alternative population growth process has been suggested by Binford (1968, 1983) which has the merit of accommodating the availability of cultural mechanisms such as contraception with the archaeological evidence for rises in population numbers.

Binford (1983) has phrased population growth in terms of a 'packing model', where new communities are formed from existing ones, thus reducing the size of the previous social groups' territories. In addition, it is argued here, new groups altering these territorial systems must have led to additional 'costs' of maintaining an extensive community social structure, as well as leading to additional costs of resource exploitation itself. Any failure of resources would have also required a level of organisation capable of transcending small family units. This sort of organisation can be phrased in terms of risk management.

1.3 RISK MANAGEMENT

Foley (1985) has suggested a number of operational currencies which include availability of time and risk minimisation as variables for explaining subsistence strategies. So far, the main application of time budgeting and risk minimisation has been in terms of lithic technology and subsistence analysis. There are numerous ethnographic examples of such risk buffering systems in operation. Wiessner (1982) has noted a continuum of risk buffering strategies that sees low latitude (for example, tropical) hunter-gatherers, such as the !Kung bushmen, relying on risk sharing strategies like reciprocity, to high latitude (for example, arctic) groups like the Inuit, who practice a high degree of storage as a way of maintaining adequate food supplies. Between these two extremes are the Northwestern Pacific Coast Indians, practising a combination of both strategies in a temperate environment quite similar to that in Europe. It is these forms of risk buffering which could have facilitated cultural mechanisms that led to the eventual adoption of farming techniques in Europe.

1.3.1 Technology

The organisation of lithic technology has been regarded by Torrence (1983 and unpublished) as a time budgeting mechanism that is also ultimately a short-term risk buffering strategy, aimed at reducing the chances of loss in the pursuit of plants or animals. Of course, this would be closely related to the seasonal distribution of resources. The time budgeting element means that technology can be used to control efficiently the time allocated to subsistence pursuits. In fact, it is possible to see the actual subsistence pattern practised as a response to time budgeting and ultimately as a risk management strategy. The level of specialist microlithic tools used during the Mesolithic can be regarded as part of the time budgeting and risk minimisation strategy.

Torrence (unpublished) has argued that mesolithic tool kits represented a high degree of uniformity and homogeneity that suggested an emphasis on reliability. Mesolithic groups were therefore responding to high levels of short-term risk due to their reliance on mobile prey and the seasonal variability in the occurrence of these resources. It is also interesting to note that in the later periods, when domesticated resources predominated, lithic material probably did not have the same risk buffering value; tool construction was more variable in quality and type, and probably had less emphasis on specialised tasks. Alternative risk buffering mechanisms were undoubtedly in operation.

1.3.2 Storage

A second risk buffering strategy that can also be detected archaeologically is storage. This can be divided into direct storage and social storage techniques.

i) Direct storage

Direct storage is where food is processed for future consumption. The success of this form of storage is of course dependent on the type of food being stored, the climate and the storage technology. Storage, however, would in most cases demand a certain degree of sedentism, a characteristic associated with most farming communities. In fact, one of the main features in the early Neolithic is the presence of pits, generally thought to have functioned as storage facilities. One additional form of storage, involving animal husbandry, is to feed animals excess crops or food unsuitable for direct human consumption and then slaughter the animals when meat is needed. Such animals might also have a primary role in enhancing an individual's social prestige (e.g. Evans-Pritchard 1940). The advantage in this is that sedentism is not a limiting factor, and mobile hunter-gatherers could therefore adopt such a storage strategy more easily.

ii) Social storage

Following on from the idea of transferring food into a more durable form, is the concept of social storage (Halstead and O'Shea 1982). This consists of reciprocal food exchanges, where a surplus accumulated by one individual or household is transferred to another in need, or what Sahlins (1972) termed 'generalised reciprocity'. An alternative social storage mechanism would involve the use of non-food 'tokens' in more extensive exchange systems. These tokens can be reconverted into food when necessary. Their collection, however, can become an object of cult associated with individual power, and their primary role could become abstracted with a new symbolic meaning. Perhaps mesolithic and early neolithic art forms such as carved bone, polished stone axes or even pottery itself had a similar symbolic role.

1.3.3 Social mechanisms

Another concept, first developed by Sahlins (1972), is that of hunter-gatherers as the 'original affluent society'. This is a rather misleading concept because much of the so-called 'spare time' stressed by Sahlins is used to maintain kinship ties and gain information about resource distribution around the foragers' territories. In other words, these activities represent risk buffering systems to avoid dietary failure. If we accept that such a strategy could be related to the fact that during the Postglacial, mesolithic groups were under great pressure to schedule the necessary time to exploit seasonal and less predictable resources than earlier Pleistocene strategies of following herds of reindeer, we can begin to discuss emerging social organisation. In order to minimise risk in subsistence pursuits, hunter-gatherers used mobility as a method for gaining information, and for balancing population numbers to available resources. If, as Binford (1983) argued, more population groups were establishing themselves in the regions during the Postglacial, the costs of exploiting small package sized (but highly productive) resources in smaller territorial ranges would have been high. Some groups would have become more sedentary, for example coastal groups like the Ertebølle (Rowley-Conwy 1983), and these high costs would have included 'social costs' of maintaining inter-community cohesion. The extensive periods of time needed to maintain various risk buffering strategies, like manufacturing specialised tools or exchanging resource information, might have made it increasingly necessary to transfer such responsibility from the individual to a more institutional or specialist level of organisation. The associated specialised information or knowledge gained from such activity, could become the object for power relations within individual groups. One way that hierarchies could evolve is through successful hunting which provided a degree of status to the individual (Mithen unpublished). Such power could be related to social storage systems that used 'tokens' as a tool for converted foodstuffs. The relationship between emerging individuals or institutions and such 'tokens' is hard to determine. Were the tokens the cause of an individual's rise to power or the effect and manifestation of an individual or institutional power base? The two must be closely related. Once such social organisations developed, the threat of large scale resource failure would have kept the system in permanent operation (cf. Minnis 1985).

It is therefore likely that an institution developed with the role of maintaining these social risk buffering activities (cf. Gamble 1986b). Perhaps it is here that we have the origins of hierarchical society. A specialised institution would be directed towards co-ordinating inter-group activity, particularly as the density of population groups would have made processing inter-group information time-consuming and highly complex for individual family units. In a paper discussing different forms of social organisation, Johnson (1982) argues that an increase in the size of human groups leads to corresponding increases in communication problems, causing disputes and conflicts. This results in a need for additional systems of organisation to maintain permanence in the society.

1.4 SEQUENTIAL AND SIMULTANEOUS HIERARCHIES

Hunter-gatherer groups have traditionally used fission, or moving away, to avoid conflict. Such a social system would rely on a high level of flexibility and has been discussed in terms of a 'sequential hierarchy' by Johnson (1982). Sequential hierarchies relate to hunter-gatherer mobility, and the organisational unit within the group changes according to the structure of the population present at the camp site. The basic organisational unit could be the extended family with populations over forty or so. At smaller camps of single families, the organisational unit would obviously be the family unit. Fission is therefore an important aspect in sequential hierarchies.

If fission is not possible because mobility is restricted, or there is some reason to maintain group size in one place (for example, a particular subsistence strategy requiring communal work), Johnson argues for the development of a 'simultaneous hierarchy' in which control is vested in a small proportion of the group, who have extensive powers over organisational units like kin groups, work groups or residential units. This must be seen in contrast to sequential hierarchies, as simultaneous hierarchies achieve organisation without the need for fission (Johnson 1982).

With respect to the population growth processes discussed previously (Binford 1983), and the need to minimise the risks involved in exploiting different resources, such organisational strategies would have been important for co-ordinating different subsistence tasks and maintaining the inter-group cohesion needed to relate population groups to available resources.

The cost of maintaining social organisation by such means as simultaneous hierarchies would have meant a certain reduction in the individual's or family's personal freedom. The institution or individual would have needed surplus food to continue its specialist work and a certain ritual or symbolic input to maintain its legitimate authority within the society.

Specialised knowledge of main food sources, and the group's reproduction network, would be two basic elements of control. The importance of kinship structures in many current small scale societies is good evidence of how a specialised knowledge maintains authority within a community (for example, the Nuer; Evans-Pritchard 1940). Anthropologists like Bender (1978, 1981) and Meillassoux (1972, 1973) have discussed the origins of such social control and its relation to early food production from a Marxist perspective. One fundamental transformation would have been the development of the concept of ownership, particularly of land and animals (e.g. Ingold 1980). It is therefore possible that the concept of land ownership had its origins in an institution that controlled access to resource information, marriage exchange and ways of maintaining social discipline (e.g. Goody 1976). How can we begin to look into the archaeological record for such evidence? I suggest that we attempt to consider early pottery, mobile art forms and human burials with high status artefacts, as representing symbols or evidence of simultaneous hierarchies.

1.5 ARCHAEOLOGICAL CORRELATES

We have now considered the role of risk management in explaining social transitions that led to the adoption of agriculture, and it is now possible to discuss some of the archaeological expectations that could be found in data sets. Undoubtedly, the major problem is that social processes (apart from colonialisation) only indirectly find their way into the archaeological record, but this is not a reason for ignoring their fundamental role in facilitating subsistence change. A further problem is the time span that needs to be considered in studying long-term trends which culminated in the adoption of agriculture. Such continuous data sets are extremely rare, and the material from northern Italy is therefore significant.

We have already noted the development of microlithic tool industries as a method of time budgeting and risk management. Their replacement by more generalised and less sophisticated assemblages in the periods when food production was more prominent suggests that bulk storage and social storage became more established as risk buffering systems during the early stages of agriculture. Keeping animals fed on surplus plant foods would also have acted as a further risk buffering system. Throughout the course of the Holocene we would expect to see evidence for a reduction in the size of foraging territories. This might be manifested in a reduction in the range of fauna exploited. Through the course of time, we would therefore expect to see a progressively smaller range of animals present in archaeological contexts. Additionally, we would expect an increase in specialised resource exploitation; for example, coastal resources or red deer.

At the same time, hierarchical social systems, capable of risk management, became more prominent. Archaeologically these could be visible in a number of ways. Evidence for long distance exchange networks and prestige items, perhaps the symbols of authority, might be found. These might represent artefacts, the manufacture of which required a certain skill or degree of time investment, such as mobile art, pottery or polished stone implements. Further evidence for social structure could be found in high status burials associated with rich grave goods.

1.6 A CASE STUDY

In order to test these expectations, a data set from the Adige valley in sub-Alpine northern Italy was analysed (Figure 1). During the Holocene the region was rich in resources, with a wide range of wild plants and animals occupying a series of ecozones. These ecozones had both a vertical and a spatial dimension and were thus relatively close together. For example, these included river and lake environments rich in fish and bird life, as well as animals like beaver and otter. Further up the hillsides occurred rich woodlands containing hazel, beech and oak (Greig 1984), ideal habitats for red deer, roe deer, pig and other herbivores. Beyond the woodland, open pastures and craggy mountain slopes provided a habitat for caprines like ibex and chamois. These different ecozones would also have been rich in edible plant species, providing further food sources for mesolithic communities. The range of potential foods can perhaps be best illustrated by examining the wild foods found on the waterlogged Bronze Age site of Fiavé. Foods that would have also been available to mesolithic groups included wild strawberries, cornelian cherries, wild apples and hazel-nuts (Jones and Rowley-Conwy 1984). However, the region, as it does today, also experienced a high degree of seasonal variation in climate, ranging from severe winters to very hot summers with long growing seasons. This would have had an effect on the settlement patterns practised throughout the year. Harsh winters would have made occupation possible only in the most sheltered of locations.

The archaeological sites studied consisted of a series of rock shelters and a cave site, located around the modern town of Trento (Bagolini 1980a; Boscato and Sala 1980; Broglio 1980). Four shelters examined contained stratified faunal and lithic material ranging from the early Mesolithic to a period some time in the early Neolithic. Ceramics, and evidence of the use of domesticated animals, were found in three of the four sites. Faunal material from an early neolithic open site four kilometres outside Trento was also studied. Open air sites from the surrounding highlands and mountains have also been considered in order to assess the size of the territories exploited. Finally, two large faunal assemblages from middle neolithic sites were also examined to add chronological depth to the study.

Although the cave site of Grotta d'Ernesto still awaits radiocarbon dating, the lithic material suggests a very late upper palaeolithic date, with microlithic material associated with the early Mesolithic (Dalmeri 1985). The fauna consisted mainly of red deer, bear and ibex. All the species, including the bear, showed clear traces of butchery. Because of the high level of faunal preservation and butchery evidence, and the associated fireplace, the site is significant in terms of postglacial settlement studies. According to the butchery evidence, it is suggested that Grotta d'Ernesto was probably

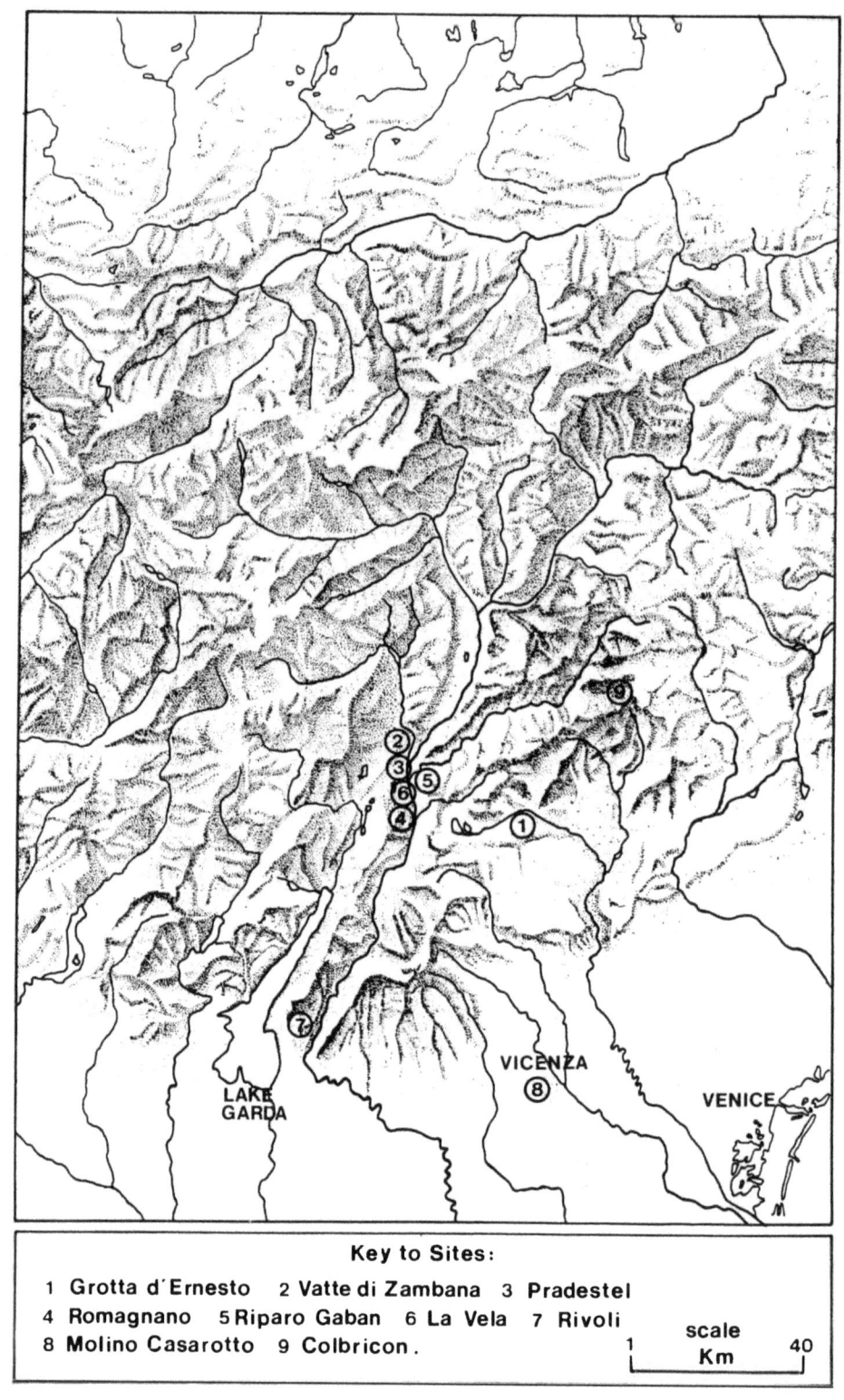

FIGURE 1 Location map showing archaeological sites mentioned in the text.

used on a short term basis for meat processing and consumption rather than for long term residential activity. Very few boneshad been split open for marrow extraction, and leg bones showed clear filleting marks. There were very few of the small bone fragments normally associated with extensive habitation.

The three rock shelters, known as Pradestel, Romagnano III, and Zambana (Figure 2) all appear to have been butchery processing sites. The shelter known as Riparo Gaban, being somewhat larger and in a more sheltered location, could have been used for more permanent residential purposes. At all four sites, butchery studies indicate intensive meat and marrow processing taking place. Practically all the red deer first and second phalanges had been split open, presumably for marrow. The majority of lower leg bones and metapodials also showed evidence of marrow extraction in their butchery. Negative evidence such as the lack of the main meat bones like femora and humeri could be used to suggest that the prime meat bearing bones were taken away from the site to more permanent locations, although there are problems in such interpretations (Clark forthcoming). These bone types are more vulnerable to taphonomic destruction due to their lower bone densities (e.g. Binford 1978, 1981). However, I would expect to be able to identify small fragments of these bones in the assemblages.

In the course of the Mesolithic there is evidence that exploitation patterns changed. The early periods show a much greater range of large mammals exploited; these included red and roe deer and pig, with higher altitude animals like ibex and chamois being particularly important at Romagnano and Pradestel. In the later stages of occupation at the sites, red deer predominate. This suggests that hunting territories became progressively smaller during the Mesolithic. This observation is also supported by fieldwork in the surrounding highlands. A mountainous area to the north east of Trento has been systematically surveyed and has revealed approximately thirty-five scatters of lithic material that typologically belong to the early, and to a limited extent middle, Mesolithic (Bagolini and Lanzinger pers. comm.). These 'off site' scatters, such as Colbricon (Figure 3), represent hunting stands and tool preparation areas (Bagolini 1980a). They occupy prominent areas regarded as too high for red deer and more than likely represent ibex or chamois hunting stands. From the later Mesolithic and Neolithic there is no evidence for such extensive areas of mountain activity. In fact it is not until the middle Bronze Age and later periods that these areas were again used. So evidence from both the rock shelters and the surrounding region suggests a gradual shift from hunting a wide range of animals to a more intensive and selective hunting strategy, probably concentrating on red deer. Red deer exploit a much smaller territory and the risk of failure to locate or kill must have been much less. It is possible that Holocene climatic factors caused an increase in forest density, and that this resulted in a concentration in lower altitude resource exploitation of red and roe deer. Animals, such as ibex, that were specifically adapted to the marginal zones of rocky mountain sides, might have moved further up the slopes, effectively making them much costlier to exploit. Pollen evidence might help to ascertain whether vegetational changes could have forced human groups during the middle Mesolithic to exploit valley systems more intensively, at the expense of higher altitude zones. Unfortunately, most pollen work conducted so far in sub-Alpine Italy lacks radiocarbon dates and any correlation with changing cultural activities is currently impossible.

In terms of emerging social hierarchies responsible for co-ordinating subsistence strategies and maintaining social cohesion, it is likely that material manifestations would consist of burial goods or other specialist made artefacts that could be used to signify status.

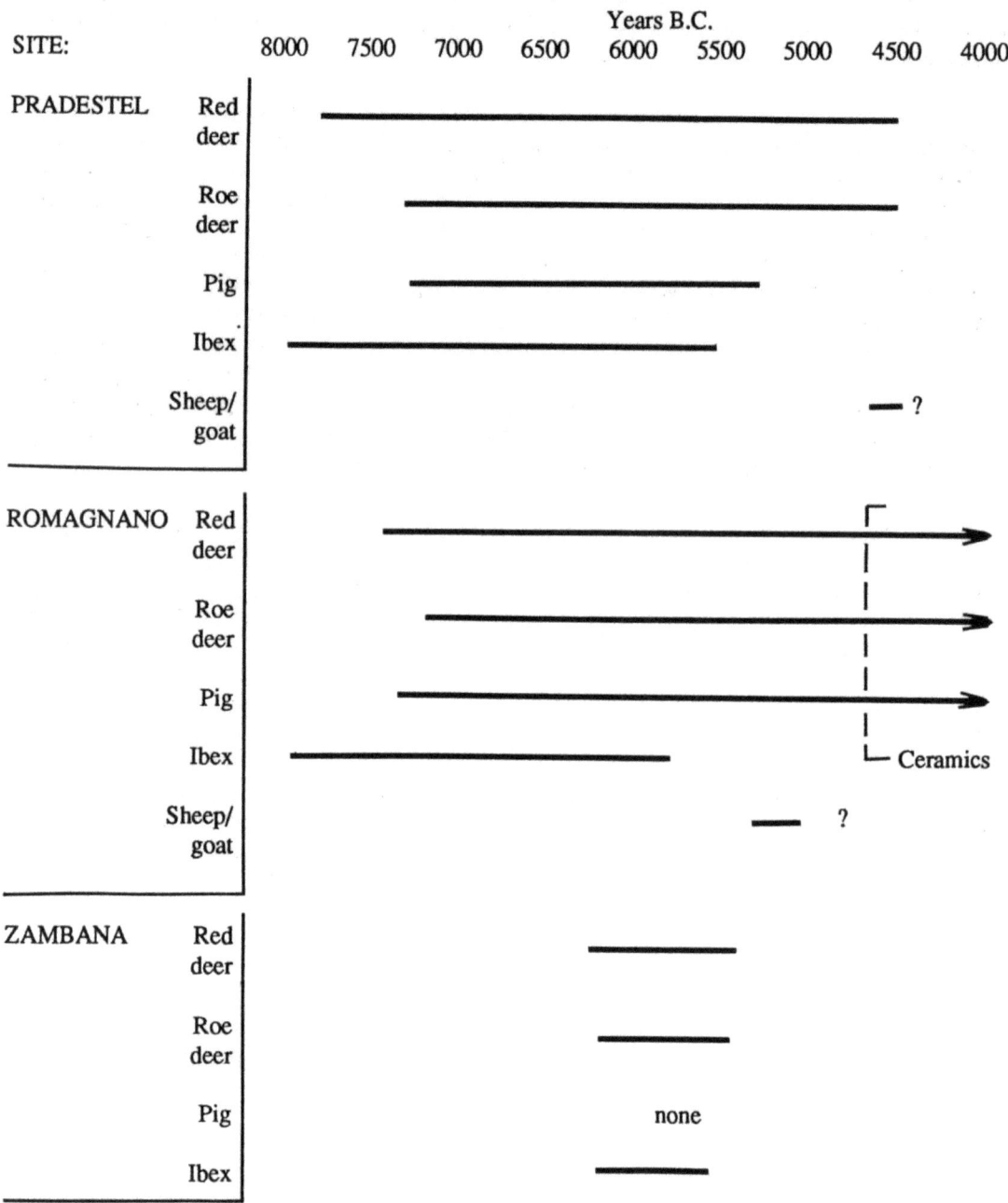

FIGURE 2 Summary chart illustrating chronological changes of large mammal exploitation from three sites in the Adige Valley in northern Italy: Pradestel, Romagnano and Vatte di Zambana (radiocarbon dates for Riparo Gaban are not available).

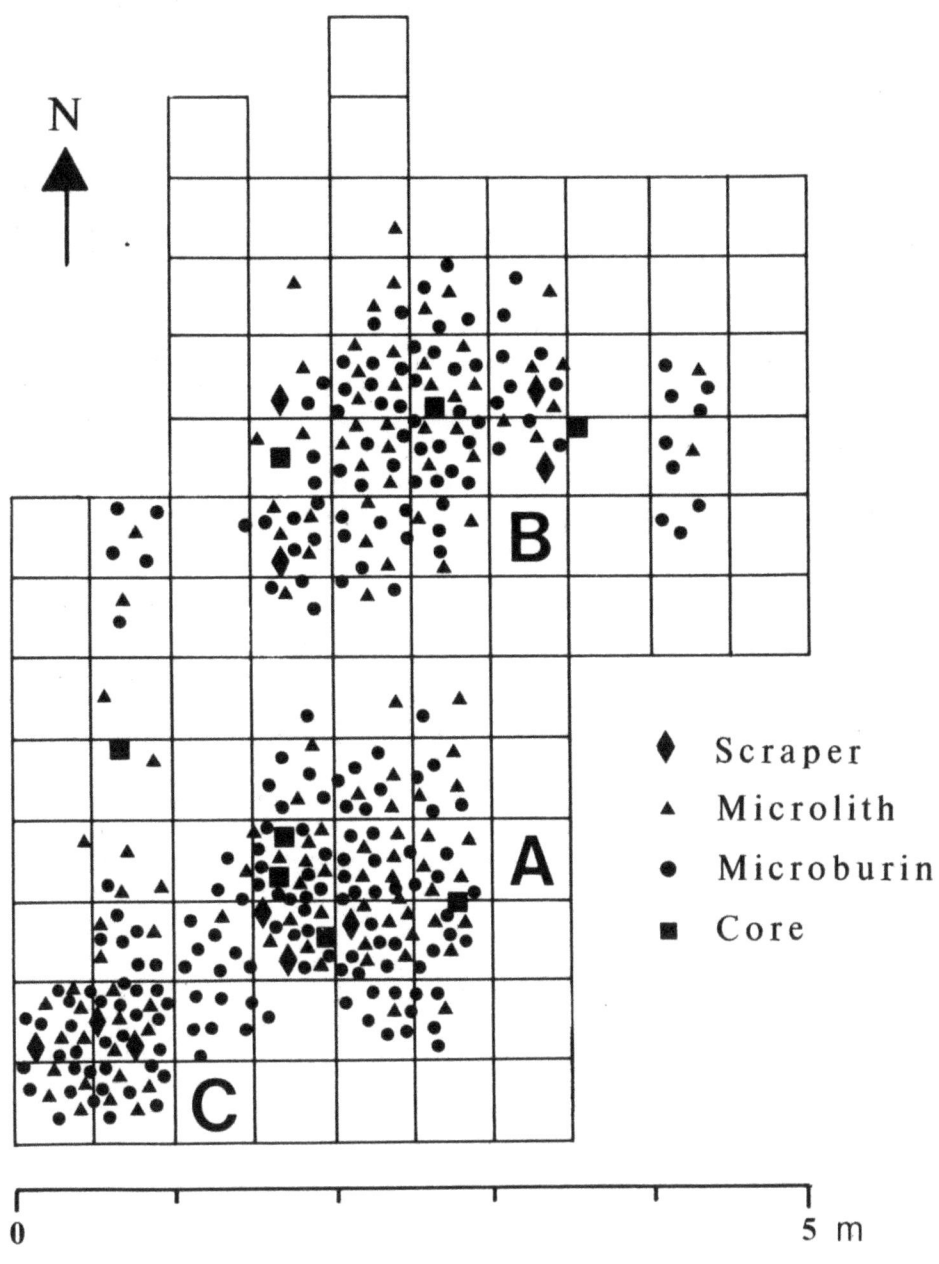

FIGURE 3 Colbricon; three flint scatters dating to *circa* 6200-6000 B.C. (2000m above sea level). Lithic material suggests that these activity areas were used to prepare tools for hunting or the primary processing of hunted animals, probably ibex (after Bagolini 1980a).

It is during the Mesolithic that we get the first substantial evidence for institutionalised status in the form of burials. Mesolithic cemeteries are invariably found in coastal regions (Vedbaek in Scandinavia) or in rich river valley systems (such as Lepenski Vir at the Iron Gates of the river Danube). Although it is easy to characterise this pattern in terms of areas of Europe where semi-sedentary communities lived in highly productive environments, cemetery evidence for the Mesolithic in the equally rich region of the Adige valley in sub-Alpine Italy is absent. Very little burial or settlement evidence is available, perhaps because the primary zone of settlement (the valley floors) have been buried under several metres of colluvium.

The late mesolithic and early neolithic levels of the rock shelter at Riparo Gaban, however, have yielded a large number of elaborately carved bone and antler items, including plaques, figurines and a hollow 'flute' made from human bone (Plates I-VI, Figure 4 and Bagolini 1980a). Furthermore, this is the period that sees the introduction of ceramics, which it is hard to consider as simply representing the remains of utilitarian pottery. Sub-Alpine Bronze Age sites like Fiavé prove that equally durable wooden vessels were skillfully manufactured and there seems no reason why mesolithic communities did not already possess such carpentry skills. The rarity of ceramics in the archaeological layers, together with their fragile nature, suggests that they might have had other functions originally, perhaps as symbols of status or part of a widespread exchange system, representing an aspect of the 'social storage' risk buffering system. It could certainly help explain why these ceramics are so uniform in shape and decoration over a very large area of northern Italy.

Is it therefore possible to suggest that those traits often defined as the 'early Neolithic' represent something more fundamental to the hunter-gatherer subsistence system? Something adopted as a status item and exchanged as part of a food risk buffering mechanism? The subsequent introduction of caprids into an otherwise hunting economy could also represent status items and not just a new food source.

The site of La Vela, near Trento, consists of an open settlement with clear evidence of storage pits and burials. The site has two main phases. The first contains a very high proportion of remains of wild animals, especially red deer and pig (Table 1). There is some evidence that domesticated sheep and cattle were also kept, but their remains are represented by just a few fragments. The second phase contains a much larger proportion of domesticated animal bones, enough to argue that farming was the main subsistence pursuit, although red deer bones were still present. This site has as yet no radiocarbon dates, but typologically the ceramics fit into the late early to middle Neolithic.

ANIMAL SPECIES	OCCUPATION 1	OCCUPATION 2
Red deer	2	79
Roe deer	1	-
Chamois	2	1
Sheep/goat	4	44
Cattle	6	70
Pig	9	70
Brown bear	-	1

TABLE 1 Provisional table of bone data (numbers of identifiable fragments) from La Vela (Trento).

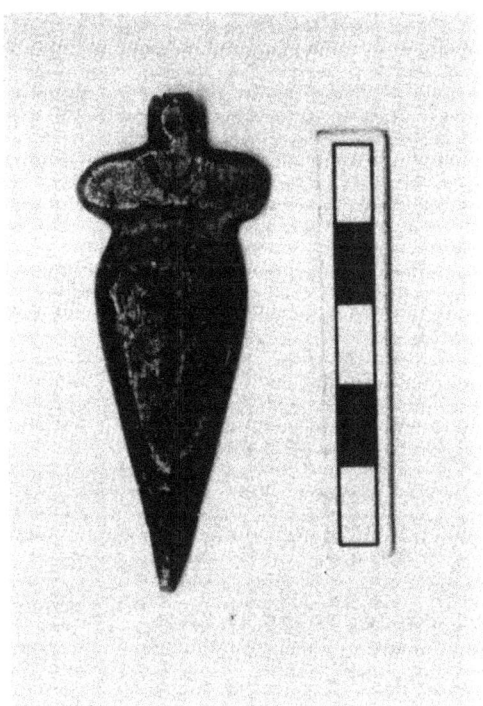

PLATES I-III Carved objects from the Riparo Gaban rock shelter (scales represent 50mm).

I Spatula carved from red deer antler. Late Mesolithic.

II Anthropomorphic figurine carved on a soft soapstone (?) material. Late mesolithic/early neolithic level.

III Female figurine carved from bone. The lower waist section revealed traces of red ochre. Late mesolithic/early neolithic level.

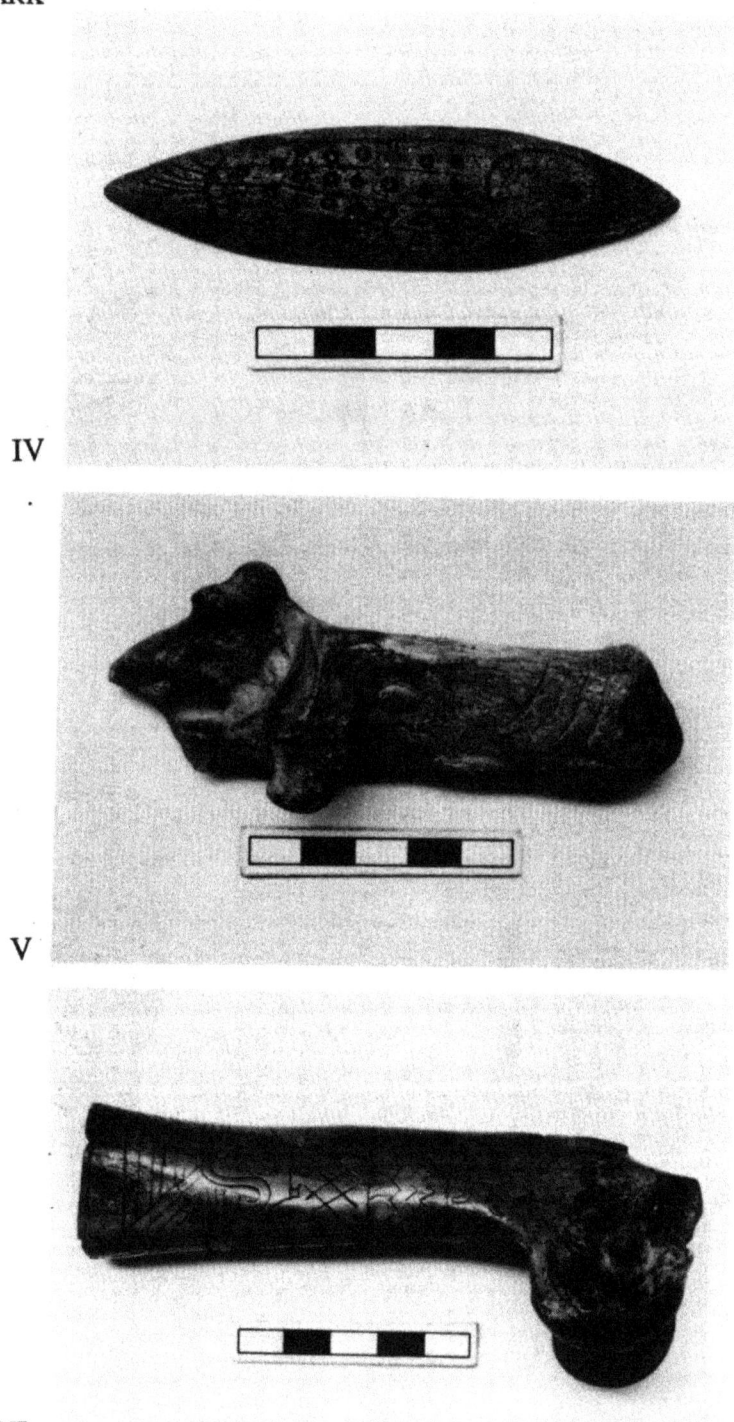

PLATES IV-VI Carved objects from the Riparo Gaban rock shelter, late mesolithic/early neolithic level (scales represent 50mm).

IV   Carved bone plaque of a fish, probably a pike.

V   Red deer calcaneum carved into the head of a red deer.

VI   Pig humerus with elaborate carved decoration. Motifs are very similar to those of the anthropomorphic figurine (Plate II).

FIGURE 4 Carved human femur from the late mesolithic/early neolithic level at the Riparo Gaban rock shelter.

Interpreted as a musical flute, but more likely of some ritual significance similar to Tsimshian shamans' 'soul catcher' as used in ritual cleansing by Northwest Coast Indians (cf. Drucker 1955).

Two additional neolithic sites, both dated to the middle Neolithic, provide further subsistence evidence. The site of Molino Casarotto, in the Berici hills south of Vicenza, has one of the biggest neolithic faunal assemblages in Europe. Practically 90% of these bones belonged to red deer and (in terms of bone size) wild pig. Using butchery evidence, I have interpreted Molino Casarotto as a specialised hunting site, located on the banks of a marshy lake. The site certainly suggests that hunting was still a major subsistence component of the middle Neolithic in northern Italy. This interpretation is supported by another middle neolithic site, that of Rivoli, located to the north of Verona on the Adige valley. Red deer and large (possibly wild) pig were also abundant species, but so too were cattle and caprines. This site was located on the spur of a steep-sided hill overlooking the Adige valley and is a site more suited as a prominent hunting stand. Most of the material was excavated from a series of pits that could have originated as storage pits. Unfortunately, settlement evidence in the form of hut remains is lacking, so the site represents something of an enigma.

The fact that hunting remained such an important component of subsistence in northern Italy suggests that the hunter-gatherer social structure continued into the Neolithic, with little of the cultural upheaval that is often associated with the transition to farming. The risks of adopting a new subsistence strategy, such as agriculture, would have been great (cf. Rindos 1984), and the maintenance of a hunting component within the subsistence system could be seen in terms of a risk management strategy. The fact that farming as a developed package arrived so late in northern Italy could be due to the high ecological diversity of the environment. The communities would have been under less pressure to adopt than in areas such as continental Europe, which contained a much more homogeneous environment in which early agriculturalists were probably more vulnerable to dietary failure and food stress. The agricultural anthropologist, Cancian (1980), has convincingly demonstrated that groups in poorer environments would have had more reason to adopt new subsistence strategies than groups living in more comfortable environments, such as a region like northern Italy. The variety of foods available to these people would have been great and who can blame them for resisting the monotony of a bread and mutton diet for so long?

1.7 SUMMARY

This paper has attempted to demonstrate that the understanding of the transition to farming can benefit from a longer time perspective than is normally used. This helps us to see various Holocene subsistence strategies, including farming, coastal exploitation and specialised mammal exploitations such as red deer hunting, as the result of long term adaptations to different environments. The need to integrate both social and ecological perspectives in such a study has always been acknowledged, but never satisfactorily applied. Although it is extremely difficult to discern social behaviour in such a sparse archaeological record, the application of risk management theory in terms of an emerging social hierarchy is a good beginning from which to integrate social and ecological theory, and it is hoped that this can be further developed in the future.

ACKNOWLEDGEMENTS

This paper derives from work currently being undertaken as part of a Ph.D. research programme and many ideas formulated in it are in the process of further refinement. I thank the British School at Rome for financing my fieldwork in Italy. A Richard Newitt Prize (awarded by the University of Southampton) is also gratefully acknowledged. My warm thanks go to many friends and colleagues in Italy who welcomed me to Trento and Vicenza, opened their museum stores and generally gave what the Italians give best, hospitality. These friends include Renato Perini, Lia and Annamaria Giovanazzi-Perini, Franco and Paola Marzatico, Michele Landzinger, Bernardino Bagolini and Antonio Dolago. Finally, I should like to thank Clive Gamble for commenting on an earlier version of this paper.

BIBLIOGRAPHY

AMMERMAN, A.J. AND CAVALLI-SFORZA, L.L. (1971) Measuring the rate of the spread of early farming in Europe. *Man* 6 674-88.

AMMERMAN, A.J. AND CAVALLI-SFORZA, L.L. (1973) A population model for the diffusion of early farming in Europe. In: *The explanation of cultural change.* (ed. Renfrew, A.C.) London: Duckworth pp. 343-57.

BAGOLINI, B. (1980a) *Il Trento Nella Preistoria Del Mondo Alpino.* Trento: Casa Editrice.

BAGOLINI, B. (1980b) *Introduzione al Neolitico.* Italy: Pordenone.

BAILEY, G.N. (ed.) (1983) *Hunter-gatherer economy in prehistory: A European perspective.* Cambridge: University Press.

BENDER, B. (1978) Gatherer-hunter to farmer: a social perspective. *World Archaeology* 10 204-22.

BENDER, B. (1981) Gatherer-hunter, intensification. In: *Economic archaeology: towards an integration of ecological and social approaches.* (eds Sheridan, A. and Bailey, G.) Oxford: British Archaeological Reports S96 pp. 149-57.

BINFORD, L.R. (1968) Post-pleistocene adaptations. In: *New perspectives in archaeology.* (eds Binford, S. and Binford, L.) Chicago: Aldine pp. 313-41.

BINFORD, L.R. (1978) *Nunamiut ethnoarchaeology.* New York: Academic Press.

BINFORD, L.R. (1981) *Bones: ancient men and modern myths.* New York: Academic Press.

BINFORD, L.R. (1983) *In pursuit of the past.* London: Thames and Hudson.

BÖKÖNYI, S. (1969) Archaeological problems and methods of recognizing animal domestication. In: *The domestication of plants and animals.* (eds Ucko, P. and Dimbleby, G.) London: Duckworth pp. 219-29.

BOSCATO, P. AND SALA, B. (1980) Dati paleontologici, paleoecologici e 3 depositi epipaleolitici in Valle dell' Adige (Trento). *Preistoria Alpina* 16 45-61.

BROGLIO, A. (1980) Culture e ambienti della fine del Paleolitico e del Mesolitico nell'Italia nord-orientale. *Preistoria Alpina* 16 7-29.

CANCIAN, F. (1980) Risk and uncertainty in agricultural decision making. In: *Agricultural decision making: anthropological contributions to rural development.* (ed. Barlett, P.F.) New York: Academic Press.

CHILDE, V.G. (1947) *The dawn of European civilisation.* (4th edition) London: Kegan Paul.

CLARK, R.H. (forthcoming) *Holocene adaptations and the beginning of agriculture: a case study from northern Italy.* Ph.D. thesis, University of Southampton.

COHEN, M.N. (1977) *The food crisis in prehistory.* New Haven: Yale University Press.

DALMERI, G. (1985) La Grotta d'Ernesto: un insediamento preistorico di grande interesse per la conoscenza del Paleolitico finale nell'area Trentino-Veneta. *Natura Alpini* **36** 31-39.

DENNELL, R.W. (1983) *European economic prehistory: a new approach.* London: Academic Press.

DENNELL, R.W. (1985) The hunter-gatherer/agricultural frontier in prehistoric temperate Europe. In: *The archaeology of frontiers and boundaries.* (eds Green, S. and Perlman, S.M.) New York: Academic Press pp. 113-39.

DUCOS, P. (1978) 'Domestication' defined and methodological approaches to its recognition in faunal assemblages. In: *Approaches to faunal analysis in the Middle East.* (eds Meadow, R.H. and Zeder, M.A.) Peabody Museum Bulletin 2. Harvard: Harvard University pp. 53-56.

DRUCKER, P. (1955) *Indians of the Northwest Coast.* USA: McGraw-Hill.

EVANS-PRITCHARD, E.E. (1940) *The Nuer.* Oxford: University Press.

FOLEY, R. (1985) Optimality theory in anthropology. *Man* **20** 222-42.

GAMBLE, C.S. (1983) Culture and society in the Upper Palaeolithic of Europe. In: *Hunter-gatherer economy in prehistory: a European perspective.* (ed. Bailey, G.) Cambridge: University Press pp. 201-11.

GAMBLE, C.S. (1986a) The Mesolithic sandwich: ecological approaches and the archaeological record of the early post-glacial. In: *Hunters in transition.* (ed. Zvelebil, M.) Cambridge: University Press pp. 33-42.

GAMBLE, C.S. (1986b) Hunter-gatherers and the origin of states. In: *States in history.* (ed. Hall, J.A.) Oxford: Blackwell pp. 22-47.

GEDDES, D.S. (1985) Mesolithic domestic sheep in west Mediterranean Europe. *Journal of Archaeological Science* **12** 23-48.

GOODY, J. (1976) *Production and reproduction: a comparative study in the domestic domain.* Cambridge: University Press.

GREIG, J. (1984) A preliminary report on the pollen diagrams and some macrofossil results from Palafitta Fiavé. In: *Scavi Archeologici nella zona Palafitticola di Fiavé-Carera: Parte 1.* (Perini, R.) Italy: Servizio Beni Culturali della Provincia Autonoma di Trento pp. 305-23.

HALSTEAD, P. AND O'SHEA, J. (1982) A friend in need is a friend indeed: social storage and the origins of social ranking. In: *Ranking, resource and exchange.* (eds Renfrew, C. and Shennan, S.J.) Cambridge: University Press pp. 92-99.

INGOLD, T. (1980) *Hunters, pastoralists and ranchers.* Cambridge: University Press.

JOHNSON, G. (1982) Organisational structure and scalar stress. In: *Theory and explanation in archaeology: the Southampton conference.* (eds Renfrew, A.C., Rowlands, M.J. and Abbott-Seagraves, B.) London: Academic Press pp. 389-421.

JONES, G. AND ROWLEY-CONWY, P. (1984) Plant remains from the north Italian lake dwellings of Fiavé (1400-1200 bc). In: *Scavi Archeologici nella zona Palafitticola di Fiavé-Carera: Parte 1.* (Perini, R.) Italy: Servizio Beni Culturali della Provincia Autonoma di Trento pp. 323-55.

LEWTHWAITE, J. (1986) The transition to food production: a Mediterranean perspective. In: *Hunters in transition.* (ed. Zvelebil, M.) Cambridge: University Press pp. 53-66.

MEILLASSOUX, C. (1972) From reproduction to production: marxist approach to economic anthropology. *Economy and Society* 1 93-105.

MEILLASSOUX, C. (1973) On the mode of production of the hunting band. In: *French perspectives in African studies.* (ed. Alexandre, P.) Oxford: University Press pp. 187-203.

MINNIS, P.E. (1985) *Social adaptation to food stress.* Chicago: University of Chicago Press.

MITHEN, S. (unpublished) *Influence, prestige and social evolution in Mesolithic society.* Paper presented at the Theoretical Archaeology Group symposium: Recent approaches to the concept of adaptation (1987 Bradford).

PRICE, T.D. AND BROWN J.A. (eds) (1985) *Prehistoric hunter-gatherers.* New York: Academic Press.

RINDOS, D. (1984) *The origins of agriculture: an evolutionary perspective.* New York: Academic Press.

ROWLEY-CONWY, P. (1983) Sedentary hunters: the Ertebølle example. In: *Hunter-gatherer economy in prehistory: a European perspective.* (ed. Bailey, G.N.) Cambridge: University Press pp. 111-26.

SAHLINS, M.D. (1972) *Stone Age economics.* London: Tavistock Publications.

TORRENCE, R. (1983) Time budgeting and hunter-gatherer technology. In: *Hunter-gatherer economy in prehistory: a European perspective.* (ed. Bailey, G.N.) Cambridge: University Press pp. 11-22.

TORRENCE, R. (unpublished) *Hunter-gatherer technology and the managment of risk.* A paper presented at the Man the Hunter conference (1986 London School of Economics).

WIESSNER, P. (1982) Beyond willow smoke and dogs' tails: a comment on Binford's analysis of hunter-gatherer settlement systems. *American Antiquity* **47** 171-80.

WINTERHALDER, B. AND SMITH, E.A. (eds) (1981) *Hunter-gatherer foraging strategies.* Chicago: University of Chicago Press.

ZVELEBIL, M. (1986) Mesolithic societies and the transition to farming: problems of time, scale and organisation. In: *Hunters in transition.* (ed. Zvelebil, M.) Cambridge: University Press pp. 167-88.

ZVELEBIL, M. AND ROWLEY-CONWY, P. (1984) Transition to farming in Northern Europe: a hunter-gatherer perspective. *Norwegian Archaeological Review* **17** 104-28.

ZVELEBIL, M. AND ROWLEY-CONWY, P. (1986) Foragers and farmers in Atlantic Europe. In: *Hunters in transition.* (ed. Zvelebil, M.) Cambridge: University Press pp. 67-93.

2

LIKE RISING DAMP? AN ECOLOGICAL APPROACH TO THE SPREAD OF FARMING IN SOUTH EAST AND CENTRAL EUROPE

Paul Halstead

ABSTRACT

This paper explores regional variability in the early farming economies of south east and central Europe. Two aspects of the contrasting regional environments of early farmers are emphasised. First, while village settlement enforced reliance on crops in south east Europe, in central Europe a dispersed settlement pattern allowed a more diverse subsistence base. Secondly, the introduction of cereal and pulse crops to central Europe, with its alien climatic regime of wet summers and cold winters, must initially have entailed a reduction in the reliability of yields and an increase in the spatial scale of failures.

The reduced reliability of crops was indeed offset by more intensive management of livestock (involving greater use of wood pasture and perhaps of dairy products) and by the exploitation of local opportunities for foraging. Equally important, however, was a change in emphasis from sharing among neighbours in south east Europe to more distant exchanges in central Europe, where local misfortunes were swamped by more widespread climatic hazards to crops.

The approach adopted is explicitly 'ecological' in recognising that both 'social' and 'economic' behaviour are integral to the survival strategies of early farmers, and that both the cultural and natural environment impose constraints upon these strategies. This offers a fuller understanding of early farming in Europe, and also demonstrates the dangers inherent in narrow palaeoeconomic or cultural interpretation of the ecofactual and artefactual record.

2.1 RISING DAMP

> 'These days, many prehistorians prefer ... to talk of the "spread of farming economy", rather than of farming peoples, to explain the origin of the European early Neolithic, but rarely specify how this spread was accomplished, or what it involved. Consequently, it is often hard to avoid the impression from much recent literature that early farming was like some kind of margarine or rising damp that eventually, and somewhat mysteriously, managed to cover most of Europe' (Dennell 1983: 155).

For much of this century, the origins of early European farmers, as of their crops and livestock, have been sought outside the continent, in the more advanced - and increasingly overcrowded - Near East (e.g. Childe 1957; Ammermann and Cavalli-

Sforza 1984). As 'the ghosts of land-hungry Asiatic farmers' begin to fade (Dennell 1983: 155), however, it is becoming increasingly clear that the inception of farming in Europe was a far from uniform process. The speed and thoroughness with which the new economy ousted the old varied both regionally and locally - as did the reasons for the change - and, in these circumstances, the distinction between immigrant farmers and native foragers may often have been as meaningless as it is archaeologically opaque (Clarke 1978; Dennell 1983; Zvelebil 1986a).

To talk of the spread of <u>farming economy</u>, however, is not simply to side-step the impenetrable 'whodunnit' problem of assigning an ethnic label to the first farmers, but rather to focus on another issue, equally important and far more soluble. Whatever their identity, the first European farmers faced a wide range of natural environments. How they coped with these diverse environments is a question of considerable interest, discussed here in the context of south east and central Europe - parts of the continent where the establishment of farming was apparently quite rapid (Zvelebil 1986b) and is relatively well documented archaeologically (e.g. Dennell in press).

2.2 AN ECOLOGICAL APPROACH

Though far from novel, the question of how early farming was adjusted to cope with the contrasting environments of south east and central Europe is not easily answered and poses some fundamental problems, both at the 'middle-range' level of reconstructing early farming economies and at the 'general theory' level of interpreting any inferred regional differences (cf. Binford 1977).

To begin with problems of reconstruction, some aspects of the farming economy (such as choice of crop species or site locations) may be inferred from the archaeological record relatively directly - some substantial taphonomic distortions notwithstanding. Farmers do not cope with their varied environments, however, simply by shuffling their repertoire of domesticates or shifting the location of their settlements. To address the question in hand, it is necessary to clarify the nature of early land use and economy in rather more detail. Was land use intensive or extensive, shifting or permanent? What was the relative importance of the arable and pastoral components of the economy? Was food production geared to productivity or security? Had farming really replaced foraging as the basis of subsistence? The reconstruction of patterns of land use or husbandry involves relatively complex modelling of human behaviour.

Local and regional differences in the behaviour of early farmers have rightly been stressed by several authors (e.g. Jarman, Bailey and Jarman 1982; Barker 1985), but the plethora of patterns of land use presented in recent studies may owe more to the divergent methods and assumptions of prehistorians than to the contrasting regional and local environments of neolithic Europe. For example, Barker's reconstruction of early agricultural land use in southern Yugoslavia emphasises the contribution to subsistence of pastoralism (Barker 1975: 95, but cf. 1985: 95), while Dennell's (1978) study of similar settlements in southern Bulgaria argues for a form of mixed farming, in which livestock were an essential but secondary complement to the arable sector. Barker relied primarily on off-site evidence (site catchment analysis), while Dennell was able to make greater use of on-site evidence (e.g. bones and seeds). In both cases, the available archaeological data were far from ideal, and it has also been suggested that Barker misinterpreted the date and arable potential of two critical soil types (Jarman, Bailey and Jarman 1982: 162). Arguably the most significant difference between the two studies, however, is that, given prevailing technology, Barker tacitly assumed the environment was used in the most productive manner possible (in terms of output per unit of land - Barker 1975), whereas Dennell assumed the most efficient use of the environment (in terms of output per unit of labour input - Dennell 1978: 100). Both viewpoints may well be valid in certain circumstances but, as *a priori* assumptions underpinning the reconstruction of past economies, they inevitably introduce circularity into subsequent consideration of how early farmers coped with environmental variability.

Turning to the interpretation of variation in early farming economies, the usual approach is to look for correlations between cultural and environmental variables. Thus the heavy reliance on ovicaprids by early farmers in southern Europe and the emphasis on cattle and/or pigs in central Europe are widely interpreted in terms of the contrast between Mediterranean and temperate climate or vegetation (e.g. Sterud 1978; Barker 1985). Conversely, Hubbard (1976) argues that the spatial distribution of early crops does not correspond to natural environmental parameters and so is governed by 'cultural preference'. Such suggested correlations are hard to evaluate, because several environmental gradients can be defined at a variety of spatial scales and with an almost infinite number of thresholds. In effect, the imaginative scholar should be able to find an environmental parameter to correspond spatially with almost any conceivable cultural variable, and *vice versa*. Clearly a more rigorous method is needed for defining relevant parameters and variables.

Some critics would go further, arguing that 'economic behaviour' is subordinate to 'social action' and that the natural environment is little more than a passive backdrop to human culture (cf. the opposing viewpoints in Sheridan and Bailey 1981). Such assumptions of social or cultural determinism also pose unavoidable problems of circular argument.

The approach followed here is adopted from evolutionary ecology (e.g. Ricklefs 1980). The environment of man is taken to include both a natural (physical and biotic) and cultural (social and material) component, and the problems of survival and reproduction posed for human communities by this natural/cultural environment are solved by a combination of economic and social behaviour. Where the various deterministic assumptions of alternative approaches serve to 'identify' relevant environmental problems and effective cultural solutions *a priori* (e.g. Higgs and Jarman 1975: 4), the present approach uses the ecological concept of the 'limiting factor' (Odum 1975: 107-8) to simplify the complexity of the real world in a manner more sensitive to particular environmental and historical circumstances.

This exercise is conducted at two temporal scales, the annual (a single 'average' year) and the interannual (a period of several years). At each scale, the social dimension of human behaviour is emphasised. At the annual scale, the size of residential groups - that is the size of the nutritional problem facing early farmers - is a basic element in the attempt to reconstruct patterns of food production and land use. At the interannual scale, patterns of residence and exchange are seen as integral components of the strategies adopted by early farmers to cope with the contrasting regional environments of south east and central Europe.

2.3 THE SPREAD OF FARMING IN SOUTH EAST AND CENTRAL EUROPE

The 'novel resources' basic to European farming spread through south east and central Europe in four broad stages (Champion, Gamble, Shennan and Whittle 1984: 117 Figure 5.5, 121 Figure 5.7), representing increasing chronological, geographical and cultural distance from Greece (Figure 1):

> (1) Greece - the proliferation of early neolithic communities, particularly in the eastern mainland, in the early sixth millennium bc.

> (2) The Balkans - the spread of Karanovo, Starcevo and Körös culture settlements through Bulgaria and Yugoslavia into south east Hungary in the latter sixth millennium bc.

> (3) Central Europe - the appearance of *Linearbandkeramik* (LBK) settlements across Hungary, Czechoslovakia, southern Poland, northern Austria, central Germany and south east Holland in the latter fifth millennium bc.

> (4) The Alpine Foreland - the appearance of Egolzwil, Older Cortaillod, Pfyn and Lutzengüetle settlements in southern Germany and northern Switzerland in the latter fourth millennium bc.

> (The extent to which these various communities were in fact supported by farming will be addressed in due course.)

Greece is a logical starting point for discussion not only because of its chronological primacy, but also because it resembles climatically (and may have been part of) the broad swathe of south west Asia in which European mixed farming developed (Higgs and Jarman 1969; Dennell 1983). Thus as farming spread out from Greece - particularly to the north - both domesticates and husbandry practices encountered a natural environment increasingly different from that to which they were already adapted.

More specifically, the climate of the Balkans and central Europe differs from that of Greece in being cooler, more continental and in having a summer rather than winter rainfall regime. Annual precipitation, on the other hand, varies relatively little with latitude (Wallén 1970, 1977). This means that, from south to north, winter cold progressively replaces summer drought as the principle climatic constraint on plant growth (e.g. Thran and Brockhuizen 1965). Livestock can adjust to such climatic differences rapidly, by virtue of their ability to move away from extreme conditions - into shelter and to alternative sources of food or water. Early cereal and pulse crops may have adjusted less easily to the progressively shorter growing season of the Balkans, central Europe and the Alpine Foreland, and to the related hazards of late frosts and destructively wet summers. Mediterranean crops commonly avoid the summer drought by early maturation, whereas winter hardiness in temperate Europe is conversely favoured by an extended vegetative period (Dantuma 1969: 134). Such changes in life cycle are likely to have been gradual.

These regional climatic trends are of course superimposed upon, and modified by, local variation in topography and soils, but local differences in the early farming economy lie outside the scope of this paper and, arguably, beyond the reach of present evidence.

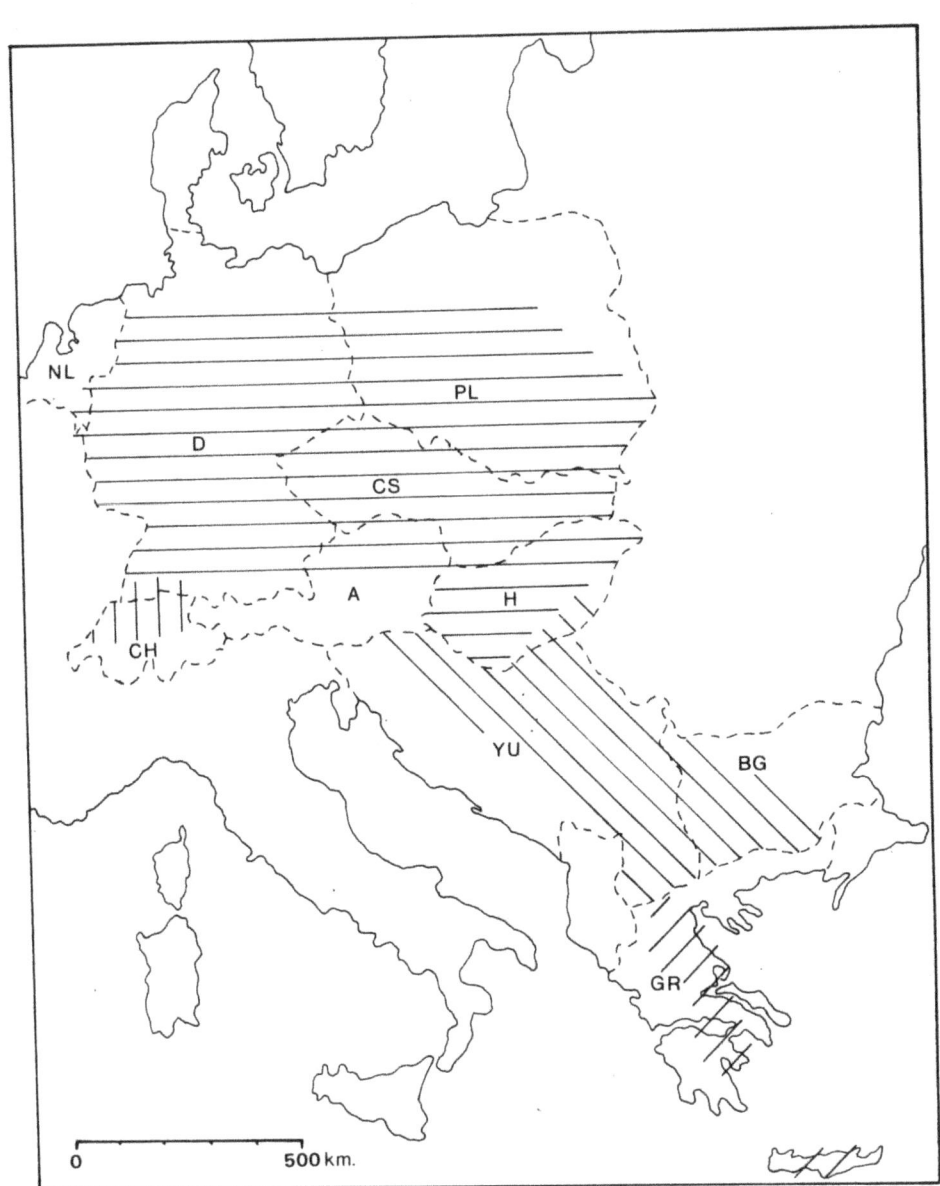

FIGURE 1 The spread of farming in south east and central Europe.

/// *circa* 6000 bc ≡ *circa* 4500 bc
\\\ *circa* 5500 bc ||| *circa* 3500 bc

GR Greece   BG Bulgaria   YU Yugoslavia
H Hungary CS Czechoslovakia PL Poland
A Austria D Germany   NL Holland
CH Switzerland

Anyway, by their locational decisions, early farmers in south east and central Europe rather minimised local variation, repeatedly 'selecting' patches of good (fertile but tractable) arable land with ready access to seasonal (particularly summer) pasture and with a favourable aspect (Jarman, Bailey and Jarman 1982).

A sharp contrast may be drawn at a regional level, however, between the relatively homogeneous landscape of the lowland basins of Greece, the southern Balkans and the LBK area, on the one hand, and the diverse environments of south east Hungary and the Alpine Foreland, on the other. The extensive seasonal wetlands of south east Hungary and the strongly accidented relief and numerous lakes of the Alpine Foreland will have offered a relatively rich variety and abundance of wild animal and plant resources (Clarke 1978).

Perhaps the most tangible feature of the natural environment, and that most directly relevant to early farming communities, was vegetation, and in this respect the parts of south east and central Europe occupied by early farmers may have been surprisingly uniform. Recent ecological and palynological research suggests that, prior to extensive interference from man and domestic animals, the fertile lowland basins of Greece were well endowed with deciduous woodland (Voliotis 1973; Horvat, Glavac and Ellenberg 1974; Bottema 1974, 1979; Greig and Turner 1974). Clearance by early farmers seems to have been neither rapid nor drastic (Bottema 1982), but the density of arboreal vegetation is problematic. Bottema (1974; van Zeist and Bottema 1982), for example, has argued that increased *Carpinus/Ostrya* pollen from the fifth millennium bc indicates higher rainfall and thicker woodland - implying arid conditions and thin woodland during the early Neolithic (sixth millennium bc). Alternatively, however, *Carpinus/Ostrya* pollen could be interpreted as an index of grazing and cutting pressure (cf. Turrill 1929), implying rather thicker woodland in the early Neolithic. Similarly, the steppe-like appearance of the Hungarian plain is now viewed as a relatively recent phenomenon (Pécsi and Jakucs 1971; Jarman, Bailey and Jarman 1982: 169). Despite local and regional differences in the composition, structure and density of vegetation, therefore, it seems clear that early farmers throughout the fertile lowland basins of south east and central Europe faced a more or less wooded landscape.

For the sake of simplicity, therefore, the ensuing discussion will concentrate on two aspects of regional variation in the natural environment - the initial climatic obstacles to crop production in temperate Europe and the relative wealth of wild food resources in south east Hungary and the Alpine fringe.

2.4 THE ANNUAL SCALE: GREECE

As the yardstick against which change in other regions will be measured, the early farming communities of Greece will be considered at some length. Within Greece, the early neolithic communities of Thessaly in the north of the country are particularly well documented archaeologically (Renfrew 1972; Theokharis 1973; Halstead 1984). Thessaly has a modified Mediterranean climate, abundant fertile soil and a network of more or less perennial watercourses. Deciduous trees grow well, where protected from man and livestock (Wace and Thompson 1912: 6-7). The early farmers built both mud-brick and post-frame houses, grouped in compact, permanent villages of *circa* 0.5-1.0ha which housed somewhere between 50 and 300 inhabitants. These villages were closely spaced (perhaps *circa* 2-4km apart) and habitation often continued on the same spot for several centuries or even millennia - long enough to form obvious settlement mounds or 'tells' (Halstead 1981a, 1984).

On an annual scale, these communities faced a variety of problems in the provision of shelter, clothing, fuel, water, and food of various types, but by far the largest and most demanding requirement - the limiting factor - was the dietary demand for energy (Halstead 1989). Because of their relatively large size, and the markedly seasonal climate of the region, there were few ways in which these village communities could meet their energy requirements - a point which may be appreciated by considering the viability of three alternative model subsistence strategies.

2.4.1 Foraging

Foraging (exploitation of naturally occurring resources) is poorly represented in on-site faunal and floral evidence, but there are acute taphonomic biases against the remains of fish, birds, small mammals and many edible plants. Until recently parts of Thessaly offered quite rich (but unreliable) seasonal opportunities for fowling and fishing (Yeoryiadhis 1880; Philippson 1950; Leake 1967) and in the distant past the same was probably true of hunting and gathering. Thessaly is not comparable with the Atlantic coast of Europe or the Pacific north west coast of America, however, and could not have supported large, permanent and closely spaced aggregations of foragers without radical modification of the natural environment (cf. Lee and DeVore 1968; Rowley-Conwy 1983). Even acorns, probably the most abundant native resource, could not have supported such villages without intensive management (Halstead 1984: 312). Cereals, pulses, sheep, goats, cattle and pigs offered far easier potential for food production on the necessary scale, and their remains predominate in excavated assemblages.

2.4.2 Pastoralism

Pastoralism, unlike foraging, supported quite a large population in Thessaly during the earlier part of this century (Sivignon 1975). Large transhumant flocks, principally of sheep, made extensive use of the lowland winter pasture (and of summer grazing in the surrounding mountains). They were intensively exploited for their secondary products - milk and, in earlier centuries, wool (Vergopoulos 1975). The recent extensive pastures are to a great extent the product of forest clearance, however, and (in the case of the lowlands) of a tenurial system (the '*chiflik*') which favoured the devotion of large areas to fallow grazed in winter (Halstead 1987a; Vergopoulos 1975). Moreover, recent pastoralists have subsisted only <u>indirectly</u> from their herds, selling animal produce in the market and buying cheaper agricultural staples (e.g. Campbell 1964).

Sheep were also the predominant livestock species in the early Neolithic, but mandibles of sheep and goat from excavations at early neolithic Prodhromos and late neolithic Ayia Sofia and Dhimini display the high levels of <u>juvenile</u> mortality (Halstead and Jones 1980; von den Driesch and Enderle 1976; Halstead 1987b) characteristic of Payne's (1973) meat production model (Figure 2). The marked <u>infant</u> mortality of Payne's milk production model might be concealed in an archaeological assemblage by poor preservation of fragile infant remains, but in neolithic Thessaly the high incidence of juvenile deaths leaves little demographic scope for further mortality among young animals. The evidence for cattle is sparse, but suggests a similar pattern (Halstead 1987b). This does not mean that meat only was produced, but that husbandry came nearer to optimising potential for meat than for milk (or wool or traction). Such a strategy is far less productive of energy per unit area than one geared to milk production (Legge 1981a). To support themselves primarily from pastoralism, therefore, the early neolithic villages of Thessaly would have needed vast flocks (perhaps 3000-18,000 sheep, or equivalent, per village - cf. Halstead 1981a: 314) and large, consolidated areas of pasture (in the order of 30-180km^2 per village; cf. le Houerou 1977) which would be hard to find even in the present open landscape - let alone in a wooded environment which obstructed grazing and herding, and made attacks by predators easy. In fact the predominance of sheep (though cattle, goats and pigs are all better suited to woodland) rather suggests that livestock were few in number.

2.4.3 Arable farming

Arable farming alone is sufficiently productive per unit area to have provided the bulk of the energy requirements of these village communities and a diet of cereal and pulse staples is compatible with available evidence for dental health in neolithic Thessaly (Poulianos 1966). Isolated caches of seed from early neolithic Sesklo (Tsountas 1908: 359) and Prodhromos (Halstead and Jones 1980) suggest separate cultivation (cf. Dennell 1976) of emmer (*Triticum dicoccum* Schübl.), pea (*Pisum sativum* L.), grass pea (*Lathyrus sativus* L.) and bitter vetch (*Vicia ervilia* (L.) Willd.). Finds elsewhere in Greece (Evans 1968; van Zeist and Bottema 1971) suggest a similar status for bread wheat (*T. aestivum* L.), einkorn (*T. monococcum* L.), six-row barley (*Hordeum vulgare* L.) and lentil (*Lens culinaris* Medik.).

Although large enough to preclude foraging and pastoralism as the basis of subsistence, communities of 50-300 people can be supported by cultivation on just a small scale (within 5-10 minutes walk from the settlement - cf. Halstead 1981a: 319 Table 11.3), and this in turn suggests the possibility of intensive husbandry (Halstead 1987a) quite unlike traditional Mediterranean agriculture (e.g. Grigg 1974: 125). For example, traction and pack animals are unnecessary in small-scale husbandry (Halstead 1987a) and are not positively attested (by faunal evidence from Pevkakia - Jordan 1975; Amberger 1979) until the Bronze Age, when the growth of large nucleated settlements made more extensive land use inevitable in some areas (Halstead 1987b). Likewise the traditional divorce between the predominant cereal crops, grown extensively, and the labour-intensive pulses, restricted to a few infield gardens, would have been an irrelevance and this may account for the relatively even representation of cereals and pulses on neolithic sites (Halstead 1984: Table 7.8). This 'horticultural' model could be further tested by examining the weed associations of early crops (an intensive form of land use has been suggested on this basis for a Bronze Age settlement in northern Greece - Jones in press).

Although crops must have been the main source of energy, the meat and perhaps milk of livestock will have introduced welcome and nutritionally valuable variety into the diet. Potentially useful secondary products included hides, wool, hair and of course manure which, probably in conjunction with cereal/pulse rotation, could have helped to maintain soil fertility. The concentration on sheep, which are particularly efficient at converting stubble and fallow fields to manure, may indicate that stock were closely integrated with arable farming.

Early neolithic settlement is less dense in most of the rest of Greece, but in other respects presents a similar picture to Thessaly. Village communities inhabited long-lived 'tell' settlements (Renfrew 1972), which have yielded only sparse traces of hunting or gathering. Cereals and pulses are more or less evenly represented among surviving crop seeds (Halstead 1981a: 316 Table 11.2), while sheep predominate among the domestic animals. Livestock mortality again conforms to the meat production model and the first evidence for work animals (faunal, artistic and archival) is from the Bronze Age (Halstead 1987b).

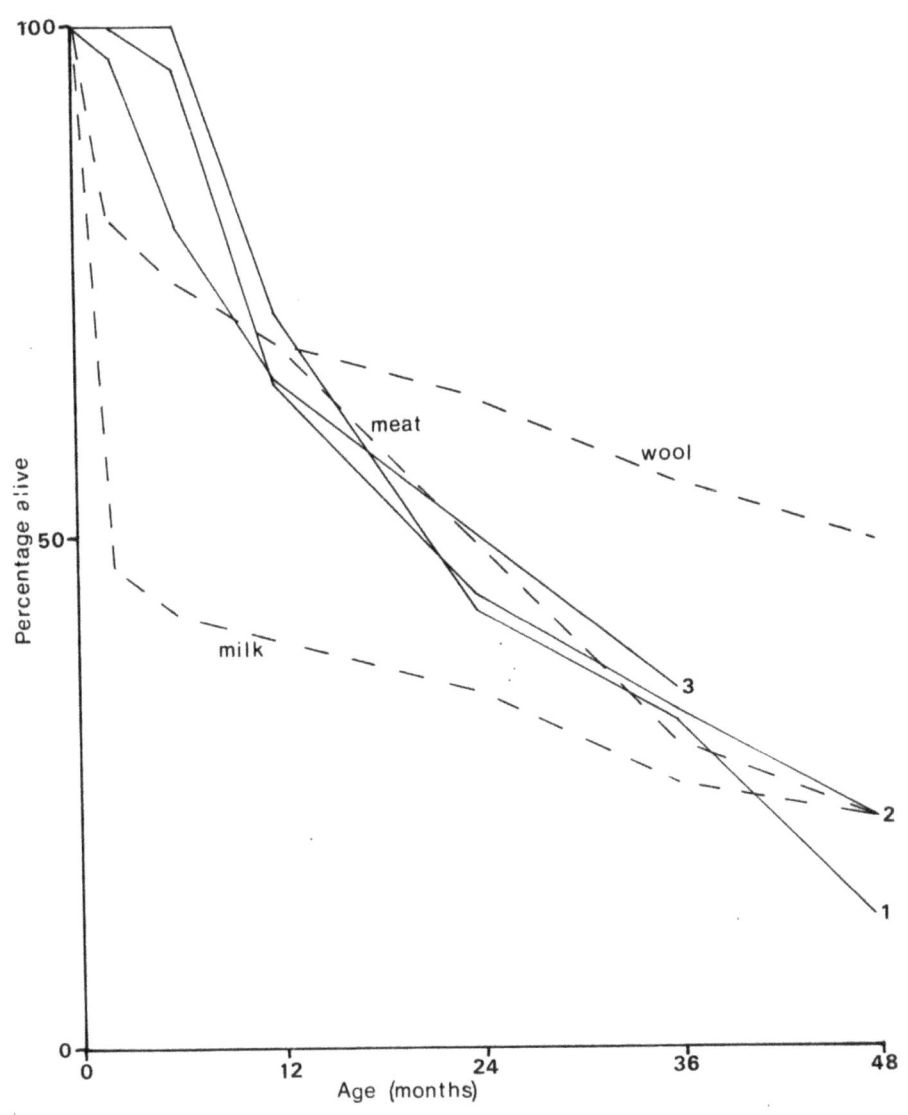

FIGURE 2 The survivorship of sheep/goat in neolithic Thessaly, Greece.

– – – – –   production models (Payne 1973)

──────   archaeozoological assemblages (each of 50+ mandibles)
  1. Early neolithic Prodhromos (Halstead and Jones 1980)
  2. Late neolithic Dhimini (Halstead 1987b)
  3. Late neolithic Ayia Sofia (von den Driesch and Enderle 1976)

N.B. Because of disparities of method, Ayia Sofia can only be compared with Payne's models up to 36 months.

Future research will doubtless document local variations in the relative contribution of hunted and gathered resources, livestock and crops to early neolithic subsistence. Arable farming must be regarded as the usual subsistence base, however, simply because the size of the early village communities rules out alternative strategies. Neither the population estimates nor the reconstruction of the natural environment suggested here are entirely unproblematic, but only drastic reappraisal would undermine this model of land use in any fundamental way.

2.5 THE ANNUAL SCALE: THE BALKANS

In the southern Balkans, Karanovo and Starcevo settlements are comparable in form, size and longevity with their Greek counterparts (e.g. Dennell and Webley 1975; Tringham 1971: 90; Whittle 1985: 50-52) but, further north, early 'tells' are rare and give way to settlements which are relatively short-lived or less concentrated on one spot (Tringham 1971: 91, 93; but cf. Kosse 1979: 129). This alternative settlement strategy is particularly clear in the Körös region of south east Hungary, on the northern margins of the first Balkan Neolithic. Here sites are often linear in form, stretching along the edge of the floodplain for up to two kilometres, with habitation occurring in discrete clusters *circa* 50m apart (Sherratt 1983). Scatters of cultural material between these concentrations may be cultivated areas subject to 'middening' or fertilising with domestic refuse. Whether these linear arrangements reflect drifting, isolated farmsteads and hamlets (of, say, up to 50 inhabitants) or longer-lived, 'exploded villages', they represent a more dispersed pattern of settlement than the compact villages of Greece and the southern Balkans.

In the southern Balkans, the compact village communities occupied an environment not dissimilar to that of northern mainland Greece and excavations have produced similar on-site evidence for subsistence. Both cereal and pulse cultigens are widely represented archaeobotanically (Renfrew 1979) and the status of emmer, six-row barley and common vetch (*Vicia sativa* L.) as separate crops has been established by detailed compositional and contextual analysis of samples from Chevdar and Kazanluk in Bulgaria - as has the casual, secondary use of some wild plant foods (Dennell 1976). In addition, separate cultivation of lentil and bitter vetch is indicated by pure caches of seed from Azmak, also in Bulgaria (Dennell 1978: 297 Table 105). In a large faunal assemblage from Anza in Yugoslavia (Bökönyi 1976) and smaller assemblages from other sites (e.g. Schwartz 1976; Dennell 1978), remains of wild animals are sparse; sheep are commoner than cattle, goats or pigs; and the available evidence for sheep at least suggests significant juvenile mortality and so husbandry geared primarily to 'meat' production (*pace* Barker 1985: 92). These communities must also have been primarily dependent on crops for subsistence and a pattern of small-scale, mixed farming again seems likely, with livestock perhaps as important for their manure as for their meat (Dennell 1978).

In the north Balkans, the more dispersed settlement pattern broadened the range of viable subsistence options. The seasonal wetlands of the Körös region also offered abundant opportunities for foraging; and the evidence for hunting, fishing, fowling and gathering, though varying greatly from site to site, is unusually rich for the south east European early Neolithic (e.g. Kosse 1979: 126-28). The same evidence, however, also demonstrates year-round occupation at several sites (Bökönyi 1972). Foraging may well have sustained sedentary habitation on the Danube, in the topographically constrained situation of the Iron Gates gorge (Bökönyi 1970), but in the Körös lowlands the extent of flooding and location of fowl, fish and game will have been highly variable and foraging cannot have been the primary subsistence base of closely spaced permanent settlements (cf. also Jarman, Bailey and Jarman 1982: 176-77).

In Körös faunal assemblages from Maroslele-Pana, Gyálarét, Röszke-Ludvár, Deszk-Olajkut and Ludas-Budzsák, livestock was again dominated by sheep (Bökönyi 1971, 1974: 22 Figures 1.1-1.3, 52 Figure 2.3) and so perhaps restricted to the areas of arable land (Kosse 1979: 130; but cf. Jarman, Bailey and Jarman 1982: 172). Age at death also seems consistent with husbandry for meat (Bökönyi 1971: 650). Arable farming remains, therefore, the most likely subsistence base and the evidence of middening again suggests small-scale, intensive husbandry. The harsher winters and summer rainfall regime of the north Balkans doubtless required some rescheduling of stock husbandry, but may have posed more serious problems in arable farming - both for the scheduling of cultivation and sowing and for the growth of crops. To some extent, both problems may have been mitigated by late sowing of millet (presumably *Panicum miliaceum* L., found in store at Nova - Kosse 1979: 128, 130, 149), in addition to cereals and pulses grown further south. Nonetheless some reduction is likely both in the area sown and in yields per unit area, as the advance of the first Balkan Neolithic ground to a halt in south east Hungary. Local opportunities for foraging may have played an important role in offsetting these problems with crop production.

2.6 THE ANNUAL SCALE: CENTRAL EUROPE

Until recently, the typical LBK settlement of central Europe was thought to have been a village, less compact than its south east European counterpart but with a broadly similar number of inhabitants (up to 150 - Soudsky and Pavlu 1972; Milisauskas 1978: 78; but cf. Kosse 1979: 133-34; Kruk 1980: 53). These estimates tend to assume that the larger LBK houses were occupied by several families - for which there is no positive evidence. Conversely, there are widespread architectural hints of some form of unidentified functional differentiation within the 'long-houses'. More problematic, however, is the very clear trend towards lateral settlement drift (e.g. Tringham 1971: 115; Milisauskas 1978: 99). The very extensive investigations on the Aldenhovener Platte in West Germany indicate gradual, long-term drift within a dispersed pattern of settlement (Kuper, Löhr, Lüning, Stehli and Zimmermann 1977; Hamond 1981) in farmsteads, hamlets or exploded villages. Thus nucleated early LBK 'villages' may be an artefact of archaeological research design - the product of excavating minute fractions of the settlement palimpsest.

Again, this dispersed pattern of settlement broadens the range of viable subsistence strategies and enhances the potential contribution of foraging and stock husbandry. Most LBK settlements are not well situated, however, to take advantage of wild resources, and the available archaeobotanical and archaeozoological evidence from excavations suggests only a minor role for foraging (e.g. Bakels 1978: 67, 77; Milisauskas 1978: 72-74 Table 5.8).

On the other hand, known sites are frequently located where they are ensured access to seasonal grazing as well as to arable land (Bakels 1978: 139, 145-46; Jarman, Bailey and Jarman 1982: 188). A contrast is afforded with south east Europe in the apparent predominance of cattle (and sometimes of pigs) over sheep in faunal assemblages from most areas of LBK settlement (Müller 1964; Clason 1968; Bökönyi 1974: 23 Figures 1.6-1.7; Milisauskas 1978: 72-74 Table 5.8; Bogucki 1982: 59 Tables 3-4, 82 Table 10, 1984: 22 Figure 3; Barker 1985: 142). Cattle are more enthusiastic browsers than sheep, can reach higher vegetation and, because of their larger body size, can tolerate a poorer quality diet (Bell 1971). The increased proportion of cattle, therefore, would have facilitated more extensive use of woodland, as well as of fallow and stubble grazing, and so permitted a greater contribution to the diet from the stock component of the economy. Unfortunately, the evidence for mortality among livestock is very limited (e.g. Bogucki 1982) and can neither confirm nor reject the suggestion, based on widespread finds of ceramic sieves suitable for cheese making (Bogucki 1984), of an upsurge in dairying.

In the absence of intensive exploitation of stock for their dairy products, crop agriculture would probably have been the major energy source and, in spite of earlier views to the contrary, this surely took the form of fixed-field rather than shifting cultivation (Willerding 1980; Rowley-Conwy 1981; Jarman, Bailey and Jarman 1982: 133-46; also Knörzer 1971; Groenman-van Waateringe 1970-71; but cf. Bakels 1978: 69). The weed species associated with LBK crop remains are quite different from those characteristic of recent cereal fields in central Europe. A number of factors may have contributed to this, including the retarded immigration of some weed species, the location of fields in small, shady clearings and the harvesting of the ears only of cereal crops (Knörzer 1971, 1979; Groenman-van Waateringe 1970-71; Bakels 1978: 68; Willerding 1980, 1981, 1983). It has also been suggested that early cereals were spring sown (e.g. Sherratt 1980; Barker 1985: 146; cf. Hillman 1981: 146-48), but at least some autumn sowing is implied by the occurrence of weeds such as *Bromus secalinus* L. and *Agrostemma githago* L. (Willerding 1980, 1983: 204-5; Hilbig and Lange 1981) in clear association with cereal crops and crop cleaning residues (Knörzer 1971; Willerding 1985: 95). A striking feature of the LBK weed flora, however, is the abundance of spring-germinating, nitrogen-loving 'Chenopodietea' species (Willerding 1980, 1983; Hilbig and Lange 1981) and this might again reflect a horticultural pattern of cereal growing, with relatively intensive manuring or middening and weeding or hoeing (Jones in press).

Evidence from more than one hundred sites includes only a few storage samples, indicating cultivation of emmer, einkorn, pea and linseed (*Linum usitatissimum* L.), but only emmer or an emmer/einkorn maslin (mixed crop) is ubiquitous (Willerding 1980, 1983). (*Bromus secalinus*, though very abundant in the Rhineland, appears to be concentrated in very weedy samples (Bakels 1978: 64-65 Figure 10) and so may just have been a waste product of crop cleaning.) The apparent narrowness of the crop repertoire, which contrasts with the diversity evident in just a handful of early neolithic samples from Greece (and, to a lesser extent, Bulgaria), will have exacerbated the unreliability of early arable farming in central Europe.

2.7 THE ANNUAL SCALE: THE ALPINE FORELAND

The evidence for early neolithic settlement in the northern Alps comes mainly from lake-side settlements, which are better preserved and have attracted more attention than their counterparts in river valleys and on higher ground. Habitation at sites away from the lakes may have been temporary, however, while lake-side sites were often demonstrably occupied both perennially and for a period of several years (Sakellaridis 1979). Most of these 'lake-villages' in fact consisted of just a few small buildings, some of which housed livestock rather than people, and so would more properly be termed 'hamlets' (Milisauskas 1978: 108; Barker 1985: 120, 123; Troels-Smith 1984).

Both this dispersed settlement pattern and lake-side habitation, offering access to seasonal grazing and to fish and fowl, enhanced the range of viable subsistence options, and the unusual organic preservation of these sites has provided a wealth of evidence for gathered plant foods (e.g. Jacomet 1987: 163) and for fishing and fowling gear (Wyss 1969: 125-26 Figures 7-8). At some sites wild mammals (particularly red deer) also account for a large proportion of the faunal assemblage (Higham 1967; Hartmann-Frick 1969: 30 Figure 15; Sakellaridis 1979; Becker 1981a; Jacomet and Schibler 1985). Of the domesticates, bones of cattle are usually most frequent, again perhaps reflecting intensive use of woodland browse, and the provision of leafy fodder for stall-feeding is documented by both pollen and macroscopic plant remains (e.g. Troels-Smith 1984). There is some evidence that both the proportion of sheep/goats (Higham 1967: 93) and the ratio of sheep to goats (Johansson 1981: 99 Figure 123) declined through the early Neolithic, which may indicate that the exploitation of woodland by cattle, pigs and goats expanded gradually.

A further indication of intensive stock husbandry is the mandibular evidence from early settlements in Switzerland for up to 40% mortality by *circa* 6 months of age among sheep/goats (Figure 3) and of up to 50% by *circa* 6-9 months among cattle (Figure 4); postcranial evidence suggests that most of the animals so culled were male (Higham 1967, 1968a and b). Higham interpreted this pattern of mortality as indicating a strategy of meat production which capitalised on the rapid growth of very young livestock to make the most effective use of the very scarce pasture available in a thickly wooded environment (Higham 1967: 93). Such a strategy is both labour intensive and risky, because of its reliance on the vulnerable categories of breeding females and their young offspring (Redding 1981; Haresign 1983), and yet would have yielded only modest amounts of meat if fodder was indeed scarce. More recently, Legge has argued that, in the case of cattle, this pattern of slaughter does not reflect the 'extravagant economics of beef production', but husbandry for dairy products (Legge 1981a: 89; also Jacomet and Schibler 1985: 129). Sheep/goat mortality falls between Payne's milk and meat models (Payne 1973), even though the archaeological assemblages fail to register the inevitable natural infant mortality (estimated at *circa* 20% by Payne). The slaughter of young lambs/kids is particularly sharp (and correspondence with the milk model closest) in the relatively early assemblages from St. Aubin IV (older Cortaillod) and Twann (classic Cortaillod), though the same is not true for cattle. Whether indicative of dairying or not, however, this high turn-over pattern of mortality coupled with stall-feeding constitutes an intensive form of livestock management which contrasts strongly with the less intensive and lower risk 'meat' strategy of early sites in Greece.

While lake-side settlement brings clear advantages for foraging and stock husbandry, it may severely constrain the availability of arable land (e.g. Jarman, Bailey and Jarman 1982: 124-25 Figures 42-43). This, coupled with the unreliable nature of early crops in the face of poor soils and a hostile climate, has led Sakellaridis to posit only a secondary subsistence role for crop production (1979: 184, 170). On the other hand, only modest areas of arable land may have been needed by the small early neolithic communities. Moreover the larger lakes have a strong ameliorating effect on the climate in their immediate vicinity, so that crop production at lake-side settlements is to a great extent cushioned from the severe problems posed elsewhere in the Alpine lowlands by a very short growing season (Higham 1967: 91; Sakellaridis 1979: 133; Jarman, Bailey and Jarman 1982: 121 Figure 41). Both Sakellaridis (1979: 164) and Jarman, Bailey and Jarman (1982: 122) agree that this favourable, arable micro-environment was strongly selected by neolithic settlement and both charred and waterlogged remains of introduced cereals, pulses and oil-plants have been found in abundance. Pfyn and Cortaillod stores of free-threshing wheat (*Triticum* cf. *aestivum* L./ *turgidum* L./ *durum* Desf.), emmer, barley and perhaps einkorn have been published

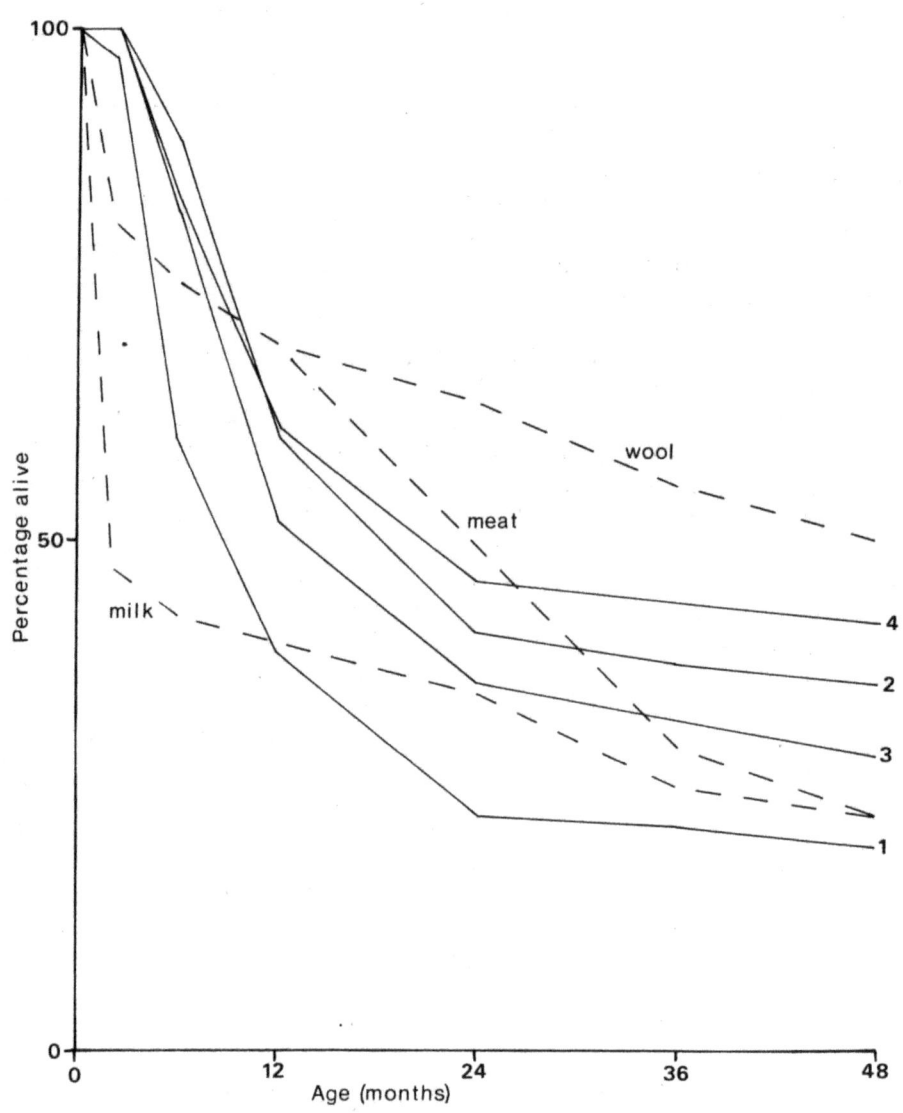

FIGURE 3 The survivorship of sheep/goat in the neolithic Alpine Foreland.

— — — —   production models (Payne 1973)

————————   archaeozoological assemblages (each of 50+ mandibles)
1. Older Cortaillod St. Aubin IV (Higham 1967)
2. Younger Cortaillod Egolzwil 2 (Higham 1967)
3. Classic Cortaillod Twann (Becker 1981a)
4. Younger Cortaillod Twann (Becker 1981b)

N.B. Because of disparities of method, the archaeozoological data can only be compared with Payne's models up to 48 months and, for Twann, the period 24-48 months is treated as one unit.

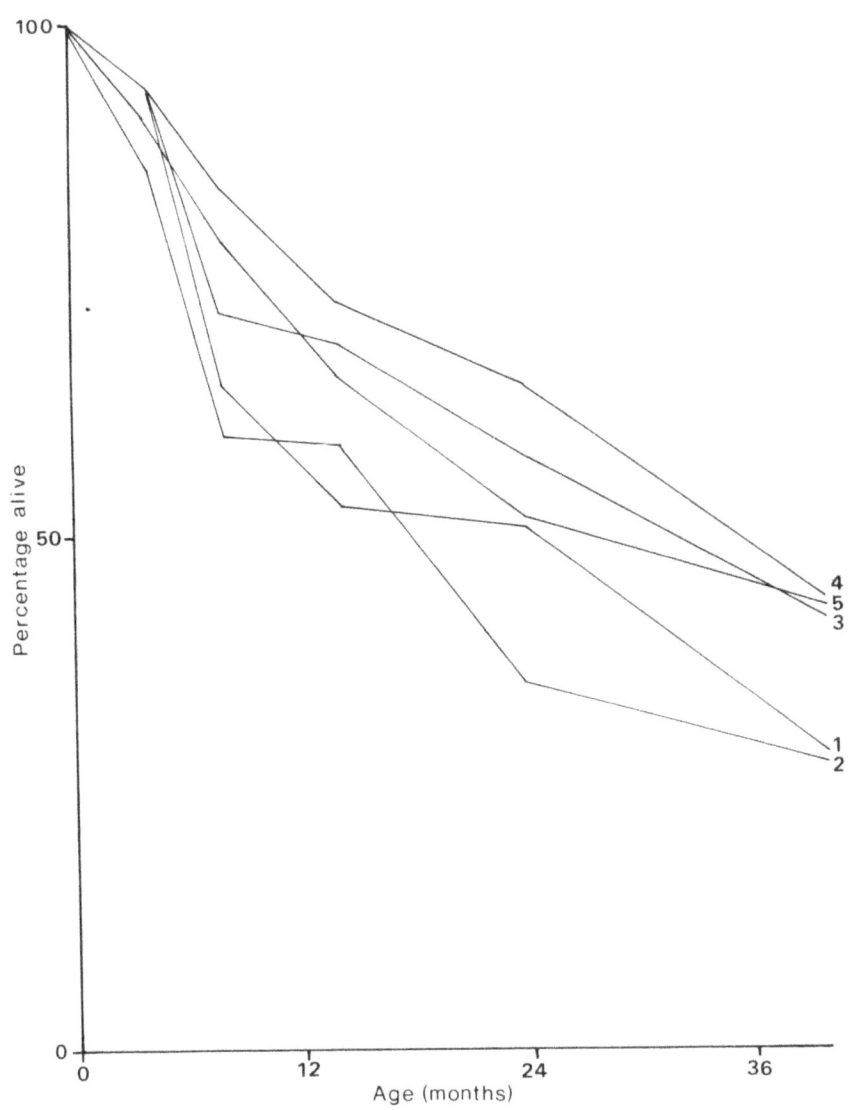

FIGURE 4 The survivorship of cattle in the neolithic Alpine Foreland.

──────── archaeological assemblages (each of 50+ mandibles)
1. Older Cortaillod St. Aubin IV (Higham 1967)
2. Younger Cortaillod Seematte-Gelfingen (Higham 1967)
3. Younger Cortaillod Egolzwil 2 (Higham 1967)
4. Younger Cortaillod Twann (Becker 1981b)
5. Pfyn Zürich (Jacomet and Schibler 1985)

N.B. Because of disparities of method, St. Aubin IV, Seematte-Gelfingen and Egolzwil 2 can only be compared with Twann and Zürich up to 40 months (and only approximately from 24 to 40 months).

recently (Jacomet 1987: 148 Table 31, 150 Table 32; Jacomet and Schlichtherle 1984: 156, 167) and earlier finds of variable contextual quality apparently include similar samples of poppy (*Papaver somniferum* L.), pea and lentil (Sakellaridis 1979: 344-45 Table 22, 358-59 Table 26, 373 Table 32, 394 Table 45). Cultivation of linseed/flax is indicated by processing waste and scraps of linen cloth (Jacomet and Schibler 1985: 133).

In the case of the lake-villages, pollen is often recovered from the immediate vicinity of the settlement, and so presumably of the cultivated area, so that the paucity of neolithic evidence for forest clearance or permanent pasture argues strongly against extensive, shifting cultivation (Sakellaridis 1979). The weeds found with early crops, which again paradoxically include species indicative of autumn sowing (e.g. *Vicia hirsuta* (L.) S.F. Gray, *V. tetrasperma* (L.) Schreber (Brombacher and Dick 1987: 209; cf. Willerding 1985: 95)) as well as a high proportion of 'Chenopodietea' (Ellenberg 1978: 812-13; Jacomet-Engel 1980: 135-40; Brombacher and Dick 1987: 209 Tables 57-58; Jacomet 1987: 149, 157-59), hint at an intensive, horticultural pattern of cultivation. Moreover, next to the Pfyn site of Thayngen Weier, there are traces of a field plot fertilised with domestic refuse and stall manure (Troels-Smith 1984).

Thus crop production, however unreliable, clearly played a significant part in the subsistence of early neolithic communities in the Alpine Foreland. In this ecologically diverse region, the dispersed nature of human settlement enabled wild resources and domestic stock at least to supplement the arable economy. The relative contribution to subsistence of foraging, stock rearing and crop production remains uncertain, but evidently varied regionally, locally and through time.

2.8 THE ANNUAL SCALE: SUMMARY

Clearly, the available evidence for subsistence in the four regions considered varies in both quality and quantity. Nonetheless, a contrast is apparent between the overwhelming dependence on cereal and pulse crops in Greece and the southern Balkans, and the indications of a greater reliance on livestock or foraging in temperate latitudes. This diversification of the subsistence base in the north Balkans, central Europe and the Alpine Foreland arguably reflects the initial unreliability of introduced cereals and pulses.

Three fundamental issues have been avoided thus far. First, the reasons for the observed switch from village to dispersed settlement, so crucial at least in enabling subsistence change, have not been considered. Secondly, the extent of interaction and interdependence within and between residential groups has been ignored. This simplification may be defended as an initial step in analysis, on the grounds that architectural evidence suggests some form of family household as the usual, basic unit of production and consumption in the cultures under consideration here (Champion, Gamble, Shennan and Whittle 1984: 140-42; Halstead 1989; Vogt 1969; cf. Flannery 1972). Thirdly, subsistence has been modelled in terms of the relatively simple problem of securing food in an 'average' year, disregarding the problem of year-to-year variation in food supply and the risk in particular of crop failure. These three related issues are discussed in the following section.

2.9 THE INTERANNUAL SCALE: GREECE

Fluctuations in production targets, labour supply and yields will all have posed problems for early crop production in Greece (Forbes 1976, 1989; Halstead 1981b, 1989). Some potential buffering mechanisms can be recognised archaeologically, including the growing of a range of staple crops and the rearing of livestock to a 'meat' strategy which enhances herd stability rather than productivity (Ellis, Jennings and Swift 1979; Redding 1981). Regular overproduction and storage of the resultant 'normal surplus' (Allan 1965), though not archaeologically detectable, can reasonably be viewed as a minimum requirement for the survival of early farming households (Halstead 1989).

In spite of these precautions, however, shortages must frequently have afflicted individual households and, from time to time, whole villages or groups of villages. In such circumstances, isolated households are only viable in the short term (Sahlins 1974) and, in neolithic Thessaly, one tangible indication of mutual help between households is the location of cooking facilities outdoors in the open yards between houses (Theokharis 1980) - an arrangement inviting the sharing at least of <u>cooked</u> food between neighbours. A further reflection perhaps of the importance of sharing and hospitality is the considerable time and skill invested in the manufacture of fine, painted pottery vessels suitable for serving and consuming food and drink (e.g. Theokharis 1973; Kotsakis 1983), and the vessels themselves may have been used as 'social storage tokens' (O'Shea 1981) in exchanges of food between less closely related households. Above all, the remarkable longevity of many early farming villages, when periodic shortage threatened the survival of individual households, clearly implies mutual help among neighbours in times of need.

In other parts of the world, village settlement has recently been maintained by the vulnerability of smaller communities to raiding (e.g. Forge 1972), but the close spacing and longevity of Thessalian sites suggests peaceful coexistence rather than endemic warfare between neighbouring villages. In fact, although most fine pottery was apparently produced and used on a very local basis, visual inspection (confirmed at least for the later Neolithic by less subjective techniques - Liritzis and Dixon 1984) indicates that a few pieces were exchanged between villages.

Visits to more distant friends or kin may also have been of vital importance on the rare occasions when drought or some other climatic extreme caused wholesale failure in one locality (cf. Richards 1939), and again there are material hints of such interaction. Obsidian from the island of Melos is widely distributed at least in coastal areas of the Aegean (Renfrew 1973). Similarities of ceramic style emphasise social interaction over rather more modest distances (up to 50-100 kilometres - e.g. French 1972; Cullen 1984; Rondiri 1985), but still on a scale far in excess of that necessary just to maintain a viable breeding population (cf. Wobst 1974).

2.10 THE INTERANNUAL SCALE: THE BALKANS, CENTRAL EUROPE AND THE ALPINE FORELAND

The southern Balkans have essentially the same winter rainfall regime as Greece and, in consequence, a similar pattern of agricultural risk and risk buffering. A range of crops was grown, livestock was probably managed for 'meat' and security, and regular overproduction of staple crops may again be assumed. External hearths (Tringham 1971: 87; Barker 1985: 91) and regional styles of fine painted pottery (Nandris 1970) again confirm the importance of assistance between neighbouring households in maintaining the solidarity and viability of long-lived village communities.

Further north, a diversified subsistence base is still apparent, for example in the nearly universal occurrence of the full complement of early livestock species (cow, sheep, goat and pig), but there is evidence of some reduction in the range of crops grown, at least by LBK farmers in central Europe. To some extent, diversity in the farming economy may have been maintained by increased emphasis on stock rearing, but the intensive, high turn-over pattern of exploitation in the Alpine region will seriously have reduced the reliability of livestock as a 'bank' of food for bad years. In south east Hungary and in the Alpine area, wild resources also made an important contribution to subsistence and the fluctuating ratio of wild to domestic animal remains through the Cortaillod period at Twann (Becker .1981a: 36 Figure 53) may reflect a greater reliance on hunting in periods of poor returns from crops and livestock.

An interesting feature of the introduction of cereal and pulse crops to the increasingly alien natural environments of temperate Europe is a radical shift in the _nature_ of agricultural risk. In Greece and the southern Balkans, early farmers selected patches of fertile soil (Jarman, Bailey and Jarman 1982: 146-52; Dennell and Webley 1975) and avoided areas of low and unreliable rainfall (Halstead 1981b) within a climatic regime to which crops were already adapted. Thus conditions were basically very favourable for the growth of cereals and pulses and problems afflicting single households (e.g. illness of a key member of the family labour force, destruction of a garden by animals or spoiling of stored grain) will have loomed large in overall agricultural risk. In these circumstances, help from near neighbours can be very effective and the observed pattern of village settlement will have brought considerable benefits.

In temperate Europe, winter cold and excessive summer rainfall replaced occasional summer drought as the principal hazards to a successful harvest and, until the development of crop strains adapted to these new conditions, losses afflicting individual households will initially have been swamped by more serious and more widespread failures caused by unfavourable weather conditions. In these circumstances, near neighbours can offer little support in bad years in return for the considerable constraints which village settlement imposes even on normal subsistence options.

Thus the shift in settlement pattern in temperate Europe must be seen in terms not only of the advantages of dispersed settlement for intensive stock husbandry and foraging, but also of the ineffectiveness of help from near neighbours as a buffer against widespread crop failure. The declining value of neighbours is confirmed by two further changes which accompany the dissolution of the village. First, fine painted pottery becomes rare in the north Starcevo and Körös areas (Tringham 1971: 79) and then disappears in the LBK and alpine regions (Tringham 1971: 121), presumably reflecting the declining importance of local sharing and hospitality. Secondly, there is a sharp decline in the stability of settlement: sites in the north Balkans, central Europe and the Alpine Foreland are characterised by relatively short-lived or drifting occupation, suggesting that the magnet for continuous occupation of the southern 'tell' sites was access to neighbours rather than to particular resource patches or to favoured spots for habitation.

Of course the corollary of declining support from neighbours was increased reliance on household-level buffering mechanisms or on more distant bonds of friendship or kinship. Household-level buffering is suggested by the increased emphasis on wild resources in south east Hungary and in the Alpine region, where the ratio of wild to domestic mammals at Cortaillod Twann varies between different parts of the settlement as well as through time (Becker 1981a: 34 Figures 48-49). The latter response will have been effective over relatively modest distances in the mountainous terrain of the Alpine Foreland (cf. Halstead 1981b), where material culture again displays marked regional diversity (Sauter and Gallay 1969; Drack 1969). In the LBK region, however, foraging was of little importance and the relative homogeneity of terrain and climate will have required very far-flung social relations as an effective buffer against scarcity. Significantly, the most distinctive feature of early LBK material culture is its extraordinary homogeneity from Hungary to Holland: pottery, for example, displays a striking uniformity of 'fabric, forms and decoration throughout the whole area' (Tringham 1971: 125; also Milisauskas 1978: 55). More subjectively perhaps, the LBK contrasts with the regionally heterogeneous cultures of the Balkans and Greece in the shift in emphasis from local production of fine painted pottery to long-distance exchange of a plethora of stone and shell objects (e.g. Sherratt 1976; Milisauskas 1978: 89 Figure 5.2; Barker 1985: 146), including the spondylus shell bracelets distributed throughout the LBK region, perhaps from sources in Greece (Shackleton and Renfrew 1970). This is not to suggest that a victim of crop failure in Holland would travel to Greece to seek help, but that LBK material culture is the product of an overlapping network of longer distance social interactions than are its more regionalised counterparts in Greece and the southern Balkans.

2.11 CONCLUSIONS

This review has traversed some fairly boggy archaeological ground, but it is clear that the damp of early farming did not rise through south east and central Europe unchanged. In Greece and the southern Balkans, the basis of subsistence was arable farming. Stock, managed to a low turn-over 'meat' strategy, will have been a reliable 'bank' of alternative food in the event of crop failure, but the dominance of sheep suggests that this livestock bank was effectively embedded in arable production. Such modest investment in stock husbandry underlines the relative reliability of early crops in this region. Delays in the spread of farming to central Europe and to the Alpine Foreland (see the sinuous distribution of early dates around the steady 'wave of advance' in Ammermann and Cavalli-Sforza 1984: 53 Figure 4.2) apparently correspond with significant readjustments - the former with increased use of woodland by cattle, the latter with more intensive stock husbandry (perhaps involving dairying) and foraging.

The trend towards increasing dependence on livestock seems to continue in Atlantic (particularly north west) Europe. Here early crops encountered the novel problem of cool, wet conditions during the summer season of ripening and harvest, and the relative abundance of nuts and fruits in the archaeobotanical record from southern England perhaps hints at a restricted role for annual crops (Moffett *et al.* this volume). Conversely, the mild winters and wet summers of western Europe ensure a perennial growing season unusually favourable for grazing animals, and the profusion of early monuments implies large tracts of open country unlikely to have been maintained by small scale arable farming. An emphasis on the livestock component of the mixed farming economy seems probable in this region (Fleming 1972: 186-87), and in Britain there are again archaeozoological hints of dairying (Legge 1981b).

That the flexible balance between crop and stock husbandry is central to the rapid dispersal of mixed farming in Europe is reinforced by comparison with North America. Here crops were not accompanied by livestock and farming spread more readily to the

Central Plains and Great Lakes, areas with a high risk of crop failure but abundant opportunities for hunting or fishing as an alternative, than to parts of the eastern United States, where crops were more reliable but alternative wild resources sparse (O'Shea 1989).

The spread of farming into temperate Europe was also marked by a shift from village settlement, with an emphasis on help between near neighbours, to dispersed settlement, with an expansion of long distance exchange. Interestingly, these radical changes in social behaviour are paralleled in the belated agricultural colonisation of marginal, arid and mountainous parts of Greece in the fifth to third and second to first millennia bc respectively. In both these latter cases, the further spread of farming was accompanied by the proliferation of small, dispersed settlements (Cherry 1979; Whitelaw 1983; Halstead 1981b, 1989, in press) and by the explosion of regional interaction (Renfrew 1972: 453 Figure 20.5; Kilian 1972, 1973) - as well as by some hints of greater reliance on hunting or stock husbandry (Halstead 1987b).

Clearly alterations in patterns of residence and social interaction are as fundamental to the spread of farming in Europe as are shifts in the balance between foraging, stock rearing and crop husbandry, adjustments to the array of domesticates or changes in the ways these are managed. Indeed the development of regional exchange systems, capable of securing communities against crop failure, may be an essential precondition of the spread of farming into some agriculturally marginal parts of Europe - or into parts of North America favourable to crop production but with only sparse opportunities for hunting and fishing (O'Shea 1989).

This review began with the question of how early farming was adjusted to cope with the contrasting environments of south east and central Europe, but its implications are rather wider. First, consideration of the structure (and, in particular, of the weaknesses) of the farming economy in different regions has led to re-evaluation of many aspects of early neolithic culture. In south east Europe, for example, village settlement, painted pottery and the predominance of sheep can no longer be viewed simply as cultural baggage inherited from the Near East. Incidentally, if patterns of early neolithic material culture are intimately related to the nature and scale of agricultural risk and risk buffering, attempts to determine the pedigree of early farmers on conventional archaeological evidence (e.g. Ammermann and Cavalli-Sforza 1984: 62) must be suspect.

Secondly, it is interesting to consider the likely effects of the development, during the course of the Neolithic and Bronze Age, of crop varieties better adapted to a temperate climate. This process cannot at present be observed directly, but increased reliability of staple crops may well be indirectly reflected in the belated appearance in temperate Europe of certain features characteristic of early farming culture in south east Europe. For example, 'tell' (i.e. long-lived village) settlement spreads to the north Balkans during the later Neolithic (Kosse 1979: 134) and there is a switch from the high turn-over 'milk' to a low turn-over 'meat' strategy of stock rearing in the Alpine Foreland by the Bronze Age (Higham 1967; Legge 1981a: 88 Figure 52).

Thirdly, the varying emphasis on arable farming, stock rearing or long-distance exchange in different parts of Europe may have contributed to the divergent regional trajectories of social change. In south east Europe, because of the importance of very local subsistence failure and exchange, variable access to the most productive arable land may always have been a powerful vehicle for social differentiation (Halstead 1981b, 1988), whereas in central Europe access to exotic exchange items may initially have been more important (cf. Shennan 1986). Similarly, competition for land is unlikely in south east or central Europe, where subsistence was dominated by small-scale arable farming, but may partly account for the profusion of early monuments in north west Europe, because of the spatially extensive nature of stock rearing (cf. Renfrew 1976; Hodder 1984). These complex issues, however, are another story.

Arguably more important than detailed arguments about the significance of village residence, dairying or painted pottery production is the overall approach adopted here, which has attempted to avoid the circular extremes of either environmental or cultural determinism. First, the size of residential groups has been as basic an ingredient as climate or vegetation in attempts to reconstruct patterns of subsistence and land use. Secondly, consideration of the weak link(s) in the strategies of early farming households has served as a basis for exploring interaction and interdependence within and between local communities, in a framework broader and more secure than that afforded by appeals to cultural rules of residence or kinship (cf. Hodder 1982). Finally, no dominant role is ascribed to either economic or social behaviour, but rather both are viewed as integral components of the survival strategies of early farmers.

ACKNOWLEDGEMENTS

Thanks to Glynis Jones and Marek Zvelebil for comments on an earlier draft of this paper, to the Sheffield undergraduates who have endured its evolution in lectures without complaint and again to Glynis Jones for drawing the figures. As referee, Geoff Bailey expressed his misgivings gently and without the cloak of anonymity. My apologies to all the above for not always heeding their advice.

BIBLIOGRAPHY

ALLAN, W. (1965) *The African husbandman*. Edinburgh: Oliver and Boyd.

AMBERGER, K.-P. (1979) *Neue Tierknochenfunde aus der Magula Pevkakia in Thessalien, 2: die Wiederkäuer*. Dissertation, University of Munich.

AMMERMANN, A.J. AND CAVALLI-SFORZA, L.L. (1984) *The neolithic transition and the genetics of populations in Europe*. Princeton: University Press.

BAKELS, C.C. (1978) *Four linearbandkeramik settlements and their environment: a paleoecological study of Sittard, Stein, Elsloo and Hienheim*. Leiden: University Press (also published as *Analecta Praehistorica Leidensia 11*).

BARKER, G. (1975) Early neolithic land use in Yugoslavia. *Proceedings of the Prehistoric Society* 41 85-104.

BARKER, G. (1985) *Prehistoric farming in Europe*. Cambridge: University Press.

BECKER, C. (1981a) *Die neolithischen Ufersiedlungen von Twann, 16: Tierknochenfunde, 3: unteres Schichtpaket (US) der Cortaillod-Kultur*. Bern: Staatlicher Lehrmittelverlag.

BECKER, C. (1981b) Besprechung der Tierarten, 1: Säugetiere. In: *Die neolithischen Ufersiedlungen von Twann, 11: Tierknochenfunde, 2: mittleres und oberes Schichtpaket (MS und OS) der Cortaillod-Kultur* (Becker, C. and Johansson, F.) Bern: Staatlicher Lehrmittelverlag pp. 35-77.

BELL, R.H.V. (1971) A grazing ecosystem in the Serengeti. *Scientific American* **225** 86-93.

BINFORD, L.R. (1977) General introduction. In: *For theory building in archaeology.* (ed. Binford, L.R.) New York: Academic Press pp. 1-10.

BOGUCKI, P.I. (1982) *Early neolithic subsistence and settlement in the Polish lowlands.* Oxford:. British Archaeological Reports S150.

BOGUCKI, P.I. (1984) Ceramic sieves of the Linear Pottery culture and their economic implications. *Oxford Journal of Archaeology* **3** (1) 15-30.

BÖKÖNYI, S. (1970) Animal remains from Lepenski Vir. *Science* **167** 1702-4.

BÖKÖNYI, S. (1971) The development and history of domestic animals in Hungary: the Neolithic through to Middle Ages. *American Anthropologist* **73** 640-74.

BÖKÖNYI, S. (1972) Zoological evidence for seasonal or permanent occupation of prehistoric settlements. In: *Man, settlement and urbanism.* (eds Ucko, P.J., Tringham, R. and Dimbleby, G.W.) London: Duckworth pp. 121-26.

BÖKÖNYI, S. (1974) *History of domestic mammals in central and eastern Europe.* Budapest: Akadémai Kiadó.

BÖKÖNYI, S. (1976) The vertebrate fauna from Anza. In: *Neolithic Macedonia.* (ed. Gimbutas, M.) Los Angeles: Institute of Archaeology, University of California pp. 313-63.

BOTTEMA, S. (1974) *Late Quaternary vegetation history of northwestern Greece.* Ph.D. thesis, University of Groningen.

BOTTEMA, S. (1979) Pollen analytical investigations in Thessaly (Greece). Palaeohistoria 21 19-40.

BOTTEMA, S. (1982) Palynological investigations in Greece with special reference to pollen as an indicator of human activity. *Palaeohistoria* **24** 257-89.

BROMBACHER, C. AND DICK, M. (1987) Die Untersuchung der botanischen Makroreste. In: *Zürich 'Mozartstrasse' neolithische und bronzezeitliche Ufersiedlungen.* (Berichte der Zürcher Denkmalpflege, Monographien 4) Zürich: Orell-Füssli Verlag pp. 198-212.

CAMPBELL, J.K. (1964) *Honour, family and patronage.* Oxford: University Press.

CHAMPION, T., GAMBLE, C., SHENNAN, S. AND WHITTLE, A. (1984) *Prehistoric Europe.* London: Academic Press.

CHERRY, J.F. (1979) Four problems in Cycladic prehistory. In: *Papers in Cycladic prehistory.* (eds Davis, J.L. and Cherry, J.F.) Institute of Archaeology monograph 14. Los Angeles: UCLA pp. 22-47.

CHILDE, V.G. (1957) *The dawn of European civilization.* (6th edition) London: Routledge and Kegan Paul.

CLARKE, D.L. (1978) *Mesolithic Europe: the economic basis.* London: Duckworth.

CLASON, A.T. (1968) The animal bones of the Bandceramic and middle age settlements near Bylany in Bohemia. *Palaeohistoria* **14** 1-17.

CULLEN, T. (1984) Social implications of ceramic style in the neolithic Peloponnese. In: *Ancient technology to modern science vol. 1.* (ed. Kingery, W.D.) Columbus: American Ceramic Society pp. 77-100.

DANTUMA, G. (1969) Remarks on the choice and the description of 160 cereal varieties. In: *Agro-ecological atlas of cereal growing in Europe, 2: atlas of the cereal-growing areas in Europe.* (ed. Brockhuizen, S.) Wageningen: Pudoc pp. 133-35.

DENNELL, R.W. (1976) The economic importance of plant resources represented on archaeological sites. *Journal of Archaeological Science* **3** 229-47.

DENNELL, R.W. (1978) *Early farming in south Bulgaria from the VI to the III millennia B.C.* Oxford: British Archaeological Reports S45.

DENNELL, R.W. (1983) *European economic prehistory: a new approach.* London: Academic Press.

DENNELL, R.W. (in press) *The origins and early development of European crop agriculture: a summary and discussion of the evidence.*

DENNELL, R.W. AND WEBLEY, D. (1975) Prehistoric settlement and land use in southern Bulgaria. In: *Palaeoeconomy.* (ed. Higgs, E.S.) London: Cambridge University Press pp. 97-109.

DRACK, W. (1969) Die frühen Kulturen mitteleuropäischer Herkunft. In: *Ur- und frühgeschichtliche Archäologie der Schweiz, 2: die Jüngere Steinzeit.* (ed. Drack, W.) Basle: Schweizerische Gesellschaft für Ur- und Frühgeschichte pp. 67-82.

DRIESCH, A. VON DEN AND ENDERLE, K. (1976) Die Tierreste aus der Agia Sofia-Magoula in Thessalien. In: *Die Deutschen Ausgrabungen auf Magulen um Larisa in Thessalien, 1966.* (Milojcic, V., Driesch, A. von den, Enderle, K., Milojcic-v. Zumbusch, J. and Kilian, K.) Bonn: Rudolf Habelt pp. 15-54.

ELLENBERG, H. (1978) *Vegetation Mitteleuropas mit den Alpen in Ökologischer Sicht.* (2nd edition) Stuttgart: Eugen Ulmer.

ELLIS, J.E., JENNINGS, C.H. AND SWIFT, D.M. (1979) A comparison of energy flow among the grazing animals of different societies. *Human Ecology* **7** (2) 135-49.

EVANS, J.D. (1968) Knossos Neolithic part 2: summary and conclusions. *Annual of the British School at Athens* **63** 267-76.

FLANNERY, K.V. (1972) The origins of the village as a settlement type in Mesoamerica and the Near East: a comparative study. In: *Man, settlement and urbanism.* (eds Ucko, P.J., Tringham, R. and Dimbleby, G.W.) London: Duckworth pp. 23-53.

FLEMING, A. (1972) The genesis of pastoralism in European prehistory. *World Archaeology* **4** 179-91.

FORBES, H.A. (1976) 'We have a little of everything': the ecological basis of some agricultural practices in Methana, Trizinia. Annals of the New York Academy of Sciences **268** 236-50.

FORBES, H. (1989) Of grandfathers and grand theories: the hierarchised ordering of responses to hazard in a modern Greek rural community. In: *Bad year economics.* (eds Halstead, P. and O'Shea, J.) Cambridge: University Press.

FORGE, A. (1972) Normative factors in the settlement size of neolithic cultivators (New Guinea). In: *Man, settlement and urbanism.* (eds Ucko, P.J., Tringham, R. and Dimbleby, G.W.) London: Duckworth pp. 363-76.

FRENCH, D.H. (1972) Notes on prehistoric pottery groups from central Greece. Athens: cyclostyled.

GREIG, J.R.A. AND TURNER, J. (1974) Some pollen diagrams from Greece and their archaeological significance. *Journal of Archaeological Science* **1** 177-94.

GRIGG, D.B. (1974) *The agricultural systems of the world: an evolutionary approach.* Cambridge: University Press.

GROENMAN-VAN WAATERINGE, W. (1970-71) Hecken im westeuropäischen Frühneolithikum. *Berichten van de Rijksdienst voor het Oudheidkundig Bodemonderzoek* **20-21** 295-99.

HALSTEAD, P. (1981a) Counting sheep in neolithic and Bronze Age Greece. *In: Pattern of the past.* (eds Hodder, I., Isaac, G. and Hammond, N.) Cambridge: University Press pp. 307-39.

HALSTEAD, P. (1981b) From determinism to uncertainty: social storage and the rise of the Minoan palace. In: *Economic archaeology: towards an integration of ecological and social approaches.* (eds Sheridan, A. and Bailey, G.) Oxford: British Archaeological Reports S96 pp. 187-213.

HALSTEAD, P. (1984) *Strategies for survival: an ecological approach to social and economic change in the early farming communities of Thessaly, N. Greece.* Ph.D. thesis, University of Cambridge.

HALSTEAD, P. (1987a) Traditional and ancient rural economy in Mediterranean Europe: plus ça change? *Journal of Hellenic Studies* **107** 77-87.

HALSTEAD, P. (1987b) Man and other animals in later Greek prehistory. *Annual of the British School at Athens* **82** 71-83.

HALSTEAD, P. (1989) The economy has a normal surplus: economic stability and social change among early farming communities of Thessaly, Greece. In: *Bad year economics.* (eds Halstead, P. and O'Shea, J.) Cambridge: University Press.

HALSTEAD, P. (in press) Simerini kai proistoriki orini oikonomia sti Pindho. In: *Proceedings of the Sixth International Congress on Aegean Prehistory. Athens.*

HALSTEAD, P. AND JONES, G. (1980) Early neolithic economy in Thessaly - some evidence from excavations at Prodromos. *Anthropoloyika* **1** 93-117.

HAMOND, F. (1981) The colonisation of Europe: the analysis of settlement processes. In: *Pattern of the past.* (eds Hodder, I., Isaac, G. and Hammond, N.) Cambridge: University Press pp. 211-48.

HARESIGN, W. (1983) *Sheep production.* London: Butterworths.

HARTMANN-FRICK, H. (1969) Die Tierwelt im neolithischen Siedlungsraum. In: *Ur- und frühgeschichtliche Archäologie der Schweiz, 2: die Jüngere Steinzeit.* (ed. Drack, W.) Basle: Schweizerische Gesellschaft für Ur- und Frühgeschichte pp. 17-32.

HIGGS, E.S. AND JARMAN, M.R. (1969) The origins of agriculture: a reconsideration. *Antiquity* **43** 31-41.

HIGGS, E.S. AND JARMAN, M.R. (1975) Palaeoeconomy. In: *Palaeoeconomy.* (ed. Higgs, E.S.) Cambridge: University Press pp. 1-8.

HIGHAM, C.F.W. (1967) Stock rearing as a cultural factor in prehistoric Europe. *Proceedings of the Prehistoric Society* **33** 84-106.

HIGHAM, C.F.W. (1968a) Trends in prehistoric European caprovine husbandry. *Man* **3** (1) 64-75.

HIGHAM, C.F.W. (1968b) Patterns of prehistoric economic exploitation on the Alpine Foreland. *Vierteljahrsschrift der Naturforschenden Gesellschaft in Zürich* **113** (1) 41-92.

HILBIG, W. AND LANGE, E. (1981) Die Entwicklung der Ackerunkrautvegetation im Gebiet des Flach- und Hügellandes der DDR. *Zeitschrift für Archäologie* **15** 41-56.

HILLMAN, G.C. (1981) Reconstructing crop husbandry practices from charred remains of crops. In: *Farming practice in British prehistory.* (ed. Mercer, R.J.) Edinburgh: University Press pp. 123-62.

HODDER, I. (1982) Sequences of structural change in the Dutch Neolithic. In: *Symbolic and structural archaeology.* (ed. Hodder, I.) Cambridge: University Press pp. 162-77.

HODDER, I. (1984) Burials, houses, women and men in the European Neolithic. In: *Ideology, power and prehistory.* (eds Miller, D. and Tilley, C.) Cambridge: University Press pp. 51-68.

HORVAT, I., GLAVAC, V. AND ELLENBERG, H. (1974) *Vegetation Südosteuropas.* Stuttgart: Gustav Fischer.

HOUEROU, H.N. LE (1977) Plant sociology and ecology applied to grazing lands research, survey and management in the Mediterranean basin. In: *Handbook of vegetation science 13: application of vegetation science to grassland husbandry.* (ed. Krause, W.) The Hague: Junk pp. 211-74.

HUBBARD, R.N.L.B. (1976) Crops and climate in prehistoric Europe. *World Archaeology* **8** (2) 159-68.

JACOMET, S. (1987) Ackerbau, Sammelwirtschaft und Umwelt der Egolzwiler und Cortaillod-Siedlungen: Ergebnisse samenanalytischer Untersuchungen. In: *Zürich 'Kleiner Hafner'.* (ed. Suter, P.J.) Berichte der Zürcher Denkmalpflege, Monographien 3. Zürich: Orell Füssli Verlag pp. 144-66.

JACOMET, S. AND SCHIBLER, J. (1985) Die Nahrungsversorgung eines jungsteinzeitlichen Pfynerdorfes am unteren Zürichsee. *Archäologie der Schweiz* **8** 125-41.

JACOMET, S. AND SCHLICHTHERLE, H. (1984) Der kleine Pfahlbauweizen Oswald Heer's - neue Untersuchungen zur Morphologie neolithischer Nacktweizen-Ähren. In: *Plants and ancient man.* (eds Zeist, W. van and Casparie, W.A.) Rotterdam: Balkema pp. 153-76.

JACOMET-ENGEL, S. (1980) Botanische Makroreste aus den neolithischen Seeufersiedlungen des Areals 'Pressehaus Ringier' in Zürich (Schweiz): stratigraphische und vegetationskundliche Auswertung. *Vierteljahrsschrift der Naturforschenden Gesellschaft in Zürich* **125** (2) 73-163.

JARMAN, M.R., BAILEY, G.N. AND JARMAN, H.N. (eds) (1982) *Early European agriculture.* Cambridge: University Press.

JOHANSSON, F. (1981) Stratigraphische Auswertung des Knochenmaterials. In: *Die neolithischen Ufersiedlungen von Twann, 11: Tierknochenfunde, 2: mittleres und oberes Schichtpaket (MS und OS) der Cortaillod-Kultur.* (Becker, C. and Johansson, F.) Bern: Staatlicher Lehrmittelverlag pp. 98-105.

JONES, G. (in press) Phytosociology and the archaeological recognition of crop husbandry practices. In: *Proceedings of the fifth OPTIMA meeting.* (eds Demiriz, H. and Phitos, D.) Istanbul.

JORDAN, B. (1975) *Tierknochenfunde aus der Magula Pevkakia in Thessalien.* Dissertation, University of Munich.

KILIAN, K. (1972) Zur mattbemalten Keramik der ausgehenden Bronzezeit und der Früheisenzeit aus Albanien. *Archäologisches Korrespondenzblatt* **2** 115-23.

KILIAN, K. (1973) Zur eisenzeitlichen Transhumanz in Nordgriechenland. *Archäologisches Korrespondenzblatt* **3** 431-35.

KNÖRZER, K.-H. (1971) Urgeschichtliche Unkräuter im Rheinland: ein Beitrag zur Entstehungsgeschichte der Segetalgesellschaften. *Vegetatio* **23** 89-111.

KNÖRZER, K.-H. (1979) Über den Wandel der angebauten Körnerfrüchte und ihrer Unkrautvegetation auf einer niederrheinischen Lössfläche seit dem Frühneolithikum. In: Festschrift Maria Hopf. (ed. Körber-Grohne, U.) *Archaeo-Physika* **8** 147-63.

KOSSE, K. (1979) *Settlement ecology of the early and middle neolithic Körös and Linear Pottery cultures in Hungary*. Oxford: British Archaeological Reports S64.

KOTSAKIS, K. (1983) *Keramiki tekhnoloyia kai keramiki dhiaforopoiisi: provlimata tis graptis keramikis tis mesis neolithikis epokhis tou Sesklou*. Ph.D. thesis, University of Thessaloniki.

KRUK, J. (1980) *The neolithic settlement of southern Poland*. Oxford: British Archaeological Reports S93.

KUPER, R., LÖHR, H., LÜNING, J., STEHLI, P. AND ZIMMERMANN, A. (1977) Struktur und Entwicklung des Siedlungsplatzes. In: *Der Bandkeramische Siedlungsplatz Langweiler 9 (Rheinische Ausgrabungen 18)*. (Kuper, R., Löhr, H., Lüning, J., Stehli, P. and Zimmermann, A.) Bonn: Rheinland Verlag pp. 304-33.

LEAKE, W.M. (1967) *Travels in northern Greece 4*. Amsterdam: Hakkert.

LEE, R.B. AND DEVORE, I. (1968) Problems in the study of hunters and gatherers. In: *Man the hunter*. (eds Lee, R.B. and DeVore, I.) Chicago: Aldine pp. 3-12.

LEGGE, A.J. (1981a) The agricultural economy. In: *Grimes Graves, Norfolk: excavations 1971-72, volume 1*. (Mercer, R.J.) Department of the Environment Research Report 11 London: Her Majesty's Stationery Office pp. 79-103.

LEGGE, A.J. (1981b) Aspects of cattle husbandry. In: *Farming practice in British prehistory*. (ed. Mercer, R.J.) Edinburgh: University Press pp. 169-81.

LIRITZIS, I. AND DIXON, J. (1984) Politistiki epikoinonia metaxi ton neolithikon oikismon Sesklou kai Dhiminiou (Thessalia). *Anthropoloyika* **5** 51-62.

MILISAUSKAS, S. (1978) *European prehistory*. New York: Academic Press.

MÜLLER, H.H. (1964) *Die Haustiere der mitteldeutscher Bandkeramiker*. Schriften der Sektion für Vor- und Frühgeschichte 17. Berlin: Deutsche Akademie der Wissenschaften zu Berlin.

NANDRIS, J. (1970) The development and relationships of the earlier Greek Neolithic. *Man* **5** 192-213.

ODUM, E.P. (1975) *Ecology*. (2nd edition) London: Holt, Rinehart and Winston.

O'SHEA, J. (1981) Coping with scarcity: exchange and social storage. In: *Economic archaeology: towards an integration of ecological and social approaches*. (eds Sheridan, A. and Bailey, G.) Oxford: British Archaeological Reports S96 pp. 167-83.

O'SHEA, J. (1989) The role of wild resources in small-scaled agricultural systems: tales from the Lakes and the Plains. In: *Bad year economics*. (eds Halstead, P. and O'Shea, J.) Cambridge: University Press.

PAYNE, S. (1973) Kill-off patterns in sheep and goats: the mandibles from Asvan Kale. *Anatolian Studies* **23** 281-303.

PÉCSI, M. AND JAKUCS, P. (1971) The natural vegetation of Hungary. In: *Hungary: geographical studies.* (eds Pécsi, M., Enyedi, G. and Marosi, S.) Budapest: Szabó Gyula pp. 109-24.

PHILIPPSON, A. (1950) *Die griechischen Landschaften, 1.1: Thessalien und die Spercheios-Senke.* Frankfurt: Klostermann.

POULIANOS, A. (1966) Palaioanthropoloyika evrimata apo ti neolithiki Thessalia. *Thessalika* **5** 71-75.

REDDING, R.W. (1981) *Decision making in subsistence herding of sheep and goats in the Middle East.* Ph.D. thesis, University of Michigan.

RENFREW, C. (1972) *The emergence of civilisation: the Cyclades and the Aegean in the third millennium B.C.* London: Methuen.

RENFREW, C. (1973) Trade and craft specialisation. In: *Neolithic Greece.* (Theokharis, D.R.) Athens: National Bank of Greece pp. 179-200.

RENFREW, C. (1976) Megaliths, territories and populations. In: *Acculturation and continuity in Atlantic Europe.* (ed. de Laet, S.J.) Brugge: De Tempel pp. 198-220.

RENFREW, J. (1979) The first farmers in southeast Europe. In: Festschrift Maria Hopf. (ed. Körber-Grohne, U.) *Archaeo-Physika* **8** 243-65.

RICHARDS, A.I. (1939) *Land, labour, and diet in northern Rhodesia: an economic study of the Bemba.* Oxford: University Press.

RICKLEFS, R.E. (1980) *Ecology.* (2nd edition) Sunbury-on-Thames: Nelson.

RONDIRI, V. (1985) Epifaniaki keramiki neolithikon theseon tis Thessalias: katanomi sto khoro. *Anthropoloyika* **8** 53-74.

ROWLEY-CONWY, P. (1981) Slash-and-burn in the temperate European Neolithic. In: *Farming practice in British prehistory.* (ed. Mercer, R.J.) Edinburgh: University Press pp. 85-96.

ROWLEY-CONWY, P. (1983) Sedentary hunters: the Ertebølle example. In: *Hunter-gatherer economy in prehistory: a European perspective.* (ed. Bailey, G.N.) Cambridge: University Press pp. 111-26.

SAHLINS, M.D. (1974) *Stone age economics.* London: Tavistock Publications.

SAKELLARIDIS, M. (1979) *The economic exploitation of the Swiss area in the mesolithic and neolithic periods.* Oxford: British Archaeological Reports S67.

SAUTER, M.-R. AND GALLAY, A. (1969) Les premières cultures d'origine méditerranéenne. In: *Ur- und frühgeschichtliche Archäologie der Schweiz, 2: die Jüngere Steinzeit.* (ed. Drack, W.) Basle: Schweizerische Gesellschaft für Ur- und Frühgeschichte pp. 47-66.

SCHWARTZ, C.A. (1976) The vertebrate fauna from Rug Bair. In: *Neolithic Macedonia.* (ed. Gimbutas, M.) Los Angeles: Institute of Archaeology, University of California pp. 364-74.

SHACKLETON, N. AND RENFREW, C. (1970) Neolithic trade routes realigned by oxygen isotope analyses. *Nature* **228** 1062-65.

SHENNAN, S. (1986) Central Europe in the third millennium B.C.: an evolutionary trajectory for the beginning of the European Bronze Age. *Journal of Anthropological Archaeology* **5** 115-46.

SHERIDAN, A. AND BAILEY, G. (eds) (1981) *Economic archaeology: towards an integration of ecological and social approaches.* Oxford: British Archaeological Reports S96.

SHERRATT, A.G. (1976) Resources, trade and technology. In: *Problems in economic and social archaeology.* (eds Sieveking, G., Longworth, I. and Wilson, K.) London: Duckworth pp. 557-81.

SHERRATT, A.G. (1980) Water, soil and seasonality in early cereal cultivation. *World Archaeology* **11** (3) 313-30.

SHERRATT, A.G. (1983) The development of neolithic and copper age settlement in the Great Hungarian Plain, part 2: site survey and settlement dynamics. *Oxford Journal of Archaeology* **2** (1) 13-41.

SIVIGNON, M. (1975) *La Thessalie: analyse géographique d' une province grecque.* Lyon: Institut des études Rhodaniennes.

SOUDSKY, B. AND PAVLU, I. (1972) The Linear Pottery culture settlement patterns of central Europe. In: *Man, settlement and urbanism.* (eds Ucko, P.J., Tringham, R. and Dimbleby, G.W.) London: Duckworth pp. 317-28.

STERUD, E.L. (1978) Prehistoric populations of the Dinaric Alps. In: *Social archeology: beyond subsistence and dating.* (eds Redman, C.L., Berman, M.J., Curtin, E.V., Langhorne, W.T., Versaggi, N.M. and Wanser, J.C.) New York: Academic Press pp. 381-408.

THEOKHARIS, D.R. (1973) *Neolithic Greece.* Athens: National Bank of Greece.

THEOKHARIS, D.R. (1980) To neolithiko spiti. *Anthropoloyika* **1** 12-14.

THRAN, P. AND BROCKHUIZEN, S. (1965) *Agro-ecological atlas of cereal growing in Europe, 1: agro-climatic atlas of Europe.* Wageningen: Pudoc.

TRINGHAM, R. (1971) *Hunters, fishers and farmers of eastern Europe 6000-3000 B.C.* London: Hutchinson.

TROELS-SMITH, J. (1984) Stall-feeding and field-manuring in Switzerland about 6000 years ago. *Tools and Tillage* **5** (1) 13-25.

TSOUNTAS, Kh. (1908) *Ai proïstorikai akropolis Dhiminiou kai Sesklou.* Athens: Athens Archaeological Society.

TURRILL, W.B. (1929) *The plant-life of the Balkan peninsula.* Oxford: University Press.

VERGOPOULOS, K. (1975) *To agrotiko zitima stin Elladha: i koinoniki ensomatosi tis yeoryias.* Athens: Exantas.

VOGT, E. (1969) Siedlungswesen. In: *Ur- und frühgeschichtliche Archäologie der Schweiz, 2: die Jüngere Steinzeit.* (ed. Drack, W.) Basle: Schweizerische Gesellschaft für Ur- und Frühgeschichte pp. 157-74.

VOLIOTIS, D. (1973) Beziehungen zwischen Klima, Boden und Vegetations-Zonen in Griechenland. *Epistimoniki Epitiris Fisiko-Mathimatikis Skholis Panepistimiou Thessalonikis* **13**.

WACE, A.J.B. AND THOMPSON, M.S. (1912) *Prehistoric Thessaly.* Cambridge: University Press.

WALLÉN, C.C. (ed.) (1970) *World survey of climatology, volume 5: climates of northern and western Europe.* Amsterdam: Elsevier.

WALLÉN, C.C. (ed.) (1977) *World survey of climatology, volume 6: climates of central and southern Europe.* Amsterdam: Elsevier.

WHITELAW, T.M. (1983) The settlement at Fournou Korifi, Myrtos and aspects of Early Minoan social organization. In: *Minoan society.* (eds Krzyszkowska, O. and Nixon, L.) Bristol: Classical Press pp. 323-45.

WHITTLE, A. (1985) *Neolithic Europe: a survey.* Cambridge: University Press.

WILLERDING, U. (1980) Zum Ackerbau der Bandkeramiker. In: *Beiträge zur Archäologie Nordwestdeutschlands und Mitteleuropas.* (eds Krüger, T. and Stephan, H.-G.) Hildesheim: August Lax pp. 423-56.

WILLERDING, U. (1981) Ur- und frühgeschichtliche sowie mittelalterliche Unkrautfunde in Mitteleuropa. *Zeitschrift für Pflanzenkrankheiten und Pflanzenschutz* Sonderheft **9** 65-74.

WILLERDING, U. (1983) Zum ältesten Ackerbau in Niedersachsen. In: *Frühe Bauernkulturen in Niedersachsen.* (ed. Wegner, G.) Archäologische Mitteilungen aus Nordwestdeutschland 1. Oldenburg: Staatliches Museum für Naturkunde und Vorgeschichte pp. 179-219.

WILLERDING, U. (1985) Zum Ackerbau der Linienbandkeramiker bei Esbeck, Stadt Schöningen, Landkreis Helmstedt. In: *Ausgrabungen in Niedersachsen: Archäologische Denkmalpflege 1979-1984.* (ed. Wilhelmi, K.) Stuttgart: Konrad Theiss Verlag pp. 92-96.

WOBST, H.M. (1974) Boundary conditions for palaeolithic social systems: a simulation approach. *American Antiquity* **39** 147-78.

WYSS, R. (1969) Wirtschaft und technik. In: *Ur- und frühgeschichtliche Archäologie der Schweiz, 2: die Jüngere Steinzeit.* (ed. Drack, W.) Basle: Schweizerische Gesellschaft für Ur- und Frühgeschichte pp. 117-38.

YEORYIADHIS, Y. (1880) *Thessalia.* Athens.

ZEIST, W. VAN AND BOTTEMA, S. (1971) Plant husbandry in early neolithic Nea Nikomedeia, Greece. *Acta Botanica Neerlandica* **20** (5) 524-38.

ZEIST, W. VAN AND BOTTEMA, S. (1982) Vegetational history of the eastern Mediterranean and the Near East during the last 20,000 years. In: *Palaeoclimates, palaeoenvironments and human communities in the eastern Mediterranean region in later prehistory.* Oxford: British Archaeological Reports S133 pp. 277-321.

ZVELEBIL, M. (ed.) (1986a) *Hunters in transition.* Cambridge: University Press.

ZVELEBIL, M. (1986b) Mesolithic societies and the transition to farming: problems of time, scale and organisation. In: *Hunters in transition.* (ed. Zvelebil, M.) Cambridge: University Press pp. 167-88.

3

HIERARCHICAL APPROACHES TO THE EVOLUTION OF COMPLEX AGRICULTURAL SYSTEMS

K.D. Thomas

ABSTRACT

Classical ecosystem models imply that homoeostatic processes are important in regulating (stabilising) natural communities. Models of the origins of agriculture, and of the ways in which past subsistence systems functioned, have often invoked feedback processes, and associated stability, to account for systems change, or maintenance. Assumptions of equilibrium in natural and human ecosystems are being increasingly questioned; non-equilibrium models are being examined as plausible alternatives to homoeostatic ones.

Some ecologists see history and chance as two important factors influencing the structure of biological communities. Others envisage structure arising in ecological systems from a hierarchy of processes which operate at different rates, or on different scales, in different levels of the hierarchy. Hierarchies within food webs can be stable under certain structural and functional conditions. In this paper I will attempt to consider some of the implications of these new ways of viewing ecosystems.

An attempt is made to apply some of the concepts of hierarchy theory to prehistoric agricultural systems in north west Pakistan. It is possible to arrange the components of agricultural systems into a hierarchy of interactive 'holons' and to show that some arrangements are potentially more stable than others. Such modelling could lead to new insights about the functioning of past agro-ecosystems. However, in a strict sense, hierarchies should consist of holons which are differentiated by rates of response; this will be difficult to apply in any detail to archaeological data sets. It is unclear how valuable hierarchy theory will prove to be in prediction (retrodiction) or explanation but it can be useful in structuring descriptions of systems. A potential advantage of hierarchical approaches is that they structure the flow of energy, materials and information in systems in ways relevant to ecologists, environmental archaeologists and cultural archaeologists.

3.1 INTRODUCTION

Agricultural systems, and human ecosystems in general, can be conceptualised as ecological systems with trophic levels, food chains (or webs) and positive and negative feedback loops which may, or may not, tend to regulate these systems. It is not clear exactly how far concepts from ecology, and in particular theoretical ecology, can be applied to the analysis and interpretation of past agricultural systems; there are clear differences between self-organised ecosystems and managed agricultural systems. Ecologists are, anyway, re-evaluating their own models and concepts concerning the organisation and function of ecosystems (Diamond and Case 1986; Kikkawa and Anderson 1986; Schulze and Zwölfer 1987) and it is far from certain whether equilibrium, disequilibrium or non-equilibrium models are the most appropriate. Some of the problems and potentials of various ecological approaches to an understanding of early agricultural systems are explored below.

3.2 COMPLEX AGRICULTURAL SYSTEMS

For the purposes of this discussion, complex agricultural systems are those in which both arable and pastoral activities yield a range of primary and secondary products and in which there are various functional relationships between the arable and pastoral subsystems. Sherratt (1981) has considered some aspects of agricultural integration in his well known paper on 'plough and pastoralism'. He envisages a secondary products package, based largely on new perceptions of the potential of domesticated animals for yielding milk, fibre and work, originating in south west Asia (and possibly south Asia too) in the fourth millennium B.C. and then spreading through Eurasia. Sherratt views these fundamentally important agricultural developments as arising from essentially historical processes (invention, innovation, perception of need, etc.) and spreading as a result of population and social factors; hence his use of the term 'the secondary products revolution' (Sherratt 1981).

In this paper I will be considering complex agricultural systems in terms of ecological theory. How did such systems function?; how were they organised or integrated?; were they inherently more stable, and therefore more successful, than simpler systems? The question of the _origins_ of complex agricultural systems will not be considered here.

The pragmatic reader may feel that this approach is likely to be highly theoretical and without much relevance to practical archaeology. In fact, the ideas discussed here originate in an empirical study of early agricultural systems of the third millennium B.C. in north west Pakistan (Thomas 1983b, 1986). The analysis of archaeological and bioarchaeological evidence led me (1983b) to infer a range of agricultural products, both primary and secondary, and their likely uses (Table 1). These elements were then used to construct models which sought to describe how the agricultural systems might have functioned (Figures 1 and 2). These models use the familiar 'box and arrow' convention, showing the possible relations between the elements and how they could have been integrated into an agro-ecosystem. It was not suggested that such systems were necessarily _stable_ but the implication was there that systems involving feedback loops were, at least, stability-seeking and potentially longer-lived than simpler, less integrated systems.

Such models are useful in providing a structural framework for diverse categories of data. They may help in the formulation of hypotheses about site or system function and also help us to understand processes of site-formation. They do not, however, explain why such systems arose or why they declined. In feedback models it is necessary either to invoke factors outside the system to explain change (such as climatic change or social shifts) or to focus on a factor within the system (such as human population growth or decline in soil fertility) which causes internal instability. Positive feedback processes, generating self-amplifying loops, have been successfully used to model changes in past subsistence and settlement systems (Harris 1977).

FOOD FOR THE HUMAN POPULATION

>Wheat, Barley, Lentils
>(*Malva, Vicia/Lathyrus*, wild fruits, such as *ber*)
>Meat of *Bos, Bubalus, Ovis, Capra*
>(Milk of these species)
>(Meat of *Equus, Sus*, freshwater fish, freshwater molluscs)

FEED FOR ANIMALS

>(Forage on the aftermath of harvest)
>(Fodder as a by-product of crop-processing)
>(Forage crops)
>(Grazing and browsing of rangeland plants)

FIBRES

>(Wool, hair, fur, tendons, intestines)
>(Straw)
>(From wild plants: Gramineae, *Calotropis, Crotalaria*)

LEATHER

>(Skins and hides)

BUILDING MATERIALS

>Wood, twigs, grasses
>(Dung)

FUEL

>Wood, (Charcoal)
>(Dung)

OTHER MATERIALS

>Bone, shell, (horn)

WORK

>Transport, (traction), (threshing)

TABLE 1 The main arable, pastoral and wild products utilised at Tarakai Qila (a settlement of the 3rd millennium B.C. in north west Pakistan (Thomas 1983a); inferred categories are in brackets. (After Thomas 1983b).

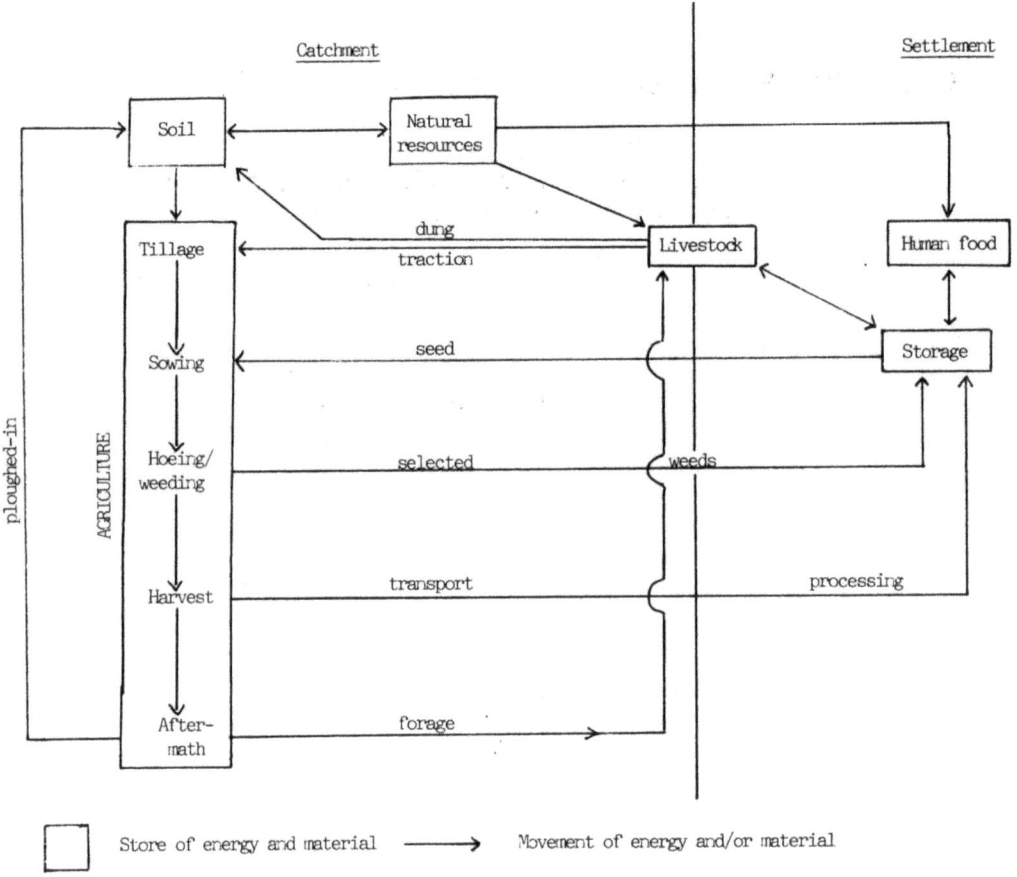

FIGURE 1 Simple model of an agricultural system of the mid 3rd millennium B.C. in north west Pakistan. Boxes represent stores of energy; arrows show the movements of energy and materials within the system. (From Thomas 1983b: Figure 13:6).

3.3 SYSTEMS AND ECOSYSTEMS

The general theory of systems elaborated by von Bertalanffy (1968) has had wide application in the natural and social sciences including, most importantly from our point of view, biology, ecology, geography, anthropology and archaeology. Notable early studies of the potential of systems approaches in archaeology are those of Binford (1965) and Flannery (1968). Watson *et al.* (1984) review the applications of systems theory in archaeology and distinguish clearly between: (a) systemic approaches to the analysis of particular sets of data and (b) the use of general systems theory to formulate ideas of how any system will function, regardless of its particular components; catastrophe theory (e.g. Renfrew 1979) is an example of such a general theory. In this paper I will be concerned with more focussed systemic approaches.

O'Neill *et al.* (1986) have reviewed the development of the ecosystem concept in ecology. Ecosystem models have been developed to permit both the organisation of complex data and the conceptualisation of how complex natural systems function. Until quite recently approaches to ecosystem analysis have invoked cybernetic concepts: the

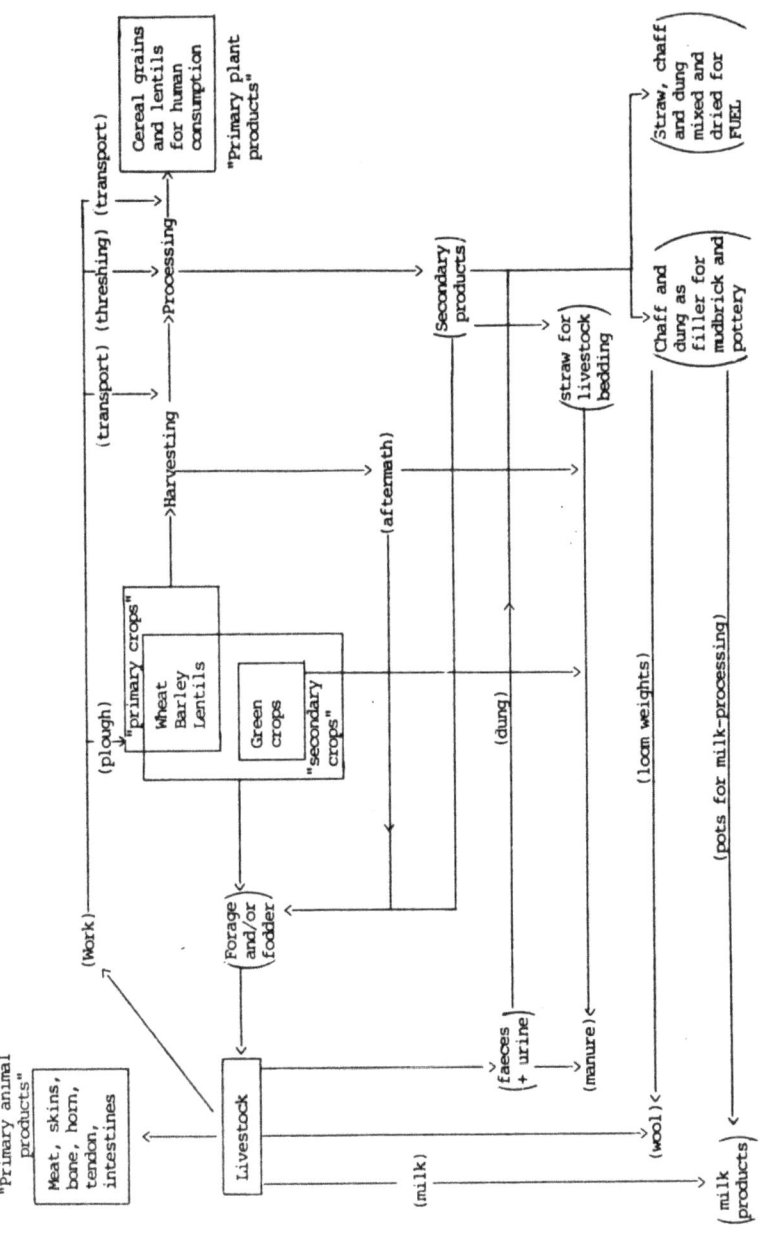

FIGURE 2 Simplified model of an agricultural system (see Figure 1), showing the main primary and secondary products (Table 1) and highlighting the role of secondary products in integrating the various elements of the system. Boxes enclose crops and other primary products, brackets enclose various categories of secondary products and arrows show pathways along which energy and materials are moved. (From Thomas 1983b: Figure 13:9).

system is maintained by transfers of energy and materials along pathways (food chains and webs) which form feed-back loops to regulate the function of the system. The energy and materials are carrying information, which controls the system. Ecosystems with more species have more complex food webs and, according to ideas from information theory, are potentially more stable than low-diversity ecosystems. These concepts, or assumptions, are increasingly being challenged by ecologists (e.g. Chessson and Case 1986; O'Neill et al. 1986; Schulze and Zwölfer 1987).

There are problems with the use of the concept of stability as applied to ecological systems. Stability is extremely difficult to demonstrate in real communities of animals and plants (Connell and Sousa 1983), mainly because ecological studies have, of necessity, been undertaken over short time scales and over small spatial scales. In theory, stability in ecological systems can be influenced by two basic factors: internal feedback processes, involving interactions between species or other components, and the degree of uncertainty in the external environment (DeAngelis and Waterhouse 1987). Some ecologists, following Sanders (1968), contend that those environments which exhibit high levels of stability, such as the deep-sea floor or the wet tropics, will develop complex, species-rich ecosystems. In such cases, ecosystem complexity is the consequence of environmental stability. Recent empirical studies, reviewed by Moore (1983), have shown that ecological systems as varied as sub-alpine meadows and tropical coral reefs are most diverse when subjected to some environmental stability; there appears to be no simple relationship between environmental stability and ecosystem complexity. The role of community complexity as a causal factor in ecosystem stability is difficult to evaluate but again there seems to be no clear relationship between them (Pimm 1984). Some ecosystems are most complex and stable when the external environment is stable; others are more complex when the external environment is subject to periodic instabilities (perhaps, in such cases, increased community complexity or diversity helps confer a higher degree of stability on the ecosystem?).

Another problem with the concept of stability, as applied to ecosystems, is distinguishing it clearly from the concept of equilibrium. Systems can exist in various types of equilibrium, some of which are stable and others unstable; unstable equilibria will not exist for long in the face of even small disturbances but stable equilibria will return to the point of stability if displaced away from it. 'Stability' in this sense is relative to the degree of perturbation which might affect the system; if the perturbation is of high magnitude then it might exceed the limits over which the equilibrium is stable, causing a new equilibrium point to be established. All ecological systems probably exist as such metastable equilibria, some requiring greater perturbations to upset them than others. Homoeostatic models of ecosystems assume that these metastable equilibria are maintained by the internal relationships between the biological components of the system; in other words, that ecosystems have equilibrium-seeking properties. A major conceptual drawback of these homoeostatic models is the difficulty of building-in parameters which allow for change.

An alternative view is that homoeostatic models are inappropriate and unecessary: ecosystems can at times exhibit properties such as stability and resilience (return to some dynamic steady-state) without the homoeostatic balance of positive and negative feedback processes (see below). Ecosystem models which do not assume the primacy of homoeostatic feedback processes have, unfortunately, been termed 'non-equilibrium models' in the literature (e.g. DeAngelis and Waterhouse 1987; Foin and Davis 1987; O'Neill et al. 1986). I say 'unfortunately' because many of these alternative models can account for stability, steady-state, equilibrium (call it what you will) in ecological systems; they are, in reality, non-feedback models. (In this paper I use the term 'non-equilibrium' in accordance with established practice but I hope that I have clarified what I mean by it).

3.4 AGRO-ECOSYSTEMS

Modern agricultural systems have been considered in terms of systems theory (e.g. Spedding 1979), and Bayliss-Smith (1982) has analysed both traditional and industrial agricultural systems as ecological systems. Indeed, ecological concepts are permeating modern agriculture to such an extent that farmers in Britain are currently evaluating their role in what they term 'the food chain', and 'diversification' is perceived as a way in which to stabilise farming enterprises in an uncertain economic climate. The differences between ecosystems and agro-ecosystems are obvious: agro-ecosystems are managed by man with the objective of sustainability and the responses which farmers may make to constraints are not necessarily determined by the environment; social factors may be of over-riding importance (Bayliss-Smith 1982).

Agro-ecosystems and ecosystems have been compared by Aldag (1987) in terms of production, energy flow, disturbance, rates of nutrient cycling and nutrient losses; people act as a major 'sink' for large amounts of energy and nutrients in agricultural systems. Aldag sees agro-ecosystems as being controlled from outside but, because people are the main primary and secondary consumers in an agro-ecosystem, it is more appropriate to consider them as an integral part of such systems.

Anthropological studies of traditional agro-ecosystems have, of necessity, included man as a focal element of the system. Such approaches have been discussed by Ellen, who has also presented an excellent critique of equilibrium models of ecosystems (Ellen 1982: 177-203). Agricultural systems have been reviewed from an archaeological perspective by Green (1980) who is less explicitly critical of equilibrium models but is clearly in sympathy with models which allow change to occur as a consequence of endogenous instabilities rather than exogenous pressures.

The first worker to use ecosystem concepts explicitly in discussing the origins and evolution of agricultural systems was Harris (1969). He suggested that the ecosystem approach was valuable in that it focussed attention on structure, function, equilibrium and change. He proposed that taxonomically diverse, generalised, agro-ecosystems would be more stable than specialised ones, but he also pointed out that increased rates of flow of nutrients (and of losses of nutrients) could have a potentially unbalancing effect on any system.

Although many ecological anthropologists and archaeologists are interested in the stability of human ecosystems (see Foin and Davis 1987 for a brief review), there have been few useful empirical studies. This is mainly because of the extremely difficult problems of defining and measuring stability. One very detailed study of a specialised subsistence system is that by Wen Dazhong and Pimentel (1986a and b), who analysed the historical records relating to organic agriculture in seventeenth century China. They showed that the agro-ecosystem was sustainable in terms of energy flows and nutrient cycles, but that its maintenance required high inputs of labour and a sophisticated appreciation of crops, manures and soil management. Simulation models of rather less sophisticated systems, shifting agriculture in the New Guinea Highlands (Foin and Davis 1987), have shown that these ecosystems were unstable and that, although feedback processes could be detected, the systems were always prevented from reaching equilibrium. Another recent study, a computer simulation of Huron swidden horticultural systems by McGlade and Allen (1986), again showed that disequilibrium models were most appropriate. One might argue for both of these last two cases, that stability in swidden systems should be viewed from the regional perspective rather than from the individual swidden plot; this was, in fact, done by Foin and Davis but not by McGlade and Allen.

3.5 HIERARCHICAL MODELS OF ECOSYSTEMS AND AGRO-ECOSYSTEMS

The rates and intensities with which ecosystem components interact with one another is highly variable. Some components interact frequently and may do so with such intensity that mutualistic coadaptations between species, as discussed by Rindos (1984), can evolve. Other components within systems may interact rarely, if at all. It is possible, therefore, to consider ecosystems as being comprised of 'holons' whose components interact with one another regularly, or at relatively high rates, but which interact with other holons in the same system at relatively low rates (O'Neill et al. 1986: 79-80). Accordingly, holons can be structured into hierarchies based on the rates and intensities of their interaction. Higher rates of interaction will be characteristic of lower levels of the hierarchy; equally, there will be low rates of interaction between holons at higher levels.

Hierarchy theory has developed over the last twenty years or so as a branch of general systems theory (Pattee 1973; Whyte et al. 1969); the application of hierarchy theory in ecology has recently been reviewed by O'Neill et al. (1986). In systems with high rates of interaction between holons (i.e. systems in which the boundaries between holons are indistinct) there is a high potential for instability because a perturbation in one holon will be transmitted rapidly to the others. Perturbations in holons of systems which are more hierarchically organised will be smoothed-out to lower frequency signals in the next level of the hierarchy, so the overall response of the system may be low. In consequence, hierarchical ecosystems can exhibit properties of stability and resilience which arise from the varying levels of interaction and connection between the components of the systems rather than from homoeostatic balances of opposing feedback loops operating throughout the system. Such systems act as damping or signal attenuating systems (O'Neill et al. 1986: 76-78). Hierarchical concepts have been applied successfully to the analysis and interpretation of natural ecosystems (e.g. McIntire and Colby 1978; Gigon 1987; see O'Neill et al. 1986 for a review). If hierarchical concepts offer a useful way of looking at ecological systems, it is worthwhile considering how they could be used in the analysis of past agro-ecosystems.

The concept of 'hierarchy' in natural and human systems is not a new one. Ecologists have long recognised 'levels of organisation' in biological systems (the organism, population, community, ecosystem, biome and biosphere). Analagously, in the context of archaeology, we can identify a hierarchy from, *inter alia*, the individual, family, village, town or city, state, and so on. These are essentially organisational hierarchies, although they also have important functional properties which can be analysed in terms of hierarchy theory. O'Neill et al. (1986) discuss the so-called 'dual hierarchy' of ecological systems, distinguishing between organisational hierarchies and process-functional hierarchies. The relationships between such dual hierarchies, for both natural and human systems (the latter being highly simplified and emphasising subsistence or economic processes), are shown in Figure 3.

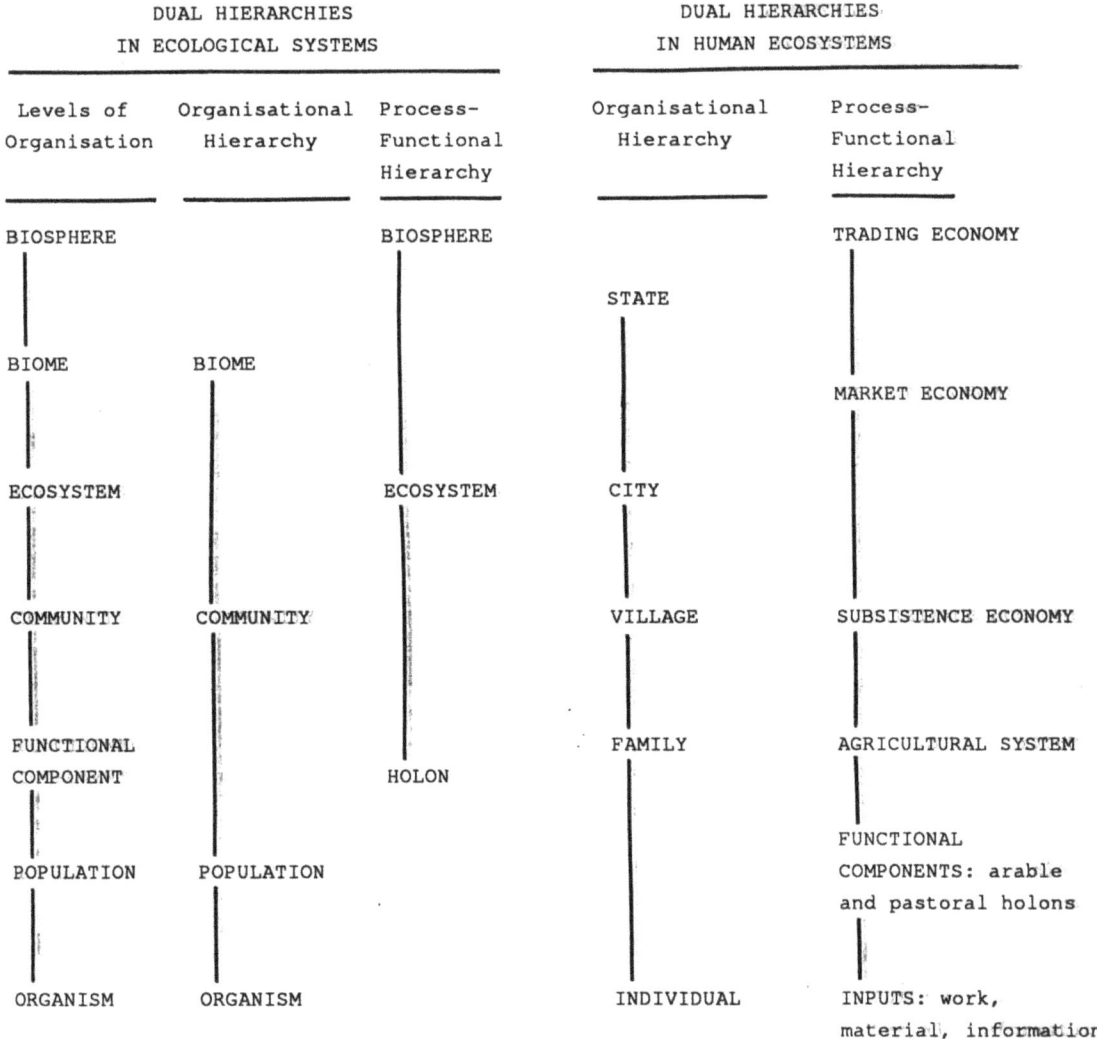

FIGURE 3 Organisational and process-functional hierarchies in ecological systems (modified from O'Neill *et al.* 1986: Figure 9:1) and human ecosystems. No equivalence is intended by the positioning of the various levels in these two types of systems.

A major difference between these natural and human systems, in addition to the purposive (or goal-oriented) nature of the latter, is that human systems involve transfers of information as well as energy and materials. Transfers of energy and materials are, in terms of cybernetic theory, transfers of information but in this paper I use 'information' in the behavioural and social sense to mean transfers (and the purposive regulation of such transfers) of concepts, ideas, news, lies, etc. Archaeologists have long been aware of the importance of information, in the sense just described, and have incorporated information flow and regulation into their systems models (Flannery and Marcus 1976), some of which have hierarchical elements (Flannery 1972; van der Leeuw 1981). Both of these authors emphasise the importance of rates of flow of information. Flannery conceptualises decision-making sub-systems as being hierarchically organised, with higher-level sub-systems regulating the output of lower-level ones; malfunctions low in the hierarchy can be corrected (cf. 'signal attenuation', discussed above) but those higher up can cause severe problems. Flannery also considers the degree to which sub-systems interact and uses the term 'hypercoherence' for systems in which there are high rates of information flow. Such hypercoherent systems are potentially unstable (as are hierarchical ecosystems in which the constituent holons are too interactive).

3.6 EARLY AGRICULTURAL SYSTEMS AS PROCESS-FUNCTIONAL HIERARCHIES

I now return to a discussion of agricultural systems of the third millennium B.C. in north west Pakistan. In an earlier paper I considered agricultural systems to be components of a nested hierarchy of systems (Thomas 1983b: 289-90), as illustrated in Figure 3 of the present paper; this hierarchical analysis was not developed further at that time. Instead, attention was focussed on particular settlements and the flow of energy and material within the settlement-resource catchment system; simple models of this are reproduced here in Figures 1 and 2.

These systems could, equally, be conceptualised as process-functional hierarchies and some possible models are shown in Figure 4. Holons are shown in separate boxes (some of which overlap) which are nested into higher hierarchical levels. In relatively simple (non-integrated?) systems, as illustrated in Figure 4A, the holons are quite distinct with (potentially) low rates of interaction between them. The situation shown in Figure 4B is a little more complex, with overlaps between the arable and pastoral holons (the animals being fed on the by-products of crop processing), but the holons remain essentially distinct. Figure 4C represents a complex agricultural system in which the arable and pastoral components are strongly integrated. In such a system the domesticated animals would be fed on crop residues, the aftermath of harvest in the fields, and possibly on forage crops. Storage of such animal feed would permit a larger population of animals to be kept alive over difficult periods of the year when rangeland productivity was low. Some of these animals would, in turn, be used to draw the plough, to manure the fields and to transport harvested crops from the fields to the settlement, and perhaps on to market. Intensive use of the rangelands, with manipulation of the natural vegetation by fire, is another possible linking feature. In this complex system, people continue to play a direct functional role in each holon (obtaining food and other products from the plant crops, domesticated animals, wild plants, wild herbivorous mammals and alternative resources, such as riverine fish and shellfish) but they also have an increasing, indirect, effect on soils (by increasing the rates of nutrient depletion and/or physical loss of soil) and on other components of the ecosystem, such as the wild predators, through modifications of the local environment.

Because people are a part of all the holons in an agro-ecosystem, there is a high level of linkage in even the simplest model outlined here (Figure 4A), implying, according to hierarchy theory, a potential for instability. In highly integrated systems (Figure 4C), with high rates of flow of energy and materials to the human component, this potential for instability is much greater. We have, therefore, the apparent paradox that a more integrated animal/plant crop production system, while being more productive, is potentially less stable than systems which are less well integrated. In fact, this is quite in accord with intuitive (or common sense) assessments of such systems: greater dependence between the components of a system will have disastrous consequences for the system as a whole if one or other of the components fails.

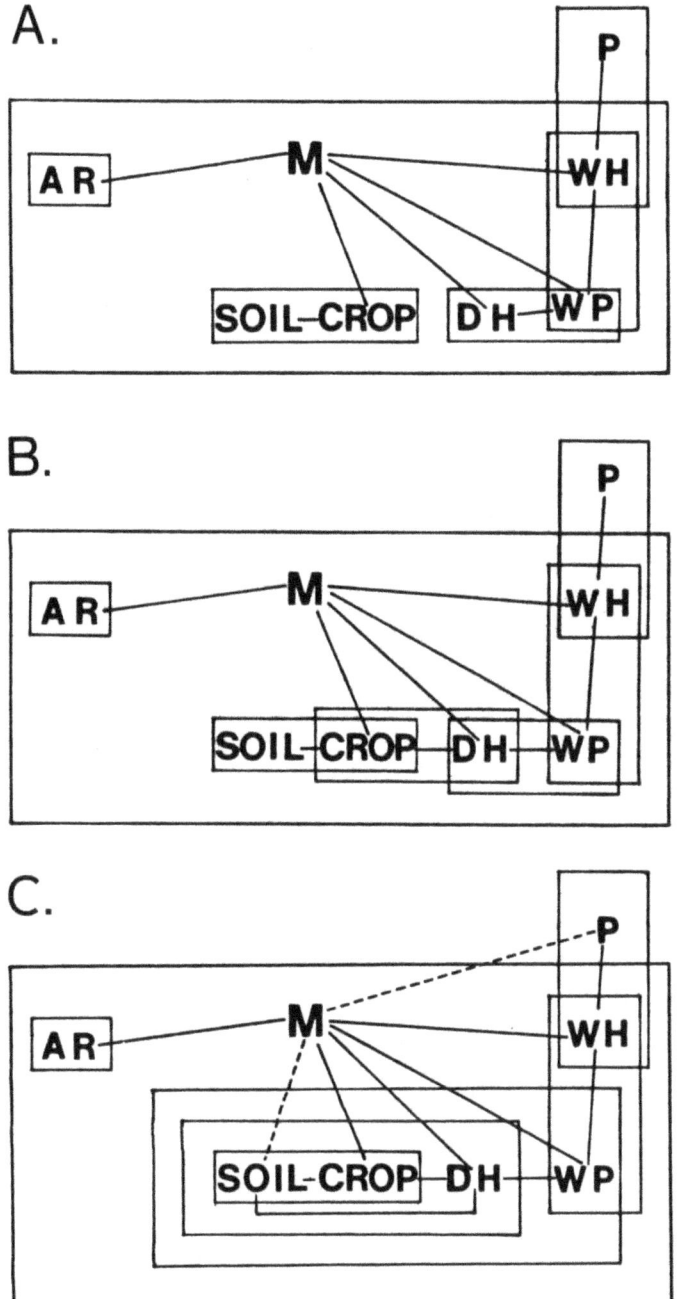

FIGURE 4 Possible process-functional hierarchies in simple prehistoric agro-ecosystems (based on the systems discussed by Thomas 1983b).

Symbols: M (the human population); DH (domesticated herbivores); WP (wild plants); WH (wild herbivores); P (wild preditors) and AR (alternative biotic resources, such as freshwater fish).

Systems with various levels of integration are shown, from those with low levels (A) to those with high levels (C). Boxes enclose holons of intensely-interacting elements. (See text for discussion.)

3.7 INTEGRATED AGRICULTURAL SYSTEMS: POSSIBLE CONSEQUENCES FOR POPULATION AND SETTLEMENT

A particularly important aspect of the hierarchical approach to agro-ecosystems is the emphasis on <u>rates</u> of interaction, production and consumption. In stable systems rates are highest within holons in the lowest levels of the hierarchy and decline both between holons at any one level and as one moves up through the levels of the hierarchy. In highly integrated agro-ecosystems, in which there are increased interactions between holons at all levels in the hierarchy, there will be increased rates of productivity (per unit time and/or unit of area) and increased rates of <u>transfer</u> of this production. These high levels of production will feed through to the human component and lead to increased population growth. Such a system will only be maintained (i.e. be stable) if there is no failure in any of the individual holons. If crop productivity were to fall (because of, for example, a shift in climate, a reduction of soil fertility or an epidemic disease affecting the animals trained to draw the plough) the existing high production/high demand system would become unstable, in which case the system would either collapse (perhaps the whole population would move elsewhere) or reorganise. In the latter case, a segment of the population could move elsewhere to found a new settlement; the rates of demand of the original settlement could therefore be lowered to a sustainable level because two new holons had been formed, which would interact at a higher hierarchical level in an expanded system. Archaeologically, both of these possible outcomes will result in an increase in the number of settlement sites in a region. It will, of course, be difficult in any group of sites to distinguish between those formed by the 'budding' process and those by the resettlement of whole populations.

It is striking to observe how the number of settlements proliferated greatly in various parts of Pakistan in the late fourth and third millennia B.C.: for example, in Baluchistan (de Cardi 1983), Sind (Flam 1987), Cholistan (Mughal 1982) and the north west (Khan *et al* 1987). Within this chronological period Flam (1982: Table 1) records an increase in both the number of sites and the median size of those sites over time. An analysis of Flam's data shows that the <u>variance</u> of site size (expressed as the coefficient of variation) also increased from the Amrian/Kot Dijian period to the Harappan period (Table 2). This increase in variance is probably associated with a diversification of site function over time: the earlier sites being mostly settlements based on local agricultural systems, and the later ones representing a broader spectrum ranging from temporary nomadic sites to large urban settlements (note: excavation is needed at more sites in order to confirm this). It is tempting to correlate these apparent changes with the development of complex, integrated, agro-ecosystems.

	Size classes of sites: areas in thousands of m^2					No. of sites	Median size	Coefficient of variation of site size
	≤5	>5	>10	>50	>200			
Amrian/ Kot Dijian	3	4	4	3	0	14	9.8	134.6
Harappan	5	4	11	5	2	27	17.0	188.6

TABLE 2 The numbers of sites in various size categories from pre-Harappan (Amrian and Kot Dijian) and Harappan periods in Sind, Pakistan. (Based on data in Flam 1982: Table 1).

As the number of contemporary settlements increases in an area, with varying levels of kin relationships between them, the development of a new hierarchical level is permitted, namely the between-site interaction sphere. The rates of interaction between such sites will be a function of both the kin relationships between the members of different settlements and of the distances between settlements. Groups of sites could therefore function as distinctive interactive holons and some ranking could emerge both within these holons and between them. Ranking, with associated regulation of the flow of information, energy and materials between settlements, could produce a stable hierarchical structure. Another way in which stable regional settlement systems could arise is by functional diversification between settlements. The development of specialised pastoral groups, and of pastoral nomadism, are examples of such diversification. Pastoralists are thought to have had an important role in Harappan socio-economic systems (Possehl 1979) and pastoral nomads have continued to be of social and economic importance in north west Pakistan up to the present day (Thomas 1983b: 309). Viewed from the perspective of systems theory, such diversification would have the effect of 'loosening-up' highly integrated systems by the creation of new, separate, process-functional holons which interact with one another at various (but higher) levels in the hierarchical system.

In both pre-Harappan and Harappan systems of the late fourth and third millennia B.C. we see the emergence of control or regulation of resources, as evidenced by clay sealings and carved steatite seals, and probably of information too. Such hierarchical control over the rates of flow of energy, materials and information was probably fundamental to the development of complex urban systems, a phenomenon seen first in an ephemeral form in the late fourth and early third millennia B.C. at sites such as Rehman Dheri (Durrani 1982), in the north west of Pakistan, and later in the mid-third millennium at Harappan sites in the Greater Indus system.

These hierarchical conceptualisations of the ways in which complex systems can become 'self organised' have analogues in van der Leeuw's (1981: 243-65) ideas of 'vortices of information flow', which also lead to the emergence of new levels of organisation in systems. Interesting though such models are, there remains a large gap between them and the archaeological data base; in the case of Pakistan archaeology, we need much more information about intra-site organisation, and inter-site relationships, in the pre-Harappan and Harappan periods. One thing is, however, clear: a hierarchical conceptualisation of complex systems provides a potentially unifying framework for the analysis of ecosystems, agro-ecosystems, economic systems and evolving social systems.

3.8 BACK DOWN TO EARTH

In practice it will prove to be very difficult to demonstrate the existence of process-functional hierarchies using archaeological data. This should not, of course, deter us from analysing archaeological data from a hierarchical viewpoint if it can be shown that hierarchy theory is likely to be relevant. How could this likelihood be assessed? One approach would be to seek archaeological evidence for levels of integration or control among the elements of past agro-ecosystems, or in human ecosystems in general. Systems in which there is a greater degree of integration of elements (i.e. those with more interactions between holons) can be regarded as being partitioned in a less hierarchical way than those in which the elements are less interactive.

For example, in the context of prehistoric Pakistan in the third millennium B.C., one could look for evidence of increased interaction between the crop-production and animal-production sub-systems. The role of animals in providing work (energy) inputs into crop production and the transport of products is shown clearly by terracotta models of ploughs (Mughal 1981) and carts (Thomas 1983b); ploughmarks have been revealed beneath the pre-Harappan levels at Kalibangan in north west India (Lal 1971). Reciprocal relationships, from crop-production systems to domesticated animals, are less easily demonstrated. The feeding of the by-products of crop processing (straw, chaff, etc.) to animals can, at present, only be inferred for these systems. The use of dung, particularly for manuring fields, is also difficult to detect in the archaeological record, although there are possibilities here. Some of the likely pathways along which dung might be moved between a settlement and its catchment are illustrated in Figure 5. A few of these pathways can be demonstrated archaeologically: charred goat faeces have been recovered from hearths, for example, and hearths have also yielded quantities of felted ashy material which could be the residue of burned dung cakes (Thomas 1983b).

Miller and Smart (1984) report that charred seeds can be recovered from dung burned as fuel. If seeds, and other parts of plants, are being introduced into sites in dung, perhaps this will be indicated by characteristic, or diagnostic, charred assemblages? Some possible pathways by which charred assemblages of plant materials could be incorporated into archaeological contexts (especially hearths) are shown in Figure 6; pathways involving the use of dung for fuel are stressed in this diagram. It is possible that the original inputs of plants, and therefore the diversity of animal feed, could be deduced from the charred assemblages. The degree of integration, if any, of crop products into the animal production system could then be inferred. This ambitious analysis has yet to be done and, as a caveat to all of this, it must be said that Gordon Hillman (pers. comm.) is not convinced that seeds in dung survive the burning process; he believes that the slow hot fire would convert them to ash. An experimental approach is clearly indicated.

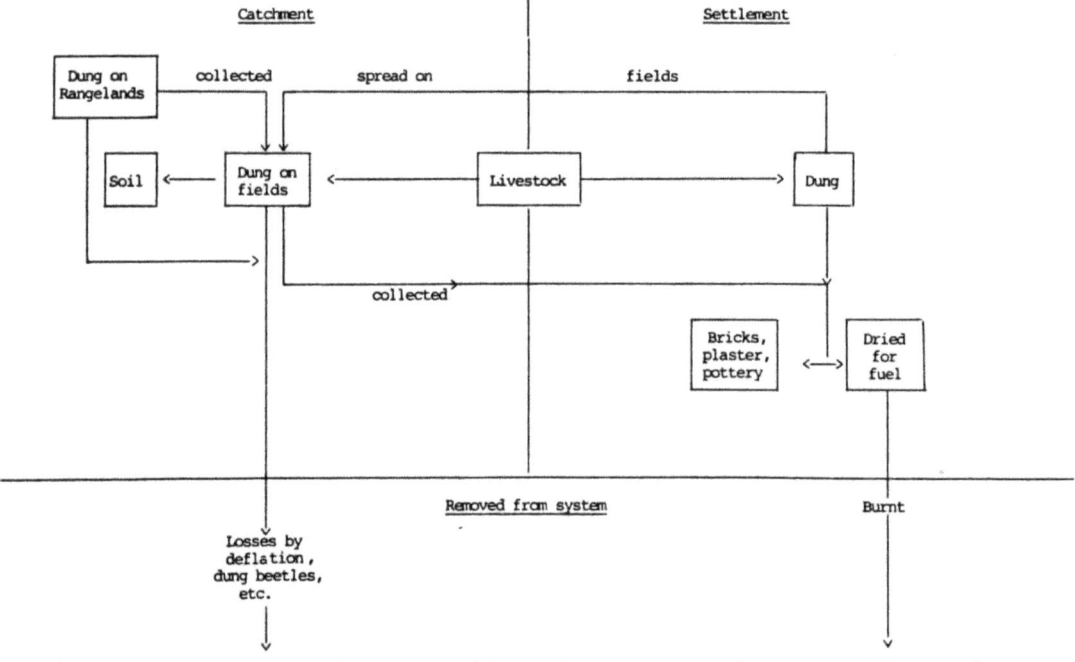

FIGURE 5 Movements and storage of dung between a settlement and its catchment or territory. (From Thomas 1983b: Figure 13:7).

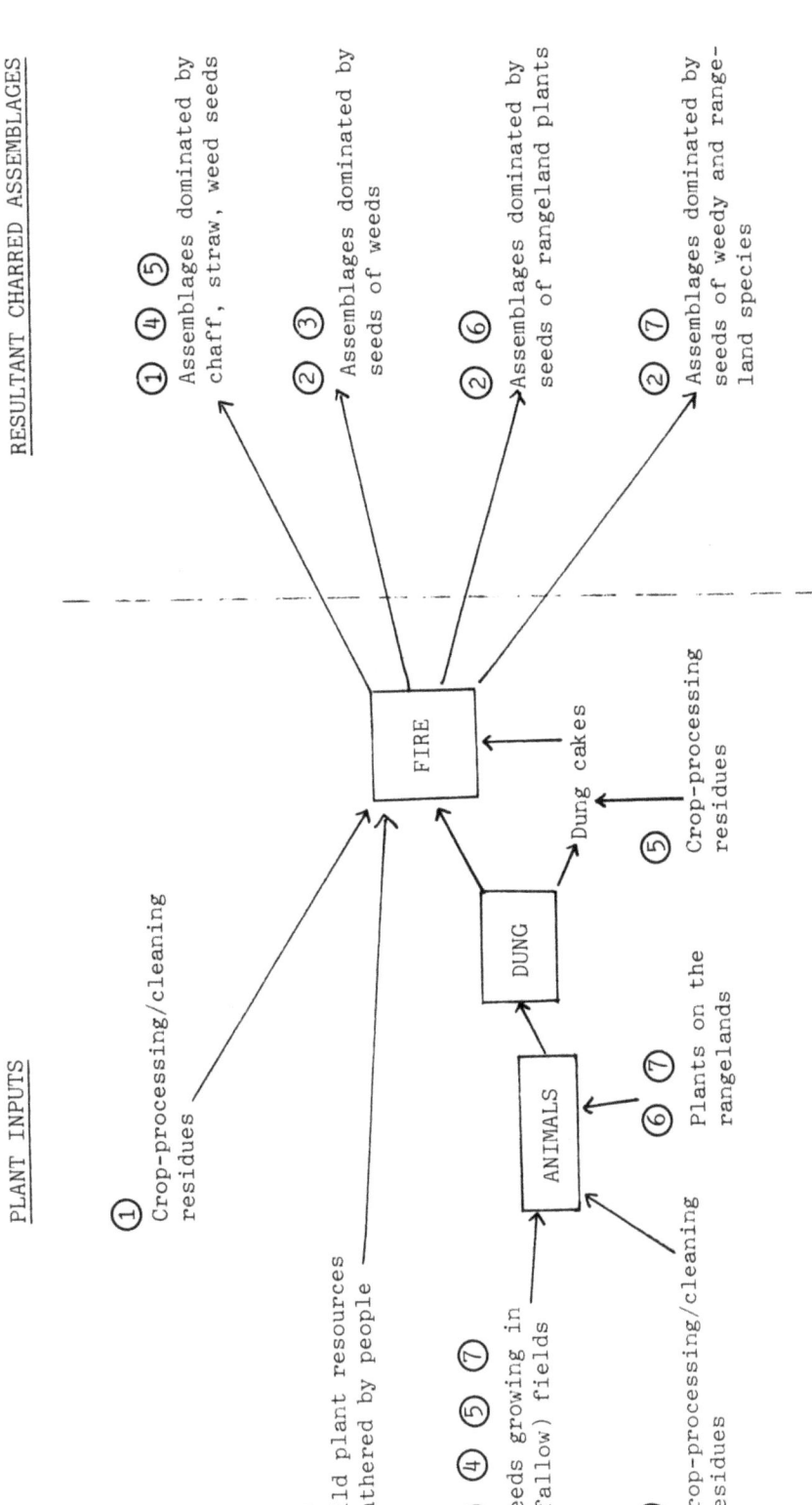

FIGURE 6 Pathways by which charred plant remains could be introduced into an archaeological site. Routes involving the use of animal dung are emphasised (see text for discussion).

3.9 CONCLUDING COMMENTS

It is important that we consider how best to handle complex sets of data. It may well be true, as David Harris suggested (not entirely seriously!) in his opening address to this conference, that theory has had its day and that we are all empiricists now. However, we have to do something with our impeccably empirical data and conceptual frameworks will still be needed to help organise it and to provide heuristic models (even if they are only descriptive ones). Systems models have proved valuable in the past and will, no doubt, do so in the future. However, those models which assume that systems are regulated by homoeostatic processes may be less realistic than alternative models, such as hierarchical ones. In strict terms, hierarchies should consist of holons which are defined by differences in the rates at which processes go on within and between them. This will be extremely difficult to demonstrate in archaeological data; it should be noted that the elements used to construct the models in Figure 4 are taxonomic or organisational concepts and therefore do not constitute genuine process-functional holons.

It is not necessary for us to demonstrate that past agro-ecosystems, or human ecosystems in general, were organised hierarchically (in the process-functional sense discussed here); in fact, it will probably be impossible to do so. However, clues to the degree of functional linkage between the elements of past systems can be gained from the archaeological record. Hierarchical approaches to past systems do not require that we construct rigid models of the sort illustrated here in Figure 4; these are strictly heuristic devices. Hierarchical conceptualisations of past systems could, however, permit the development of new hypotheses concerning the function and evolution of complex systems, and possibly provide an explanatory framework. A great advantage of hierarchical approaches is that they structure the flow of energy, materials and information in systems in ways relevant not only to ecologists and environmental archaeologists, but also to cultural archaeologists.

ACKNOWLEDGEMENTS

I thank my colleagues, Robert Knox and Farid Khan, for valuable discussion. Harry Kenward gave valuable positive and negative feedback which helped clarify both my thoughts and the ways in which I express them.

BIBLIOGRAPHY

ALDAG, R. (1987) Simple and diversified crop rotations - approach and insight into agroecosystems. In: *Potentials and limitations of ecosystem analysis.* (eds Schulze, E.-D. and Zwölfer, H.) Ecological Studies No. 61. Berlin and Heidelberg: Springer Verlag pp. 100-14.

BAYLISS-SMITH, T.P. (1982) *The ecology of agricultural systems.* Cambridge: University Press.

BERTALANFFY, L. VON (1968) *General systems theory.* New York: Braziller.

BINFORD, L.R. (1965) Archaeological systematics and the study of cultural process. *American Antiquity* **31** 203-10.

CHESSON, P.L. AND CASE, T.J. (1986) Overview: nonequilibrium community theories: chance, variability, history and coexistence. In: *Community ecology.* (eds Diamond, J. and Case, T.J.) New York: Harper and Row pp. 229-39.

CONNELL, J.H. AND SOUSA, W.P. (1983) On the evidence needed to judge ecological stability or persistence. *American Naturalist* **121** 789-824.

DEANGELIS, D.L. AND WATERHOUSE, J.C. (1987) Equilibrium and nonequilibrium concepts in ecological models. *Ecological Monographs* **57** 1-21.

DE CARDI, B. (1983) *Archaeological surveys in Baluchistan, 1948 and 1957.* Institute of Archaeology Occasional Publication No. 8. London: Institute of Archaeology.

DIAMOND, J. AND CASE, T.J. (eds) (1986) *Community Ecology.* New York: Harper and Row.

DURRANI, F.A. (1982) Rehman Dheri and the birth of civilization in Pakistan. *Bulletin of the (University of London) Institute of Archaeology* **18** (for 1981) 191-207.

ELLEN, R. (1982) *Environment, subsistence and system. The ecology of small-scale social formations.* Cambridge: University Press.

FLAM, L. (1982) Suggested archaeological evidence for complex social organizations in prehistoric Sind. In: *Anthropology in Pakistan: recent socio-cultural and archaeological perspectives.* (eds Pastner, S. and Flam, L.) South Asian Occasional Papers and Theses No. 8. Ithaca, New York: Cornell University pp. 219-30.

FLAM, L. (1987) Recent explorations in Sind: palaeogeography, regional ecology and prehistoric settlement patterns (*circa* 4000-2000 B.C.). In: *Studies in the archaeology of India and Pakistan.* (ed. Jacobson, J.) Warminster, England: Aris and Phillips.

FLANNERY, K.V. (1968) Archaeological systems theory and early Mesoamerica. In: *Anthropological archaeology in the Americas.* (ed. Meggers, B.) Washington, D.C.: Anthropological Society of Washington pp. 67-87.

FLANNERY, K.V. (1972) The cultural evolution of civilizations. *Annual Review of Ecology and Systematics* **3** 399-426.

FLANNERY, K.V. AND MARCUS, J. (1976) Formative Oaxaca and the Zapotec cosmos. *American Scientist* **64** 374-83.

FOIN, T.C. AND DAVIS, W.G. (1987) Equilibrium and nonequilibrium models in ecological anthropology: an evaluation of 'stability' in Maring ecosystems in New Guinea. *American Anthropologist* **89** 9-31.

GIGON, A. (1987) A hierarchic approach in causal ecosystem analysis: the calcifuge-calcicole problem in alpine grassland. In: *Potentials and limitations of ecosystem analysis.* (eds Schulze, E.-D. and Zwölfer, H.) Ecological Studies No. 61. Berlin and Heidelberg: Springer Verlag pp. 228-44.

GREEN, S.W. (1980) Toward a general model of agricultural systems. *Advances in Archaeological Method and Theory* **3** 311-55.

HARRIS, D.R. (1969) Agricultural systems, ecosystems and the origins of agriculture. In: *The domestication and exploitation of plants and animals.* (eds Ucko, P.J. and Dimbleby, G.W.) London: Duckworth pp. 3-15.

HARRIS, D.R. (1977) Settling down: an evolutionary model for the transformation of mobile bands into sedentary communities. In: *The evolution of social systems.* (eds Friedman, J. and Rowlands, M.J.) London: Duckworth pp. 401-17.

KHAN, F., KNOX, J.R. AND THOMAS, K.D. (1987) The Bannu archaeological project: a study of prehistoric settlement in Bannu District, Pakistan. *South Asian Studies* **3** 83-90.

KIKKAWA, J. AND ANDERSON, D.J. (eds) (1986) *Community Ecology: Pattern and Process.* Oxford: Blackwell Scientific Publications.

LAL, B.B. (1971) Perhaps the earliest ploughed field so far excavated anywhere in the world. *Puratattva* **4** 1-3.

LEEUW, S.E. VAN DER (1981) Information flows, flow structures and the explanation of change in human institutions. In: *Archaeological approaches to the study of complexity.* (ed. Leeuw, S.E. van der) Amsterdam: University of Amsterdam pp. 228-329.

McGLADE, J. AND ALLEN, P.M. (1986) Fluctuation, instability and stress: understanding the evolution of a swidden horticultural system. *Science and Archaeology* **28** 44-50.

McINTIRE, C.D. AND COLBY, J.A. (1978) A hierarchical model of lotic ecosystems. *Ecological Monographs* **48** 167-90.

MILLER, N.F. AND SMART, T.L. (1984) Intentional burning of dung as fuel: a mechanism for the incorporation of charred seeds into the archaeological record. *Journal of Ethnobiology* **4** 15-28.

MOORE, P.D. (1983) Ecological diversity and stress. *Nature* **306** 17.

MUGHAL, M.R. (1981) New archaeological evidence from Bahawalpur. In: *Indus Civilisation: new perspectives.* (ed. Dani, A.H.) Islamabad: Centre for the Study of the Civilization of Central Asia, Quaid-i-Azam University pp. 33-41.

MUGHAL, M.R. (1982) Recent archaeological research in the Cholistan Desert. In: *Harappan Civilization: a contemporary perspective.* (ed. Possehl, G.L.) Warminster, England: Aris and Phillips pp. 85-95.

O'NEILL, R.V., DEANGELIS, D.L., WAIDE, J.B. AND ALLEN, T.F.H. (1986) *A hierarchical concept of ecosystems.* Monographs in Population Biology 23. Princeton: University Press.

PATTEE, H.H. (ed.) (1973) *Hierarchy theory.* New York: Braziller.

PIMM, S.L. (1984) The complexity and stability of ecosystems. *Nature* **307** 321-26.

POSSEHL, G.L. (1979) Pastoral nomadism in the Indus civilization: an hypothesis. In: *South Asian archaeology 1977.* (ed. Taddei, M.) Naples: Istituto Universitario Orientale Seminario di Studi Asiatici Series Minor VI pp. 537-51.

RENFREW, C. (1979) Systems collapse as social transformation: catastrophe and anastrophe in early state societies. In: *Transformations: mathematical approaches to cultural change.* (eds Renfrew, C. and Cooke, K.L.) London: Academic Press pp. 481-506.

RINDOS, D. (1984) *The origins of agriculture. An evolutionary perspective.* London: Academic Press.

SANDERS, H.L. (1968) Marine benthic diversity: a comparative study. *American Naturalist* **102** 243-82.

SHERRATT, A.G. (1981) Plough and pastoralism: aspects of the secondary products revolution. In: *Pattern of the past: studies in honour of David Clarke.* (eds Hodder, I., Isaac, G. and Hammond, N.) Cambridge: University Press pp. 261-305.

SCHULZE, E.-D. AND ZWÖLFER, H. (eds) (1987) *Potentials and limitations of ecosystem analysis.* Ecological Studies No. 61. Berlin and Heidelberg: Springer Verlag.

SPEDDING, C.R.W. (1979) *An introduction to agricultural systems.* London: Applied Science Publishers.

THOMAS, K.D. (1983a) Tarakai Qila: site, economy and environment. In: *Site, environment and economy.* (ed. Proudfoot, B.) Oxford: British Archaeological Reports S173 pp. 127-44.

THOMAS, K.D. (1983b) Agricultural and subsistence systems of the third millennium B.C. in north-west Pakistan: a speculative outline. In: *Integrating the subsistence economy.* (ed. Jones, M.) Oxford: British Archaeological Reports S181 pp. 279-314.

THOMAS, K.D. (1986) Environment and subsistence in the Bannu Basin. In; *Lewan and the Bannu Basin.* (eds Allchin, F.R., Allchin, B., Durrani, F.A. and Khan, F.) Oxford: British Archaeological Reports S310 pp. 13-33.

WATSON, P.J., LEBLANC, S.A. AND REDMAN, C.L. (1984) *Archeological explanation. The scientific method in archeology.* New York: Columbia University Press.

WEN DAZHONG AND PIMENTEL, D. (1986a) Seventeenth century organic agriculture in China: I. Cropping systems in Jiaxing region. *Human Ecology* **14** 1-14.

WEN DAZHONG AND PIMENTEL, D. (1986b) Seventeenth century organic agriculture in China: II. Energy flows through an agroecosystem in Jiaxing region. *Human Ecology* **14** 15-28.

WHYTE, L.L., WILSON, A.G. AND WILSON, D. (eds) (1969) *Hierarchical structures.* New York: Elsevier.

II

DOMESTICATION

4

SIZE AND SEX: EVIDENCE FOR THE DOMESTICATION OF CATTLE IN THE NEAR EAST

Caroline Grigson

4.1 INTRODUCTION

It hardly needs saying that one of the most fundamental changes in human development has been the change from hunting and gathering to the utilisation of domestic plants and animals, particularly ungulates. Although most people know what is meant by the term domestication there is little consensus on its definition. As far as ungulates go one can say that animals have been domesticated when a breeding group has been genetically isolated from the parent population by people who interfere with its breeding, food supplies and movements. The five main criteria used to establish whether the animal remains from an archaeological site are from wild or domesticated animals are:

(1) The presence of a particular species outside the geographical range of the wild forbear.

(2) A sudden diachronic increase of a species within a faunal spectrum (Clutton-Brock 1981; Davis 1982).

(3) An age structure which differs from that found in the wild. Ducos (1968, 1969, 1978b) considers that a difference *per se* is enough to indicate domestication; Perkins (1964), Hesse (1978, 1982) and others claim a predominance of young animals. On the other hand Bökönyi (1976, 1977) states categorically that the bones of old animals predominate in an assemblage of domesticates.

(4) An age and sex structure which differs from that found in the wild (Hesse 1984).

(5) A reduction in size (Møhl 1957; Boessneck 1958; Jewell 1962; Degerbøl 1963; von den Driesch and Boessneck 1976; Grigson 1982; and many others).

Of all these criteria the first is probably the most reliable, but unfortunately it is of little use in most archaeological contexts as the wild ancestors of all the common domestic animals were very widely distributed. The second criterion may be valid, but has in fact only been used to support the idea of domestication in the replacement of gazelles by goats in the Near East.

The problems with demographic interpretations are that different authorities use different criteria, we usually do not know what the age structure of the local wild population was, and the implication that hunters hunted randomly is probably untrue (Binford 1978). More convincingly, however, Hesse (1984) claims domestication of the goat at Ganj Dareh in the Zagros in the ninth/eighth millennia bc, because although equal numbers of each sex were found almost all the males were killed before, and the females after, the age of two and a half years (note Ganj Dareh is now generally considered to be eighth-seventh millennia bc; Gebel 1984). Goats, cattle and other ungulates tend to form separate nursery herds composed of females and their young for much of the year. Thus such a demographic structure could simply represent predation on a nursery herd, which would have been far easier than hunting solitary males, but it may also be that the people appropriated such herds on the hoof. This could be a scenario for the initial domestication of an ungulate species.

Although the last criterion, diminution in size, does not necessarily follow domestication the fact is that modern domestic ungulates are smaller than their wild forebears. Thus modern domestic cattle (*Bos taurus* L.) are smaller than their wild ancestor, the aurochs (*B. primigenius* Boj.), although in other respects their anatomy is very similar (Grigson 1978). When this size difference is traced through time in European sites there is a marked diminution which invariably occurs at the beginning of the Neolithic. In Britain and Switzerland and probably elsewhere in Europe the size change in cattle is accompanied by a predominance of females (Grigson 1982; Legge 1981; Sakellaridis 1979), a size reduction in pigs and the presence of exotic domesticates (sheep and goats). There is little doubt that the whole package represents a fundamental change in man-animal relationships, and that that change was the husbandry of domestic ungulates. Nevertheless it must be remembered that the wild aurochs persisted into historical time; although it may have undergone a diminution over time this was probably much more gradual than that documented in the archaeological record which is, by definition, the material remains of human activity.

It is generally assumed that the large cattle of the Natufian (tenth-ninth millennia bc) and earlier were wild, and it is known from pictorial evidence showing milking (Grigson 1987) that the small cattle of the fourth millennium bc were domesticated. We need to know what happened between the tenth and fourth millennia bc.

The present work concentrates on the diminution of cattle as an indicator of domestication in the Near East, but also uses size in an attempt to analyse one aspect of demography: the relative numbers of the two sexes.

In assessing the size of *Bos* in the Levant in the late Pleistocene and early Holocene, Davis (1981) came to the conclusion that cattle had probably undergone two stages of diminution. The first stage, about which he was uncertain, occurred at the end of the Pleistocene (in the Natufian period of the tenth and ninth millennia bc); and the second stage occurred between the Pottery Neolithic B and the Pottery Neolithic (seventh and sixth or fifth millennia bc) at Jericho. He interpreted the first diminution as post-Pleistocene dwarfing related to climatic change and the second as the effect of domestication. It should be mentioned that the Pottery Neolithic sample comprised a single bone.

Jarman (1969) discussed size change in cattle in the eastern Mediterranean area and, though he was very disparaging about the idea that the Holocene diminution was the result of domestication, his data do show that the diminution did happen.

In the same year I pointed out that we know that modern domestic cattle are smaller than the ancestral wild aurochs (Grigson 1969). I argued that diminution could be general, for example, the known decrease in size of the aurochs between the Pleistocene and the early Holocene throughout northern Europe, or might occur in isolated populations. I postulated that only one isolating mechanism is known to be associated with size reduction in cattle in mainland areas: domestication; where the presence of people can be assumed archaeologically that is the most likely explanation for the presence of cattle that are smaller than the local aurochs. Wendorf and Schild (1984) criticised me on the grounds that I wrote in my 1969 paper that people are the only factor involved in the diminution of cattle, but clearly they misunderstood my arguments about local as opposed to general diminution in a large area.

The present paper investigates size change from about 10,000 bc to 2000 B.C. on the basis of the more copious data now available from other parts of the Near East as well as the Levant. Care has been taken to assess the evidence on the basis of known size ranges and sexual bimodality, rather than individual measurements (Davis 1981), means (Jarman 1969), or size indices with unknown sample sizes and degrees of variation (Kurtén 1959).

4.2 MATERIAL

Sites from which cattle measurements have been utilised are listed in the Appendix together with the periods and dates and names of the authors of the published reports. I have added unpublished data of my own from fifth and fourth millennia bc sites in the Levant and Dr Davis has kindly provided details of the measurements that he used in his paper on size (Davis 1981) and some from his as yet unpublished sites in Iran.

Dating the sites has been a major problem. Even when radiocarbon dates are available, not all authorities agree about the validity of particular dates and even if a date is accepted, problems of stratigraphy mean that it does not necessarily provide an accurate date for other material in the same context. Equally archaeologists do not always agree about the dating of particular types of pottery and do not always specify exactly what pottery was present. The division of time into intervals is bound to be arbitrary, and there will always be difficulties about dates that fall at the boundary between one period and another. Some archaeozoological reports give only the vaguest information about the dates of their material. In assigning dates I have made much use of Mellaart's classic work on the Neolithic (Mellaart 1975) and also of Mellaart (1970), Burney (1977) and Gebel (1984).

An additional problem with dating is the use of calendar dates and calibrated or uncalibrated carbon dates. As uncalibrated dates are generally invoked for the earlier periods and calendar dates for the late fourth and third millennia, in the present paper third millennium dates are calendar years (B.C.), tenth-fourth millennia dates are uncalibrated radiocarbon years (bc) and later fourth millennium dates B.C. are included with the radiocarbon fourth millennium.

The geographical positions of the sites are shown in Figure 1. It was found that the data fell into two major groups: those from the western part of the Near East - the Levant, northern Syria and central and western Turkey; and those from the east - Iraq, Iran and Turkestan.

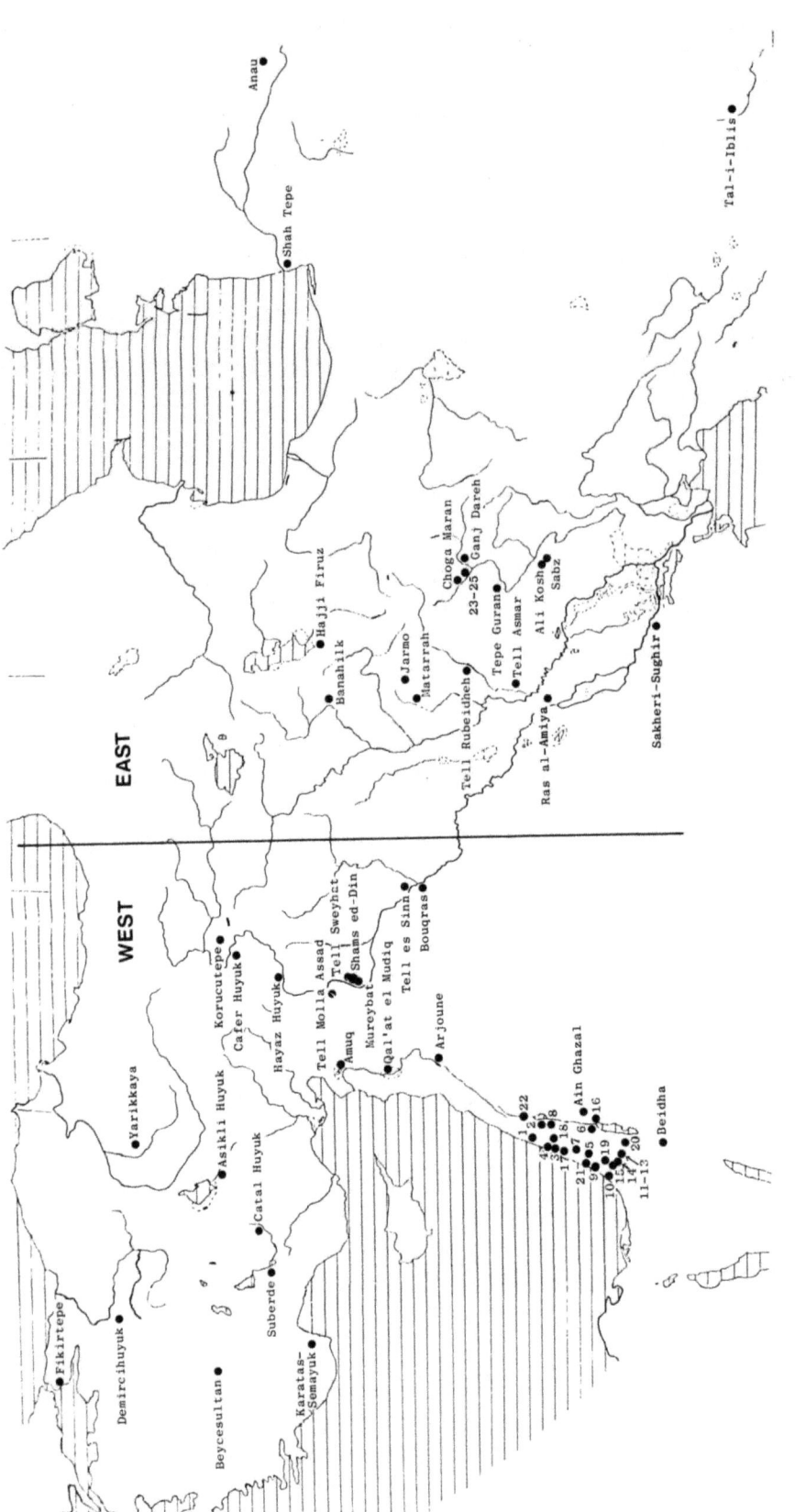

FIGURE 1 Map of the Near East showing the sites mentioned in the text, and the division between the western and eastern areas utilised in the present paper.

1. Hayonim. 2. Mallaha. 3. Kebara. 4. El Wad. 5. Hatoula. 6. Jericho. 7. Shukbah. 8. Tel Eli. 9. Ashkalon. 10. Qatif Y2 and Y3. 11. Neve Noy. 12. Bir es-Safadi. 13. Shiqmim. 14. Gilat. 15. Grar. 16. Teleilat Ghassoul. 17. Metzer. 18. Megiddo Tombs and En Shadud. 19. Tell Gat. 20. Arad. 21. Tel Aviv. 22. Tel Dan. 23. Tepe Sarab. 24. Siahbid. 25. Dehsavar.

lower third molar, length at base of crown, 48.8	naviculo-cuboid GB 67
humerus Bd 97*	metatarsal Bp 62
humerus Bt 89	metatarsal Bd 68
radius Bp 100	anterior proximal phalanx Bp 39
metacarpal Bp 74	posterior proximal phalanx Bp 35.5
metacarpal Bd 73	anterior middle phalanx Bp 36
tibia Bd 78	posterior middle phalanx Bp 34
calcaneum Gl 165	
astragalus GLl 83	

TABLE 1 Selected dimensions, in millimetres, of the female *Bos primigenius* from Ullerslev, Denmark, measured by Degerbøl (Degerbøl and Fredskild 1970). Abbreviations follow von den Driesch (1976).

* Only those dimensions that Degerbøl seems to have measured from the same points as those recommended by von den Driesch have been reproduced here and to this end Tove Hatting of the Universitets Zoologiske Museum, Copenhagen, has kindly remeasured the distal breadth of the humerus (Bd).

4.3 METHOD

The problem most often encountered in archaeozoological comparisons is paucity of material. In making size comparisons ideally one should use the absolute size of various bone elements. This method is both simple and suitable for statistical tests. However, as in most inter-site archaeozoological comparisons, there are not enough data to do this in the Near East for any but a few sites. I therefore follow Meadow (1981, 1983, 1984) and Buitenhuis (1985) in making logarithmic comparisons with a standard skeleton. Like Meadow, but unlike Buitenhuis who used natural logarithms (base e), I use common logarithms (base 10). This allows dimensions of different bone elements to be plotted in the same histogram. The skeleton chosen by Buitenhuis is a female *Bos primigenius* from Ullerslev in Denmark, dated by pollen analysis to the early Boreal. It is described, with detailed measurements, by Degerbøl (Degerbøl and Fredskild 1970).

Although von den Driesch's measurement guide (1976) appeared after Degerbøl's work, like her he based his measurement positions on Duerst (1930) and it is likely that they are identical. However, it should be noted that the distal breadth of the humerus which he gives as 92mm should in fact be 97mm (I am indebted to Tove Hatting for checking this measurement for me). The dimensions used here are listed in Table 1.

Degerbøl's work has established, without any doubt, the wide sexual dimorphism of *Bos primigenius* and also the high degree of overlap in size between wild cows and prehistoric domestic bulls. It is important to establish, not only the range of variation of the measurements, but also the degree of bimodality. This has been done by plotting the measurements that Degerbøl took from complete and partial skeletons of known sex and from Danish mesolithic sites (Figure 2). It has been assumed that any bimodality in the Near Eastern archaeological material will be the result of sexual differences. Such sexual differences furnish information which is at least as valuable as overall size change, and for this reason only dimensions that are sexually dimorphic have been chosen; so despite its relative frequency the lower third molar (M3) length has not been utilised. Although proximal and middle phalanges are the most commonly occurring elements they are not suitable for comparisons of this kind unless they have been accurately assigned to the fore or hind foot, which is rarely attempted. However, the Ullerslev M3 lengths, and proximal breadths of the phalanges have been included in Table 1 for the sake of completeness, and because they have some utility for general size comparisons.

4.4 RESULTS

4.4.1 Wild cattle in the Near East

Histograms of the log differences from Ullerslev of all the available measurements of *Bos primigenius* from millennium to millennium from archaeological sites in the early Holocene of the Middle East show that from the Natufian period (tenth and ninth millennia bc) to the Pre-ceramic and Pre-pottery Neolithic B (PPNB) (seventh millennium bc) there was no difference in size of *Bos* from the various areas and no diminution over time (Figure 3A-E). All the larger samples showed a marked bimodality which, on the basis of what is known about sexual dimorphism in wild cattle (see above) must be due to sexual differences. Equal numbers of each sex appear to have been taken.

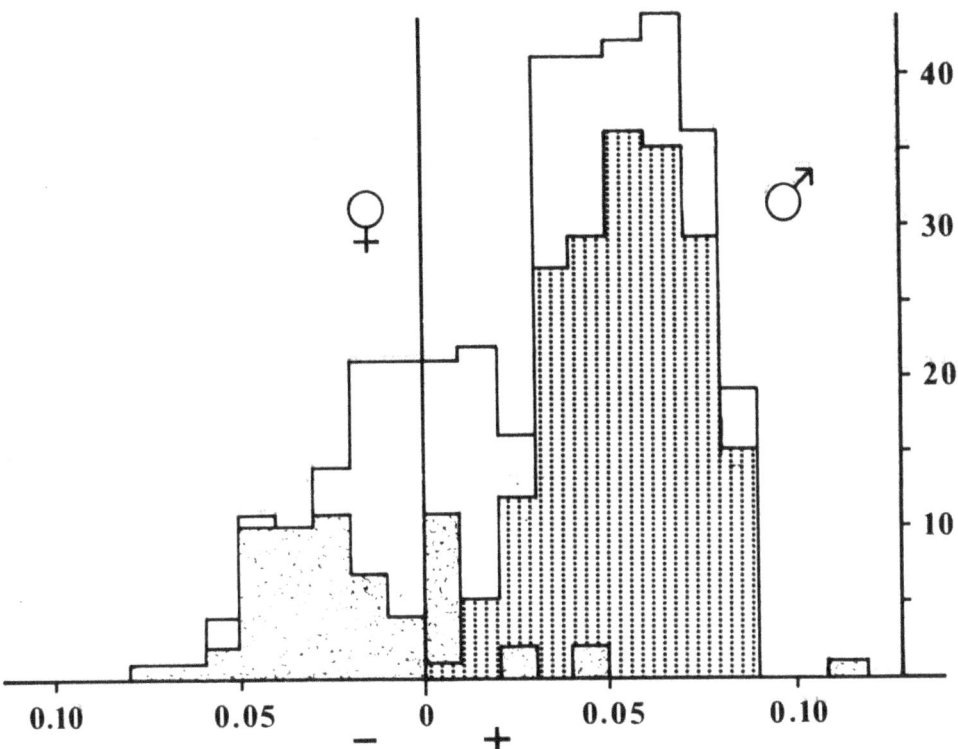

FIGURE 2 Size and sexual dimorphism of *Bos primigenius* in the Danish Mesolithic. Various dimensions (defined in Table 1) of complete skeletons of known sex and of single specimens from archaeological sites compared on a $\log_{10}$ scale with a standard animal (the Ullerslev cow).

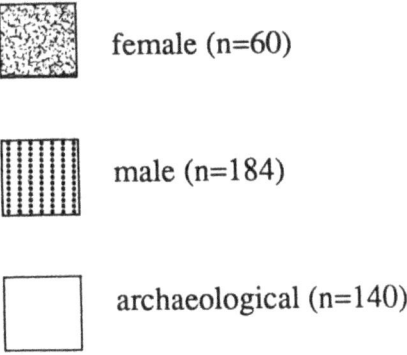

Data from Degerbøl and Fredskild 1970.

FIGURE 3 The size of *Bos* in the Middle East in the early Holocene from the tenth to the sixth millennium bc.

Note the similarity in size of all except the sixth millennium cattle in the western part of the region (3G), which are smaller.

- 3A   Kebara 1931 B, Hayonim B, Shukbah, Hatoula (Natufian), El Wad B, Jericho (Proto-neolithic); n=24.

- 3B   Ganj Dareh; n=5

- 3C   Mureybet, Jericho (PPNA), Hatoula (PPNA); n=26.

- 3D   Beidha, Jericho (PPNB), Tell Eli (IV), Ain Ghazal, Tell Molla Assad, Tell es Sinn, Bouqras, Suberde, Cafer Hüyük, Asikli Hüyük; n=35. The smallest dimension is of a distal humerus from Beidha identified by Perkins (in Hecker 1975) as *Bos*, but probably from *Alcelaphus*.

- 3E   Ali Kosh; n=10.

- 3F   Matarrah, Tepe Guran, Jarmo, Hajji Firuz, Sarab; n=40.

- 3G   Ashkalon, Jericho (Pott. Neo. A), Qalat el Mudiq, Amuq (A and B), Fikirtepe; n=136.

For sources of data see Appendix.

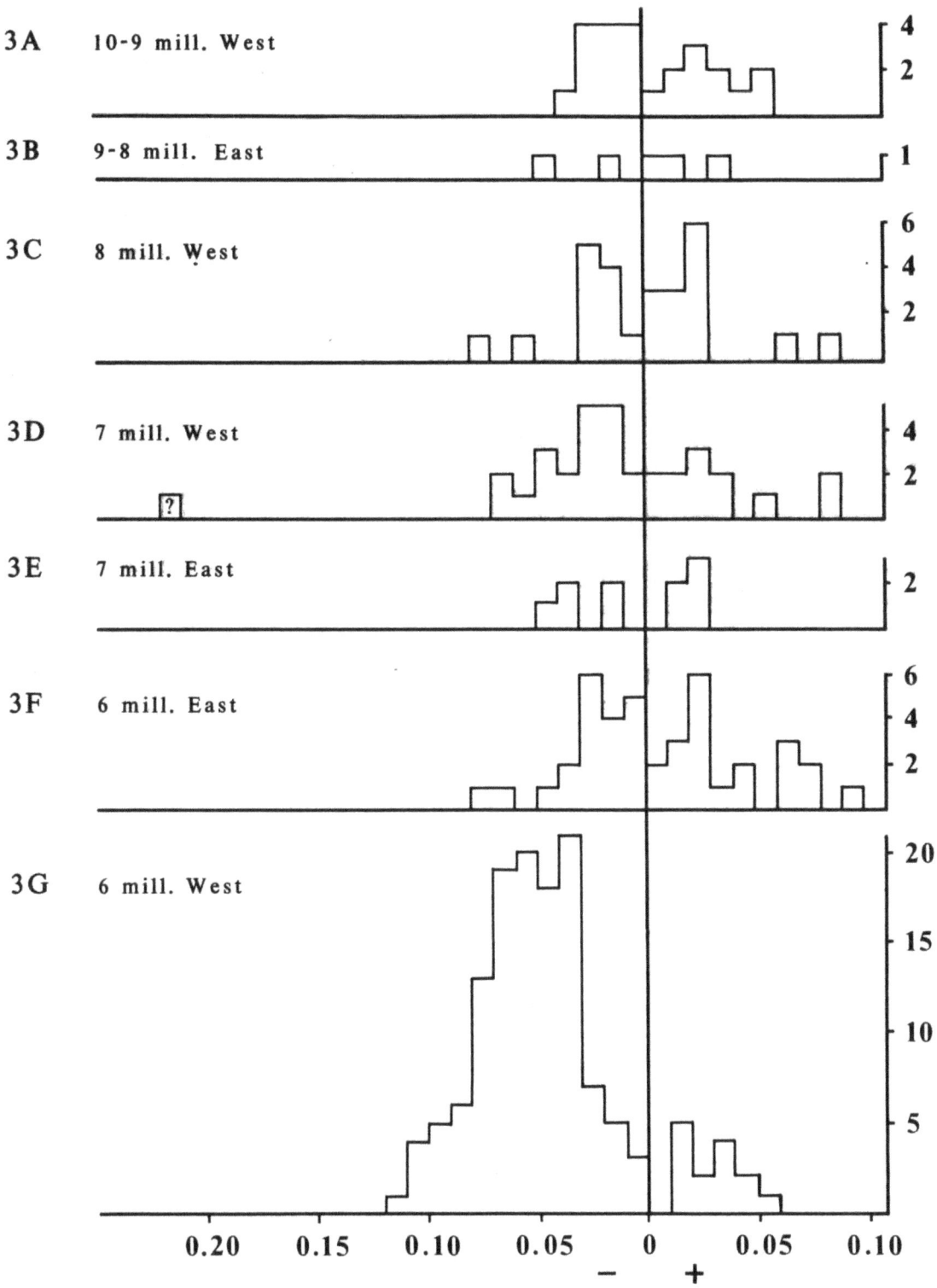

Bouchud (1987) has recently claimed that, in addition to *Bos primigenius* the Natufian site of Mallaha in the Jordan Valley had a second, smaller species of *Bos*. He calls this *Bos taurus* and maintains that it was wild. Certainly some of the measurements that he presents (and some of those published by Ducos (1968, 1978c) in an earlier work on the same material) are too small for *Bos primigenius*. It seems extremely unlikely that there was a second species of *Bos* present in the Near East in the Natufian and that it was confined to Mallaha. Is it possible that these bones are intrusive? As well as *Bos*, red deer (*Cervus elaphus* L.), fallow deer (*Dama mesopotamica* Pallas), and hartebeest (*Alcelaphus buselaphus* Brooke) are all represented at Mallaha. All of these species are sexually dimorphic and all overlap in size, so it must have been extremely difficult to identify material, particularly when fragmentary. One wonders whether the so-called *Bos taurus* bones have been misidentified - could they be male *Alcelaphus*? An additional point is that the taxon *Bos taurus* is incorrect for a wild *Bos* species. *Bos taurus* L. is the binomial of unhumped, domestic cattle.

The earliest period for which any degree of domestication of cattle has been claimed is the Pre-pottery Neolithic A (PPNA) (eighth millennium bc), in particular at Mureybet in north eastern Syria, where Ducos and Helmer (1982) claim that the cattle show signs of 'protoélevage' (incipient, or pre-domestication) because of the ageing pattern based on tooth wear, criterion three above (see also Ducos 1978c). However, the histogram of size in the eighth millennium bc (Figure 3C) shows that cattle were of the same size as in the preceding millennia, so the morphometric data do not support the claim for domestication in this period.

Full domestication of cattle has been claimed for the subsequent (seventh) millennium bc at the PPNB sites of Bouqras and Tell es Sinn in north eastern Syria by Clason (1980) and at Hayaz Hüyük, a little to the north in modern Turkey, by Buitenhuis (1985). Some of the measurements from Bouqras and Tell es Sinn were considered by Clason to be below the lower size limit of wild cattle. However, Figure 3D indicates that all the dimensions published so far are actually within the size range of wild cattle in the Near East, so the claim for domestication of cattle in the seventh millennium bc cannot be substantiated. Nevertheless the histogram of all the PPNB measurements together does suggest a slight emphasis on females.

Hayaz Hüyük is more difficult to evaluate. As already mentioned it was Buitenhuis (1985) who first used the Ullerslev cow as the standard *Bos primigenius*; from his comparison of its dimensions with those from Hayaz he concluded that some of the cattle at Hayaz were domesticated. However, he only gives the natural logarithms (base e) equivalents of his measurements without the raw data, he does not say which dimension each represents, and he does not compare them with the range of wild cattle. If his data are translated into common logarithms (base 10) and compared with the ranges used in the present work, there is a slight indication of a shift to the left, which could be interpreted either as an emphasis on females or as a slight reduction in size. However, as his data may include measurements that are unsuitable for the present comparisons, this cannot be substantiated until the raw data are published.

Although Bökönyi (1976, 1978) has claimed that domesticated cattle were present in the seventh millennium bc faunal samples that he was studying from Labweh (Lebanon), no morphometric data have been published to validate this claim. Similarly there is no published evidence to support his view that cattle were probably domestic at Tell Ramad (southern Syria) in the late seventh millennium bc, as well as at pre-pottery Jericho and aceramic Hacilar (south western Turkey). The author of the faunal reports on Jericho (Clutton-Brock 1971, 1979) states on the basis of size that the cattle in the pre-pottery levels were wild, and no measurements of cattle, nor any other evidence of domestication have been published for either Hacilar (Perkins 1973; Westley 1970), or Tell Ramad (Hooijer 1966; Contenson 1966).

It should be noted that the data given by Perkins (1969) for cattle size in the seventh millennium bc levels at Çatal Hüyük (also in Turkey) are all well within the wild range, despite his widely accepted claim (Bökönyi 1976; Clutton-Brock 1981) that they were domesticated (see Figure 5).

Continuation of the comparison into the sixth millennium bc in the east of the area (Iran and Iraq) is bedevilled with problems of dating. The most important site is Jarmo, in the Zagros, which has pre-ceramic levels underlying ceramic levels. Opinions differ as to the dates of the levels, but they are usually considered to be just before and just after 6000 bc respectively. However in the tables of measurements in the faunal report, Stampfli (1983) does not distinguish between the pre-ceramic and ceramic levels. As more animal remains were retrieved from the ceramic than the pre-ceramic levels, and as their range of variation is consistent with the idea, the cattle measurements are treated here as representing a single population of about, or just after, 6000 bc.

All the bones from Jarmo are in the wild range; so are all of those from other early ceramic sites of the same timespan in Iran and Iraq - Tepe Guran (Flannery pers. comm.), Sarab (Bökönyi 1977) and Matarrah (Stampfli 1983); and those from the later sixth millennium bc site of Hajji Firuz (Meadow 1983). When data from all these sites are plotted together (Figure 3F), the histogram shows bimodality and equal numbers of the two sexes. Bökönyi's (1976, 1978) claim that the cattle from the sixth millennium bc at Tell-es-Sawwan (Iraq) were domesticated cannot be substantiated in the absence of published data. However, the very small fraction of the fauna that cattle represent (0.07% and 1.30% respectively) argues against their domestic status.

The situation in the sixth millennium bc in the western part of the Near East is very different and is discussed in detail below. The similarity in size of cattle from the tenth to the sixth millennia bc in the east (Figure 4B), to that of the tenth to the seventh millennia bc (Figure 4A) in the west allows all the measurements to be combined in a single histogram (Figure 4C). This is a summary of the size of wild cattle in the Middle East in the early Holocene and can be used as a basis of comparison with cattle remains from subsequent sites to assess whether they had undergone diminution. One would expect the width of the range and the degree of sexual difference to be about the same. Comparison of this histogram (Figure 4C) with that of the dimensions of skeletons of known sex, as well as of archaeological material from the early Holocene of Denmark (shown in Figure 1) suggests that the range of variation is much the same. Both ranges are markedly bimodal and, as the bimodality in the Danish sample is demonstrably due to sexual dimorphism, that must also be true of the Near Eastern sample. The Danish sample is not suitable for statistical comparison, because it is not random and it contains sets of dimensions from the same individuals. However, on a visual basis the position of the two peaks and of the valley between them is about 0.020 less on the log scale in the Near Eastern than in the Danish sample, suggesting that they were very roughly 4.50% smaller.

This can be tested by comparing the absolute size of the astragalus, which is one of the elements that occurs most commonly in archaeological material. Table 2 shows differences in the maximum length of the astragalus in the early Holocene from north

FIGURE 4 The size of *Bos primigenius* in the Middle East in the early Holocene, grouped geographically, but excluding the data from the west during the sixth millennium.

4A   The sites in the western part of the area from 10th-7th millennium. (Figures 3A, 3C, 3D combined); n=84.

4B   The sites in the eastern part of the area from 10th-6th millennium (Figures 3B, 3E, 3F combined); n=55.

4C   All the data (except Figure 3G) combined: the size, range and sexual dimorphism of *Bos primigenius* in the Middle East in the early Holocene. Because of the similarity of the curve to that of *Bos primigenius* of known sex in Denmark (Figure 2) it is assumed that the two peaks represent females and males respectively; n=139.

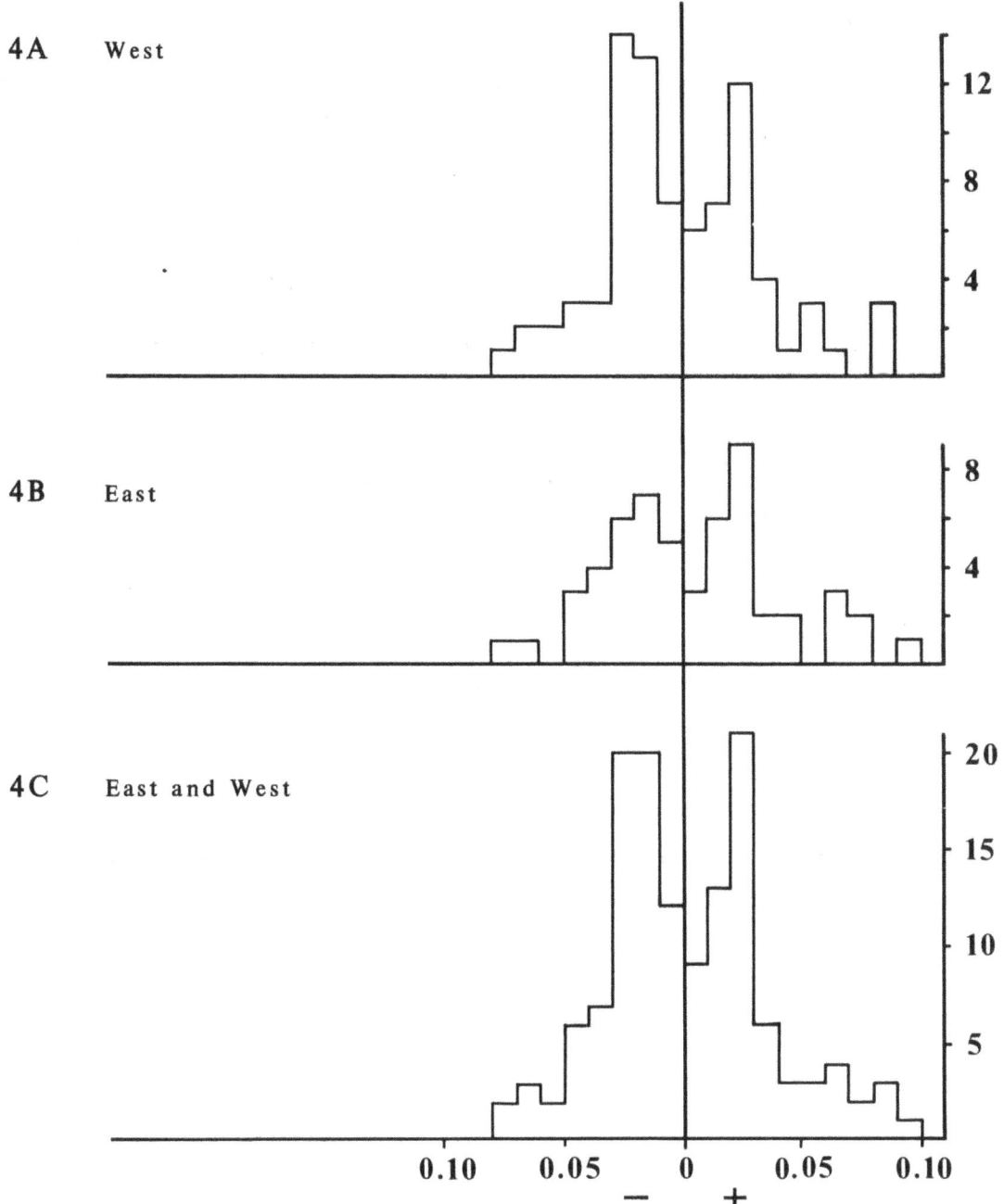

western to south eastern Europe to the Near East, the mean of the Near Eastern *Bos* being about 4% less than that of northern Europe. The difference is highly significant (p = <0.001). This confirms the view of Grigson (1969), Saxon (1974) and von den Driesch and Boessneck (1976) that there was a north-south size cline in cattle size in the early Holocene.

Histograms of cattle size in subsequent millennia (e.g. Figure 7C) suggest that wild cattle survived in the Near East throughout the period under discussion, that is until at least the third millennium B.C.

	n	range	mean	s	V
N. Europe	81	77-97	86.31	4.43	5.13
Hungary	96	77-95	84.12	4.04	4.80
Near East	51	73-90	83.05	4.01	4.82

TABLE 2 *Bos primigenius* (wild ox), astragalus maximum length (GL) in different areas in the early Holocene.

The mean for the Near Eastern sample is 3.80% less than that for northern Europe and the difference is highly significant (p = <0.001).

North European measurements from Grigson (1969) and Degerbøl (in Degerbøl and Fredskild 1973 - excluding his kitchen midden material); Hungarian from Bökönyi (1962); Near Eastern from Hayonim, El Wad B, Kebara B, Hatoula, Jericho, Bouqras, Asikli Hüyük, Mureybet, Ali Kosh, Jarmo, Sarab and Matarrah.

For sources of data see Appendix.

4.4.2 Domestic cattle of the sixth millennium bc

Continuation of the comparison of cattle size on a millennium by millennium basis shows that in the sixth millennium bc in the western part of the Near East (Figures 5B and 3G) cattle were, in contrast to those of the same period in the east (Figure 3F), considerably smaller than wild cattle (Figure 5A). The inference is that those in the west had been domesticated. The histogram shows not only that there was a marked concentration on females, although this alone would be enough to account for a reduction in the mean size, but also that the females and the few males were considerably smaller than wild females and males. One or two of the larger bones from Fikirtepe in Turkey were identified by Boessneck and von den Driesch (1979) as wild cattle.

In the west as in the east in the sixth millennium bc there are problems with dating. The site from which the majority of the data are derived is Fikirtepe in Eastern Turkey (Boessneck and von den Driesch 1979). Although the authors give the date as about 5000 bc, other authorities (e.g. Mellaart 1970) suggest a late sixth millennium bc date. Even so, the few data from the other sixth millennium bc sites (in the western part of the Near East) are all within the Fikirtepe range.

It should be stated that at any site in which domestication was actually happening, one would expect a wider range of sizes than is normally encountered in a wild population. Although Perkins (1969) claimed such a situation in the ceramic levels at Çatal Hüyük in Turkey, and indeed the value of the coefficient of variation (13.8) calculated from his figures is high, nevertheless a comparison of his data (Figure 5C) with those from contemporary sites (Figure 5B) shows that those at the small end of the range of the humerus distal breadth measurements (which are the only dimensions that he presents) are in fact far too small for sixth millennium bc *Bos*. Either they were misidentified (deer?), or they represent contamination from a much later period.

A confusion with deer seems most likely as Perkins had already misidentified the deer remains from Shanidar as *Bos* (Perkins 1960; Reed 1959), a fact which to give him credit he soon admitted (Perkins 1964; Reed 1960). Nevertheless his problems with identification seem to have persisted because in footnote number 7 in the Çatal Hüyük report (Perkins 1969) he wrote '... *Bos primigenius*, which has elements in its skeleton that cannot be distinguished from *Cervus elaphus*.' These problems seem to have continued as he was responsible for assigning the very small bones from Beidha (Hecker 1975) and Karatas-Semayük (Hesse and Perkins 1974) to *Bos*, though the histograms (Figures 3D and 7C respectively) show that this must have been incorrect.

As Perkins gave the statistical parameters but not the individual measurements at Çatal Hüyük, the distribution of size within the range cannot be assessed in detail. However the eccentric position of the mean shows that the curve must have been skewed to the right, so probably only one or two measurements were outside the range of *Bos*. It is possible that the remainder represent domestic animals or wild cows, but the manner of their presentation makes this impossible to verify.

Although Perkins' data for both the seventh and sixth millennia bc were obviously inadequate to support his assertion that they constituted evidence for the earliest domesticated cattle in the Near East, that assertion has been accepted and widely quoted by archaeologists and archaeozoologists, despite the lack of any more published data of the rest of the osteological material from Çatal Hüyük. This is an example of the unfortunate tendency among archaeozoologists working on Near Eastern material to publish the more exciting results of their work in advance of, or even in place of, the data that might provide the evidence. In the case of Çatal Hüyük the identifications cannot be checked because, as Perkins himself told me when I visited him in 1975, the entire collection had been lost. I understand that Pierre Ducos will soon remedy the situation by publishing the data that he managed to obtain before the loss of the collection. The measurements that he has already published in summary form (Ducos 1978a and c) all fall in the area of overlap between wild and domestic cattle so they cannot be used as evidence either way.

4.4.3 The fifth millennium bc

Comparison of the two fifth millennium bc samples (Figure 6B and C) with those of the domestic cattle of the sixth millennium bc in the Levant (Figure 6A) suggests a further, slight reduction in size. This is less marked in the sample from Iran and Iraq (Figure 6C), though this may be due to a higher proportion of males, as there is some suggestion of bimodality.

FIGURE 5A AND B The size of *Bos* in the sixth millennium in the western part of the Near East (5B), compared with the size of wild cattle (5A). Because of their diminution and the apparent stress on females the sixth millennium cattle are presumed to be domestic. The largest dimension (dotted screen) is a proximal metacarpal which the authors of the original report (Boessneck and von den Driesch 1979) identified as wild ox.

Data from Figures 3G and 4C respectively.

FIGURE 5C The size of cattle at Catal Hüyük - the range and the mean ± one standard deviation for the distal breadth (Bd) of the humerus. Those from the seventh millennium (n=12) were clearly wild: some of those from the sixth millennium (n=33) are smaller than the lower end of the size range of contemporary domestic cattle (5B) suggesting misidentification; neither the seventh nor the sixth millennium data constitute evidence for the presence of domestic cattle at Catal Hüyük.

Data from Perkins (1969).

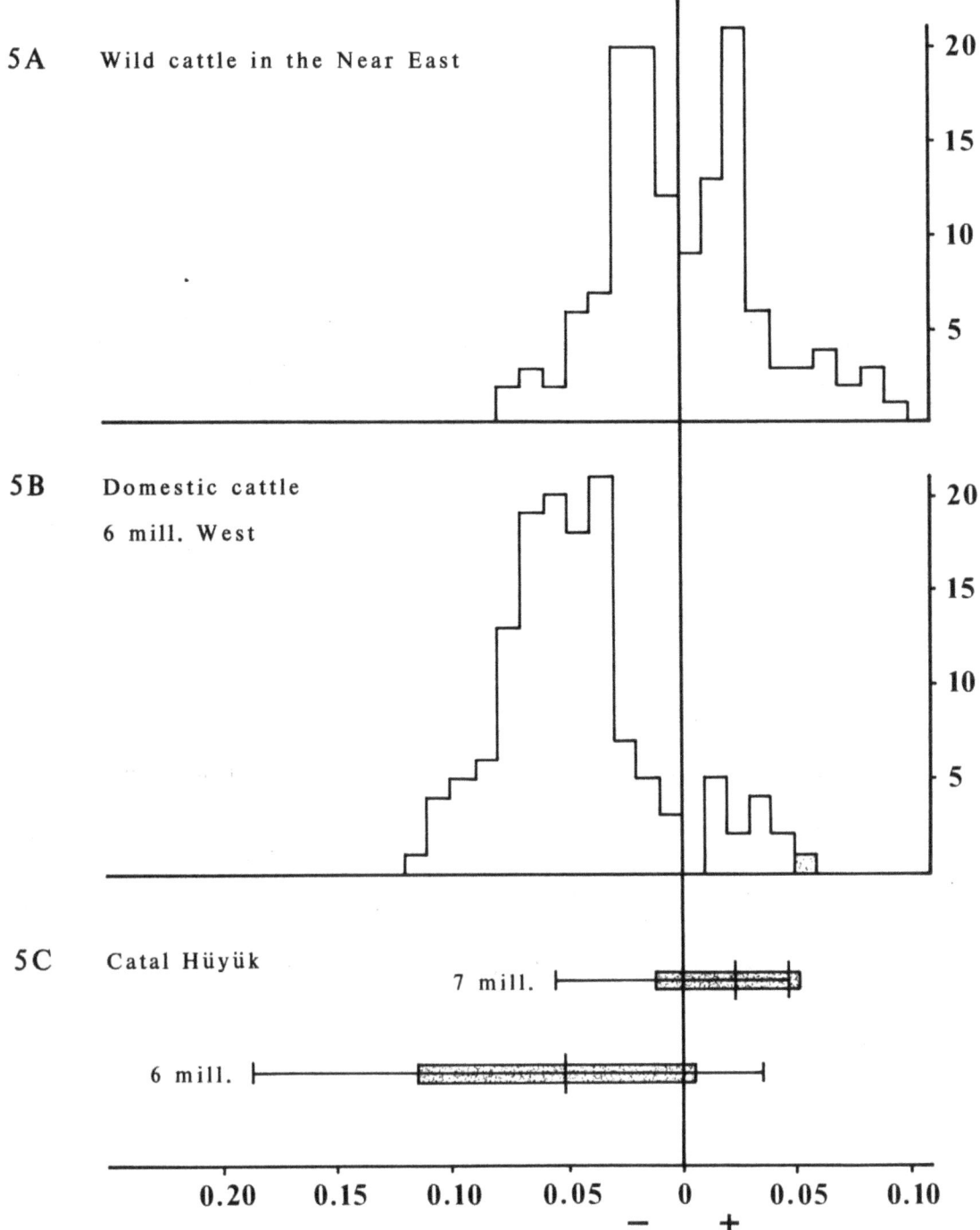

FIGURE 6 Diminution of domestic cattle from the sixth to the fourth millennium bc. None of the histograms of subsequent millennia (6B-6E) show the same emphasis on females as in the west in the sixth millennium (6A), but in the fourth millennium in the west (6E) there is an indication of trimodality, possibly related to the presence of castrates.

6A Data from Figure 3G.

6B Tell Dan (Neolithic), Shams ed Din, Amuq (C, D and E), Arjoune (V) and (VII); n=34.

6C Banahilk, Ras al Amiyah, Siahbid, Tepe Sabz, Anau (IA, IB and II), Tal-i-Iblis I; n=63.

6D Tell Rubeidheh, Choga Maran, Dehsavar, Tal-i-Iblis II; n=16.
The largest dimension is the distal breadth of a tibia from Tal-i-Iblis which the author of the original report (Bökönyi 1967) identified as wild ox.

6E Bir es-Safadi, Neve Noy, Shiqmim, Qatif Y2, Qatif Y3, Grar, Gilat, Metzer, Teleilat Ghassoul, Megiddo Tombs, En Shadud, Tel Eli I, Tel Aviv, Arjoune VI, Qa'lat el Mudiq, Beycesultan; n=243.
The largest dimension is the distal breadth of a complete metacarpal from the Chalcolithic level of the Megiddo Tombs. Its shape (i.e. the ratio of breadth/length) shows that it is of a domestic bull.

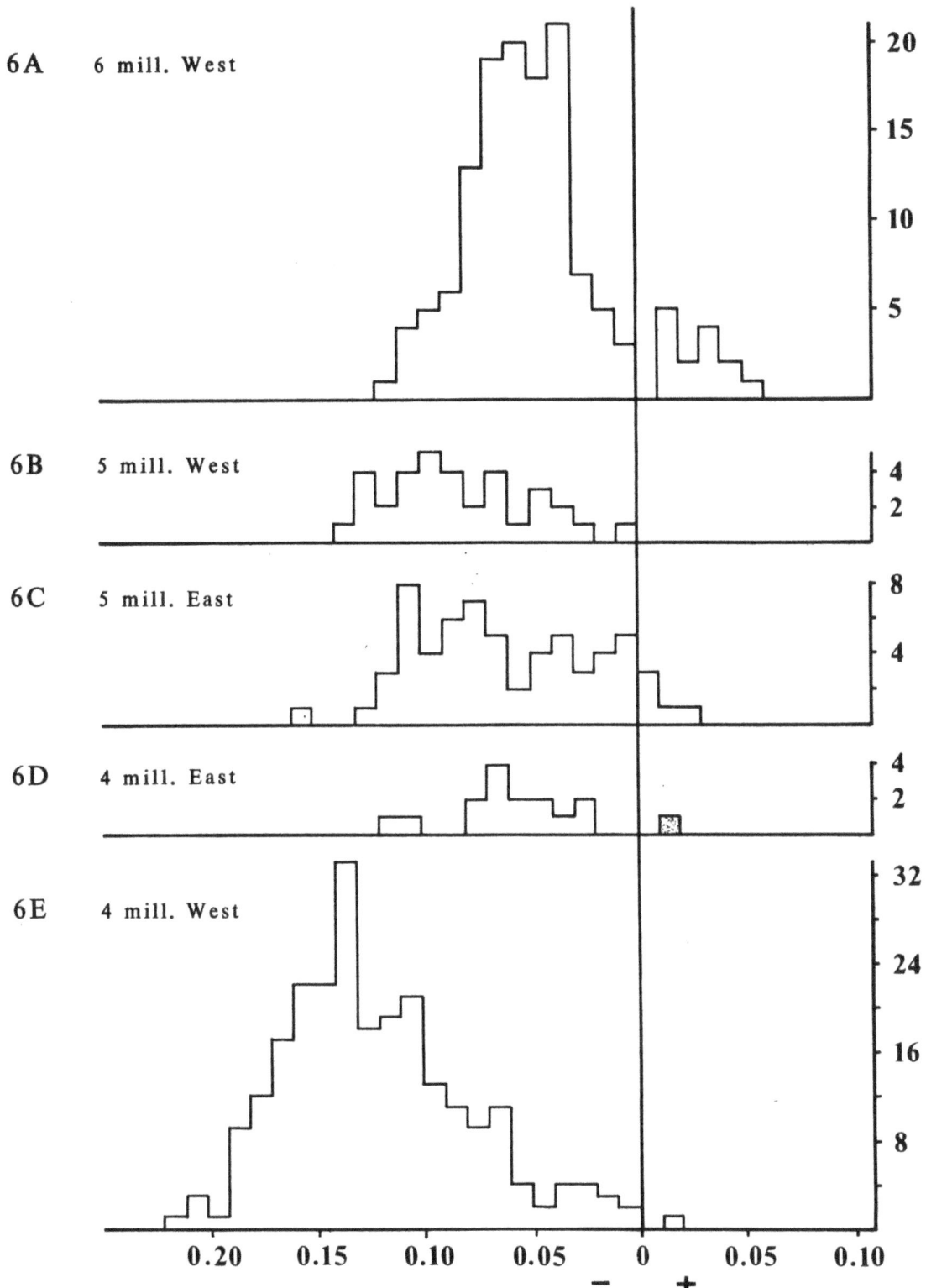

FIGURE 7 Cattle size in the third millennium B.C. compared to that in the fourth millennium.

Although the mean has clearly shifted to the right in the third millennium (7C) compared with the fourth millennium (7A) in the west, the range remains the same, suggesting a greater emphasis on males, probably including castrates. The smallest dimension in 7C is the proximal breadth of a metacarpal from Karatas Semayuk identified by Perkins (in Hesse and Perkins 1974) as *Bos*, but more probably *Cervus*.

There are too few data from the East (7B and D) for any firm conclusions to be drawn.

- 7A   data from Figure 6E

- 7B   data from Figure 6D.

- 7C   Arad, Tell Gat, Qa'lat el Mudiq, Sweyhat, Korucutepe, Yarikkaya, Karatas Semayuk; n=118.

  Dotted screen: Demircihüyük (*Bos primigenius?*); n=9.

- 7D   Sakheri Sughir, Tell Asmar, Shah Tepe III; n=23.

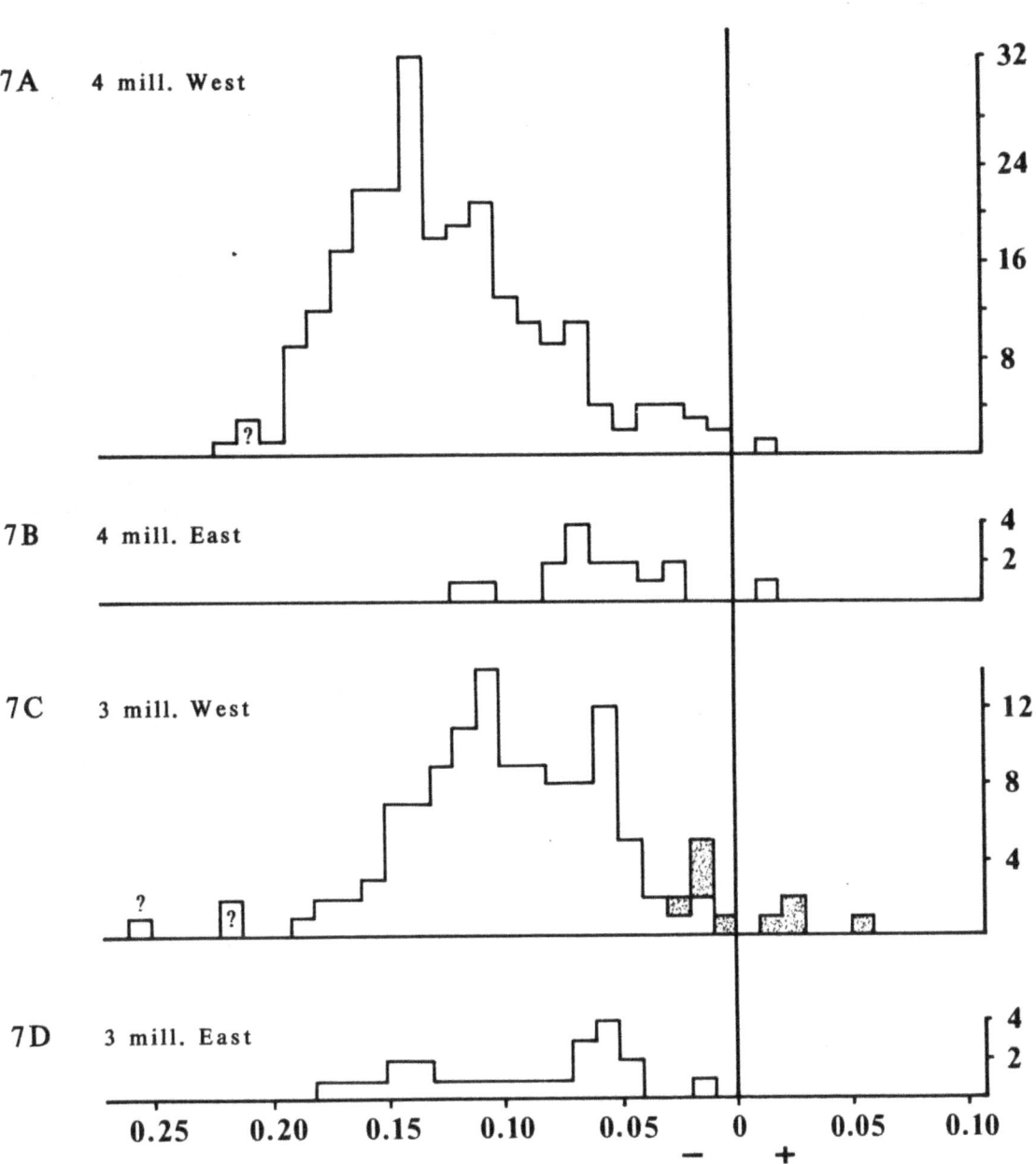

7A 4 mill. West

7B 4 mill. East

7C 3 mill. West

7D 3 mill. East

Neither of the fifth millennium bc histograms has the same strong emphasis on cows that characterises the sample from the sixth millennium bc in the western part of the area.

4.4.4 The fourth millennium bc

The histogram of the size of cattle in the fourth millennium bc in the Levant and Turkey (Figure 6E) shows a further marked diminution. It is likely that the animals, represented by the small peak at the small end of the range, are actually not *Bos*. They are from sites in the southern Levant and were identified (Grigson n.d.) as small bovid (?*Bos*, ?*Alcelaphus*). The broad spread of the remaining measurements and the slight suggestion of trimodality could be due to the presence of castrates, whose breadth measurements tend to be intermediate between those of cows and entire bulls (Fock 1966). Perhaps castration had been introduced as a way of keeping males docile so that they could be used for ploughing and other types of draught? The small peak at the large end of the range includes some metapodials whose distal ends were unusually expanded. Such expansion is thought to indicate the use of cattle as draught or plough animals (Ryder 1959; Grigson 1987).

The very small sample from Iran and Iraq (Figure 6D) is suggestive of the use of males. It includes one bone from the late Ubaid level at Tal-i-Iblis on the Iranian plateau which the author of the report (Bökönyi 1967) suggested was from an aurochs and this may indeed be true.

4.4.5 The third millennium B.C.

A comparison of the histogram of the large third millennium B.C. sample from the Levant and Turkey (Figure 7C) with that from the fourth millennium bc (Figure 7A) suggests an increase in size. However the ranges of the two samples are almost identical, even though the mean has clearly moved to the right. This pattern of the distribution of size might be due to the presence of a larger number of castrates and entire males. The small scatter of large bones from Demircihüyük in Turkey are those identified by Rauh (1981) as wild cattle.

Although too small for any firm conclusions the sample from Iran and Iraq (Figure 7D) suggests bimodality.

4.5 CONCLUSIONS

Analysis of the morphometric data published so far suggests that domestic cattle first appeared in the Middle East in the sixth millennium bc at sites in the western part of the area (western Turkey, the Levant, northern Syria and adjacent parts of central Turkey). In sites of the sixth millennium bc in Iraq and Iran on the other hand cattle seem to be wild. It is possible that this difference is illusory, as the eastern sites with published data happen to fall earlier in the sixth millennium bc than the western sites.

However although the cattle of the seventh millennium bc sites were of the wild size, there is a suggestion that in the western part of the area there was a stress on females; this could be interpreted as predation upon nursery herds and possibly as a first step towards domestication. Whether one can call this 'protoélevage' or 'incipient domestication' depends on a much more detailed analysis of the demographic data. The archaeozoology of island sites may be relevant here; despite the presence of domestic sheep, goats and pigs, all imported into aceramic Khirokitia on Cyprus early in the sixth millennium bc, cattle are absent (Davis 1984b). Domestic cattle are however present in the early sixth millennium bc at Knossos on Crete (Jarman and Jarman 1968), though Cyprus presumably received its domesticates from south east Europe, which is not included in the present study.

The stress on domestic cows in the sixth millennium bc in the western part of the Near East, particularly Fikirtepe in western Turkey, is interesting. Presumably most of the males were slaughtered when young, their remains being too fragile to survive in the archaeological record. Such a scenario would be a logical way of maintaining the herd whilst minimising the hazards of dealing with adult bulls. Whether such an imbalance of the sexes is indicative of milking is an open question. Certainly it could be the follow-on of a possible early stage of domestication in which people might have appropriated nursery herds, rather than killing individuals from the general population.

By the fifth millennium bc cattle in the east were clearly of domestic size, and though the sample size is small there are indications of bimodality, suggesting that more use was made of adult males. In both the east and the west cattle are slightly smaller than the domestic cattle of the preceding millennium.

The fourth and third millennia bc/B.C. samples from the east are very small, probably too small to draw conclusions from, except that whilst all are in the domestic size, those of the fourth millennium bc are surprisingly large. Perhaps by chance only bulls' bones have been reported upon.

The pattern of size distribution within the large samples of the fourth and third millennia bc/B.C. from the west are more confused; the explanation may be that castration had been introduced and that the presence of animals whose bones were likely to have been intermediate in width between those of cows and bulls, but overlapping with both, has obscured the usual bimodality. Indeed there is a suggestion in each of an intermediate peak.

The results of the present work have been expressed in rather coarse, large-scale entities. It is to be hoped that the same method will be applied to smaller and smaller chronological and spatial units and eventually to single levels of single sites.

Although more and more detailed information should become available as more sites are excavated in the future, it is clear that far more data could be extracted from material that has already been excavated. In some cases the data have been published, not always in adequate detail, and in others the data exist, but the appalling backlog of publication, coupled with a resistance to the publication of large numbers of measurements, has seriously hampered any advance in our knowledge of animal domestication in the Near East.

ACKNOWLEDGEMENTS

I am very grateful to Juliet Clutton-Brock for allowing me to remeasure the Jericho material and for discussions on the provenance of particular measured cattle bones; to Simon Davis, Ilse Köhler-Rollefson, Kent Flannery and Sebastian Payne for providing me with unpublished cattle bone measurements; and to Dr Tove Hatting for checking the measurements of the Ullerslev skeleton in the Universiteits Zoologiske Museum, Copenhagen.

APPENDIX

Middle Eastern animal bone reports that include measurements of cattle bones and teeth suitable for the present study.

Note on dating: 10th-4th millennium dates are in radiocarbon years (bc); late 4th millennium calendar dates (B.C.) are included with the radiocarbon 4th millennium; 3rd millennium dates are calendar dates (B.C.). Sites marked 6* are of around 6000 bc.

SITE AND LEVEL	PERIOD	MILLENNIUM	REFERENCE
Ain Ghazal	PPNB	7	Köhler-Rollefson pers. comm.
Ali Kosh	Aceram. Neo.	7	Hole, Flannery and Neely 1969; Flannery pers. comm.
Amuq A & B	early Pott. Neo.	6	Stampfli 1983
Amuq C, D & E	late Pott. Neo.	5	Stampfli 1983
Anau Ia,Ib,II	Early Chalco.	5	Duerst 1908
Arad	EBA II	3	Davis 1976; Lernau 1978
Arjoune V & VII	late Pott. Neo.	5	Grigson forthcoming
Arjoune VI	Chalco.	4	Grigson forthcoming
Ashkalon	Early Pott. Neo.	6	Ducos 1968
Asikli Hüyük	Aceram. Neo.	7	Payne 1985
Banahilk	Early Chalco.	5	Laffer 1983
Beidha	PPNB	7	Hecker 1975
Beycesultan	Late Chalco.	4	Ducos 1965
Bir es-Safadi	Chalco.	4	Ducos 1968; Grigson forthcoming
Bouqras	PPNB	7	Clason 1980
Cafer Hüyük	Aceram. Neo.	7	Helmer 1985
Catal Hüyük XII-X	Earliest Pott. Neo.	7	Perkins 1969
Catal Hüyük VI	Early Pott. Neo.	6	Perkins 1969
Choga Maran	Middle Chalco.	4	Davis 1984a, pers. comm.
Dehsavar	Late Chalco.	4	Bökönyi 1977
Demircihüyük	EBA/MBA	3	Boessneck and von den Driesch 1977/8; Rauh 1981
El Wad B	Natufian	10-9	Bate 1937; Jarman 1969

En Shadud	EBA Ia	4	Kolska-Horwitz 1985
Fikirtepe	Early Pott. Neo.	6	Boessneck and von den Driesch 1979
Ganj Dareh	Pott. Neo.	8-7	Hesse 1978
Gilat	Chalco.	4	Grigson no date
Grar	Chalco.	4	Grigson forthcoming
Hajji Firuz	Pott. Neo.	6	Meadow 1983
Hatoula 4	Natufian	10-9	Davis 1985
Hatoula 3 & 2	PPNA	8	Davis 1985
Hayaz Hüyük	PPNB	7	Buitenhuis 1985
Hayonim Cave B	Natufian	10	Davis 1981, pers. comm; Bouchud 1987
Jarmo	Aceram./Pott. Neo.	6*	Stampfli 1983
Jericho	Natufian	9	Clutton-Brock 1979; Grigson no date
Jericho	PPNA	8	Clutton-Brock 1979; Grigson no date
Jericho	PPNB	7	Clutton-Brock 1979; Grigson no date
Jericho	Pott. Neo. A	6	Clutton-Brock 1979; Grigson no date
Karatas-Semayuk	EBA I & II	3	Hesse and Perkins 1974
Kebara 1931 B	Natufian	10	Jarman 1969; Saxon 1974
Korucutepe	EBA II	3	Boessneck and von den Driesch 1975
Matarrah	Pott. Neo.	6	Stampfli 1983
Megiddo Tombs	Chalco.	4	Bate 1938
Metzer	Chalco.	4	Ducos 1968
Mureybet	PPNA	8	Ducos 1978a
Neve Noy	Chalco.	4	Grigson forthcoming
Qa'lat el Mudiq	Early Pott. Neo.	6	Gautier 1977
Qa'lat el Mudiq	Chalco.	4	Gautier 1977
Qa'lat el Mudiq	EBA	3	Gautier 1977
Qatif Y3	late Pott. Neo.	4	Grigson forthcoming
Qatif Y2	Chalco.	4	Grigson forthcoming
Ras-al-Amiya	Early Chalco.	5	Hole, Flannery and Neely 1969
Sakheri Sughir	EDI	3	Bökönyi and Flannery 1969
Shah Tepe III	EBA	3	Amschler 1939
Shams ed Din	Late Pott. Neo.	5	Uerpmann 1982
Shiqmim	Chalco.	4	Grigson forthcoming
Shukbah B	Natufian	10-9	Bate 1942; Jarman 1969
Siahbid 1960	Early Chalco.	5	Bökönyi 1977
Siahbid 1978	Early Chalco.	5	Davis 1984a, pers. comm.
Suberde	Aceram. Neo.	7	Perkins 1969; Perkins and Daly 1968
Tal-i-Iblis I	Early Chalco.	5	Bökönyi 1967
Tal-i-Iblis II	Middle Chalco.	4	Bökönyi 1967

Jabotinsky St., Tel Aviv	Chalco.	4	Ducos 1968, pers.comm.
Teleilat Ghassoul	Chalco.	4	Grigson no date
Tel Eli IV	PPNB	7	Jarman 1974
Tel Eli Ia-Ic	Chalco.	4	Jarman 1974
Tell Asmar	ED	3	Hilzheimer 1941
Tell Dan	late Pott. Neo.	5	Kolska-Horwitz 1987
Tell Gat	EBA	3	Ducos 1968
Tell es Sinn	PPNB	7	Clason 1980
Tell Molla Assad	PPNB	7	Clutton-Brock 1985
Tell Rubeidheh	Late Chalco.	4	Payne forthcoming
Tell Sweyhat	EBA	3	Buitenhuis 1985
Tepe Guran	Pott. Neo.	6*	Flannery pers. comm.
Tepe Sabz	Early Chalco.	5	Hole, Flannery and Neely 1969; Flannery pers. comm.
Tepe Sarab	Pott. Neo.	6*	Bökönyi 1977
Yarikkaya	EBA	3	Boessneck and Wiedermann 1977

BIBLIOGRAPHY

AMSCHLER, J.W. (1939) Tierreste des Ausgrabungen von dem 'Grossen Konigshugel' Shah Tepe, in Nord-Iran. *Report of the scientific expedition to the NW Provinces of China under the leadership of Dr Sven Hedin. Publication 9, VII Archaeology* 4. pp. 35-129.

BATE, D.M.A. (1937) Palaeontology: the fossil fauna of the Wady el-Mughara caves. In: *The stone age of Mount Carmel. Volume 1.* (eds Garrod, D.A.E. and Bate, D.M.A.) Oxford: Clarendon Press pp. 139-240.

BATE, D.M.A. (1938) *Animal remains from the Meggido Tombs.* University of Chicago Oriental Institute Publication 32. pp. 209-13.

BATE, D.M.A. (1942) The fossil mammals of Shukbah. *Proceedings of the Prehistoric Society* **8** 15-20.

BINFORD, L.R. (1978) *Nunamiut ethnoarchaeology.* New York: Academic Press.

BOESSNECK, J. (1958) Zur Entwicklung vor- und frühgeschichtliche Haus- und Wildtier Bayerns im Rahmen der gleichzeitigen Tierwelt Mitteleuropas. *Studien an vor- und frühgeschichtlichen Tierresten Bayerns.* Munich: Institut für Palaeoanatomie, Domestikationsforschung und Geschichte der Tiermedizin der Universität München.

BOESSNECK, J. AND DRIESCH, A. VON DEN (1975) Tierknochenfunde vom Korucutepe bei Elazig in Ostanatolien. In: *Studies in ancient civilization, Korucutepe. Volume 1.* (ed. Van Loon, M.N.) Amsterdam pp. 1-216.

BOESSNECK, J. AND DRIESCH, A. VON DEN (1977/8) Vorlaufiger Bericht uber die Untersuchungen Knochenfunden vom Demirçihüyük. *Istanbuler Mitteilungen* **27/28** 54-59.

BOESSNECK, J. AND DRIESCH, A. VON DEN (1979) *Die Tierknochenfunde aus der Neolithischen Siedlung auf dem Fikirtepe bei Kadikoy am Marmarameer.* Munich: Institut für Palaeoanatomie, Domestikationsforschung und Geschichte der Tiermedizin der Universität München.

BOESSNECK, J. AND WIEDEMANN, U. (1977) Tierknochen aus Yarikkaya bei Bogazkoy, Anatolien. *Archaeologie und Naturwissenschaften* **1** 106-28.

BÖKÖNYI, S. (1962) Zur Naturgeschichte des Ures in Ungarn und das Problem der Domestikation des Hausrindes. *Acta Archaeologica Hungarica* **14** 175-214.

BÖKÖNYI, S. (1967) The prehistoric vertebrate fauna of Tal-i-Iblis. In: *Investigation at Tal-i-Iblis.* (ed. Caldwell, J.R.) Springfield: Illinois State Museum pp. 309-17.

BÖKÖNYI, S. (1976) Development of early stockrearing in the Near East. *Nature* **264** 19-23.

BÖKÖNYI, S. (1977) *Animal remains from four sites in the Kermanshah Valley, Iran.* Oxford: British Archaeological Reports S34.

BÖKÖNYI, S. (1978) Environmental and cultural differences as reflected in the animal bone samples from five early Neolithic sites in southwest Asia. In: *Approaches to faunal analysis in the Middle East.* (eds Meadow, R.H. and Zeder, M.A.) Peabody Museum Bulletin, Harvard University 2 pp. 57-62.

BÖKÖNYI, S. AND FLANNERY, K.V. (1969) Faunal remains from Sakheri Sughir. In: *The administration of rural production in an early Mesopotamian town.* (ed. Wright, H.T.) Anthropological Papers Museum of Anthropology, University of Michigan 38 143-49.

BOUCHUD, J. (1987) *La faune du gisement Natoufien de Mallaha (Eynan) Israel.* Mémoires et travaux du centre de recherche français de Jerusalem no. 4. Paris: Association Paléorient.

BUITENHUIS, H. (1983) The animal remains of Tell Sweyhat, Syria. *Palaeohistoria* 25 131-44.

BUITENHUIS, H. (1985) Preliminary report on the faunal remains of Hayaz Hüyük from the 1979-1983 seasons. *Anatolica* 12 61-74.

BURNEY, C. (1977) *From village to empire: an introduction to Near Eastern archaeology.* Oxford: Phaidon.

CLASON, A.T. (1980) The animal remains from Tell es Sinn compared with those from Bouqras. *Anatolica* 7 35-53.

CLUTTON-BROCK, J. (1971) The primary food animals of the Jericho Tell from the Proto-Neolithic to the Byzantine Period. *Levant* 3 41-55.

CLUTTON-BROCK, J. (1979) The mammalian remains from the Jericho Tell. *Proceedings of the Prehistoric Society* 45 135-57.

CLUTTON-BROCK, J. (1981) *Domesticated animals from early times.* London: Heinemann and British Museum (Natural History).

CLUTTON-BROCK, J. (1985) Mammalian remains from Tell Molla Assad, Syria. In: *Holocene settlement in North Syria.* (ed. Sanlaville, P.) Oxford: British Archaeological Reports S238 pp. 163-66.

CONTENSON, H. DE (1966) Découvertes récentes dans le domaine du Néolithique en Syrie. *L'Anthropologie* 70 388-91.

DAVIS, S.J.M. (1976) Mammal bones from the Early Bronze Age city of Arad, northern Negev, Israel. *Journal of Archaeological Science* 3 153-64.

DAVIS, S.J.M. (1981) The effect of temperature change and domestication on the body size of Late Pleistocene to Holocene mammals of Israel. *Paleobiology* 7 101-14.

DAVIS, S.J.M. (1982) Climatic change and the advent of domestication: the succession of ruminant artiodactyls in the Late Pleistocene-Holocene in the Israel region. *Paléorient* 8 5-15.

DAVIS, S.J.M. (1984a) The advent of milk and wool production in western Iran: some speculations. In: *Animals and archaeology 3: Early herders and their flocks.* (eds Clutton-Brock, J. and Grigson, C.) Oxford: British Archaeological Reports S202 pp. 265-78.

DAVIS, S.J.M. (1984b) Khirokitia and its mammal remains, a Neolithic Noah's Ark. In: *Fouilles recentes a Khirokitia (Chypre) 1977-1981.* (ed. Le Brun, A.) Etude Néolithique Editions Recherche sur les Civilizations. Paris: A.D.P.F. Memoir **41** (1) 147-62, (2) 163-79.

DAVIS, S.J.M. (1985) A preliminary report of the fauna from Hatoula: a Natufian-Khiamian (PPNA) site near Latroun, Israel. In: *Le site Natufien-Khiamian de Hatoula près de Latroun, Israel.* CRJF Jerusalem Annexe B 71-118.

DEGERBØL, M. (1963) Prehistoric cattle in Denmark and adjacent areas. In: *Man and Cattle.* (eds Mourant, A.E. and Zeuner, F.E.) Royal Anthropological Institute Occasional Paper 18 pp. 69-79.

DEGERBØL, M. AND FREDSKILD, J. (1970) The urus (*Bos primigenius* Bojanus) and neolithic domesticated cattle (*Bos taurus domesticus* Linn) in Denmark. *Kongelige Danske Videnskabernes Selskabo Biologica Skripta* **17** 1-177.

DRIESCH, A. VON DEN (1976) *The measurement of animal bones from archaeological sites.* Peabody Museum Bulletin 1. Harvard: Harvard University.

DRIESCH, A. VON DEN AND BOESSNECK, J. (1976) Die Tierknochenfunde vom Castro do Zambujal. *Studien uber Frühe Tierknochenfunde von der Iberischen Halbinsel* **5** 1-129.

DUCOS, P. (1965) La faune de Beycesultan. *Occasional Papers of the British Institute at Ankara* **8** 139-54.

DUCOS, P. (1968) *L'origine des animaux domestiques de Palestine.* Publications de l'Institut de Préhistoire de l'Université de Bordeaux 6.

DUCOS, P. (1969) Methodology and results of the study of the earliest domesticated animals in the Near East (Palestine). In: *The Domestication and exploitation of plants and animals.* (eds Ucko, P.J. and Dimbleby, G.W.) London: Duckworth pp. 265-75.

DUCOS, P. (1978a) *Tell-Mureybet (Syrie, IX-VII millennaires) étude archaeozoologique et problèmes d'écologie humaine Volume 1.* Paris: CNRS.

DUCOS, P. (1978b) 'Domestication' defined and methodological approaches to its recognition in faunal assemblages. In: *Approaches to faunal analysis in the Middle East.* (eds Meadow, R.H. and Zeder, M.A.) Peabody Museum Bulletin 2. Harvard: Harvard University 2 pp. 53-56.

DUCOS, P. (1978c) La faune de Beisamoun dans les collections du Musée préhistoriques de la Vallée du Houleh. In: *Abou Ghosh et Beisamoun.* (ed. Lechevalier, M.) Paris: Association Paléorient pp. 257-68.

DUCOS, P. AND HELMER, D. (1982) Le point actuel sur l'apparition de la domestication dans le Levant. In: *Préhistoire du Levant* (eds Cauvin, J. and Sanlaville, P.) Actes du Colloque International Centre Nationale de la Recherche Scientifique **598** 523-28.

DUERST, J.U. (1908) Animal remains from the excavations at Anau. In: *Excavations in Turkestan. Expedition of 1904. Volume 2.* (ed. Pumpelly, R.) Washington, D.C.: Carnegie Institute pp. 341-442.

DUERST, J.U. (1930) Vergleichende Untersuchungsmethoden am Skelett bei Saugern. *Abderhaldens Handbuch der Biologischen Arbeitsmethoden* **7** 125-530.

FOCK, J. (1966) *Metrische Untersuchungen an Metapodien einiger europaischer Rinderrassen.* Dissertation, Munich: Institut für Palaeoanatomie, Domestikationsforschung und Geschichte der Tiermedizin der Universität München.

GAUTIER, A. (1977) Sondage dans le tell d'Apamée (1974). 2. Etude des restes osseaux animaux. *Bulletin de la Societé Royale Belge Anthropologie et Préhistoire* **88** 77-93.

GEBEL, H.-G. (1984) Das akeramische Neolitikum Vordasiens. *Beihefte zum Tübingen Atlas des Vorderen Orients Reihe B* 52.

GRIGSON, C. (1969) The uses and limitations of differences in absolute size in the distinction between the bones of aurochs (*Bos primigenius*) and domestic cattle (*Bos taurus*). In: *The domestication and exploitation of plants and animals.* (eds Ucko, P.J. and Dimbleby, G.W.) London: Duckworth pp. 277-94.

GRIGSON, C. (1978) The craniology and relationships of four species of *Bos*. 4. The relationship between *Bos primigenius* Boj. and *Bos taurus* L. and its implications for the phylogeny of the domestic breeds. *Journal of Archaeological Science* **5** 123-52.

GRIGSON, C. (1982) Sexing neolithic domestic cattle skulls and horn cores. In: *Ageing and sexing animal bones from archaeological sites.* (eds Wilson, B., Grigson, C. and Payne, S.) Oxford: British Archaeological Reports 109 pp. 25-35.

GRIGSON, C. (1987) Shiqmim: pastoralism and other aspects of animal management in the Chalcolithic of the Northern Negev. In: *Shiqmim 1. Studies concerning Chalcolithic societies in the Northern Negev Desert (1982-1984).* (ed. Levy, T.E.) Oxford: British Archaeological Reports S356 pp. 219-41.

HECKER, H. (1975) *The faunal analysis of the primary food animals from pre-pottery Neolithic Beidha (Jordan).* Ann Arbor, Michigan and London: University Microfilms International.

HELMER, D. (1985) Etude préliminaire de la faune de Cafer Hoyük (Malatya-Turquie). *Cahier de l'Euphrate* **4** 117-20.

HESSE, B. (1978) *Evidence for husbandry from the early Neolithic site of Ganj Dereh in western Iran.* Ann Arbor, Michigan and London: University Microfilms International.

HESSE, B. (1982) Slaughter patterns and domestication: the beginnings of pastoralism in western Iran. *Man* (N.S.) **17** 403-17.

HESSE, B. (1984) These are our goats: the origins of herding in west central Iran. In: *Animals and archaeology 3: Early herders and their flocks.* (eds Clutton-Brock, J. and Grigson, C.) Oxford: British Archaeological Reports S202 pp. 243-64.

HESSE, B. AND PERKINS, D. (1974) Faunal remains from Karatas-Semayük in southwest Anatolia. *Journal of Field Archaeology* **1** 149-60.

HILZHEIMER, M. (1941) *Animal remains from Tell Asmar.* Oriental Institute of the University of Chicago, Studies in Ancient Oriental Civilisation 20.

HOOIJER, D.A. (1966) Animal remains found at Bouqras and Ramad in 1965. *Annales Archéologiqes Arabes* **16** 193-95.

HOLE, F., FLANNERY, K.V. AND NEELY, J.A. (1969) *Prehistory and human ecology of the Deh Luran Plain.* Memoirs of the Museum of Anthropology, University of Michigan 1.

JARMAN, M.R. (1969) The prehistory of upper Pleistocene and recent cattle. Part 1. East Mediterranean, with reference to north west Europe. *Proceedings of the Prehistoric Society* **35** 236-66.

JARMAN, M.R. (1974) The fauna and economy of Tel Eli. *Mitekufat Haeven* **12** 50-70.

JARMAN, M.R. AND JARMAN, H.N. (1968) The fauna and economy of early Neolithic Knossos. *Annual of the British School at Athens* **63** 241-64.

JEWELL, P.A. (1962) Changes in size and type of cattle from prehistoric to medieval times in Britain. *Zeitscrift für Tierzüchtung und Züchtungsbiologie* **77** 159-67.

KOLSKA-HORWITZ, L. (1985) The En Shadud faunal remains. In: *En Shadud, salvage excavations at a farming community in the Jezreel Valley, Israel.* (ed. Braun, E.) Oxford: British Archaeological Reports S249 pp. 168-77.

KOLSKA-HORWITZ, L. (1987) Animal remains from the Pottery Neolithic levels at Tel Dan. *Mitekufat Haeven* **20** 114-18.

KURTÉN, B. (1959) Rates of evolution in fossil mammals. *Cold Spring Harbour Symposium on Quantitative Biology* **24** 205-15.

LAFFER, J.P. (1983) The faunal remains from Banahilk. In: *Prehistoric archaeology along the Zagros Flanks.* (ed. Braidwood, L.S.) University of Chicago, Oriental Institute Publication 105 pp. 629-47.

LEGGE, A.J. (1981) Aspects of cattle husbandry. In: *Farming practice in British prehistory.* (ed. Mercer, R.J.) Edinburgh: University Press pp. 169-81.

LERNAU, H. (1978) Faunal remains, strata III-I. In: *Early Arad.* (ed. Amiran, R.) Jerusalem: Israel Exploration Society pp. 83-113.

MEADOW, R.H. (1981) Early animal domestication in South Asia: a first report of the faunal remains from Mehrgarh, Pakistan. In: *South Asian Archaeology 1979*. (ed. Hartel, H.) Berlin: Dietrich Reimer pp. 143-79.

MEADOW, R.H. (1983) The vertebrate faunal remains from Hasanlu Period X at Hajji Firuz. In: *Hasanlu excavation reports 1: Hajji Firuz Tepe, Iran*. (ed. Voigt, M.M.) Philadelphia University Museum Monograph 50 pp. 369-422.

MEADOW, R.H. (1984) Animal domestication in the Middle East: a view from the eastern margin. In: *Animals and archaeology 3: Early herders and their flocks*. (eds Clutton-Brock, J. and Grigson, C.) Oxford: British Archaeological Reports S202 pp. 309-37.

MELLAART, J. (1970) (a) The earliest settlements in western Asia from the ninth to the end of the fifth millennium B.C. (b) Anatolia before 4000 B.C. In: *Cambridge Ancient History. Volume 1:i*. (eds Edwards, I.E.S., Gadd, C.J. and Hammond, N.G.L.) Cambridge: University Press pp. 248-326.

MELLAART, J. (1975) *The Neolithic of the Near East*. London: Thames and Hudson.

MØHL, U. (1957) Zoologisk gennemgang af knoglematerialet fra Jernaldersbopladserne Dalshoj og Sorte Mulde. Bornholm. In: *Bornholm i Folkvanderings- tiden*. (ed. Klint-Jensen, O.) Nationalmuseets Skrifter, storre Beretninger II.

PAYNE, S. (1985) Animal bones from Asikli Hüyük. *Anatolian Studies* **35** 109-22.

PERKINS, D. (1960) The faunal remains of Shanidar Cave and Zawi Chemi Shanidar: 1960 season. *Sumer* **16** 77-79.

PERKINS, D. (1964) Prehistoric fauna from Shanidar, Iraq. *Science* **144** 1565-66.

PERKINS, D. (1969) Fauna of Çatal Hüyük: evidence for early cattle domestication in Anatolia. *Science* **164** 177-79.

PERKINS, D. (1973) Chronologies in Old World archaeology. *American Journal of Archaeology* **77** 279-82.

PERKINS, D. AND DALY, P. (1968) Suberde. *Scientific American* **219** 96-106.

RAUH, H. (1981) *Knochenfunde von Saugetieren aus dem Demircihüyük (Nordwest-Anatolien)*. Dissertation, Munich: Institut für Palaeoanatomie, Domestikationsforschung und Geschichte der Tiermedizin der Universität München.

REED, C.A. (1959) Animal domestication in the prehistoric Near East. *Science* **130** 1629-39.

REED, C.A. (1960) A review of archaeological evidence on animal domestication in the prehistoric Near East. In: *Prehistoric investigations in Iraqi Kurdestan*. Oriental Institute of the University of Chicago, Studies in Ancient Oriental Civilisation **31** 119-45.

RYDER, M.L. (1959) The animal remains found at Kirkstall Abbey. *Agricultural History Review* **7** 1-5.

SAKELLARIDIS, M. (1979) *The economic exploitation of the Swiss area in the Mesolithic and Neolithic periods*. Oxford: British Archaeological Reports S67.

SAXON, E.C. (1974) The mobile herding economy of Kebarah Cave, Mt Carmel: an economic analysis of the faunal remains. *Journal of Archaeological Science* **1** 27-45.

STAMPFLI, H.R. (1983) The fauna of Jarmo with notes on animal bones from Matarrah, the 'Amuq and Karim Shahir. In: *Prehistoric archaeology along the Zagros Flanks*. (ed. Braidwood, L.S.) University of Chicago, Oriental Institute Publication 105 pp. 629-47.

UERPMANN, H.-P. (1982) Faunal remains from Shams ed-Din Tannira, a Halafian site in northern Syria. *Berytus* **30** 3-52.

WENDORF, F. AND SCHILD, R. (1984) Conclusions. In: *Cattle-keepers of the Eastern Sahara: the Neolithic of Bir Kiseiba*. (eds Wendorf, F., Schild, R. and Close, A.) Dallas: Southern Methodist University pp. 404-28.

WESTLEY, B. (1970) The mammalian fauna. In: *Excavations at Hacilar*. (ed. Mellaart, J.) Edinburgh: University Press pp. 245-47.

III

THE EARLIEST AGRICULTURE IN CONTINENTAL EUROPE: THEORY, METHODOLOGY AND AGRICULTURAL PRACTICE

5

THE CEREAL POLLEN RECORD AND EARLY AGRICULTURE

Kevin J. Edwards

ABSTRACT

The detection of cereal pollen provides evidence of agricultural activity, even in the absence of archaeological data. The confidence attached to cereal pollen identifications assumes special importance when the earliest phases of apparent cereal cultivation in an area are detected. Problems of identification are discussed as a prelude to considerations of pollen statistics, dispersal, species exclusivity, the use of supporting palaeoecological evidence and stratigraphic integrity. The number of sites in the peripheral areas of north west Europe which have produced cereal-type grains in pre-elm decline contexts is increasing and while this may be viewed positively, caution in the interpretation of such finds is urged.

5.1 INTRODUCTION

The detection of cereal pollen in the pollen record provides the most convincing *prima facie* evidence for cultivation and for human activity in the microfossil record. Such evidence is clearly of interest to palynologists attempting to interpret their palaeoecological data. It is also of interest to prehistorians, since it not only corroborates archaeological evidence but can also be used to confirm conjectured primary agriculture or to reveal the presence of early agriculturalists when the archaeological data are absent. Problems arise, however, over the confidence attached to cereal pollen identification and the significance of positive finds. Such matters assume great importance when the beginnings of agriculture in an area are concerned although this temporal focus is not critical to the identification debate in palynological terms - the correct detection of cereal grains in later contexts is equally desirable. In this paper, the question of identification is discussed as a prelude to a wider examination of the utilisation of cereal pollen and supplementary data. Much of the discussion relates to the evidence obtained from lake and peat deposits rather than mineral soils, though the observations frequently apply to any polleniferous medium. This decision was taken not because soils are of little use in this context - Dimbleby (1985) makes a series of pertinent comments regarding the usefulness of Cerealia pollen incorporated in soils - but rather because they frequently pose additional problems relating to the stratigraphic integrity of samples and such pedological considerations are outside the scope of this paper. Most examples are drawn from Europe since much of the research on the topic has been carried out there, though the contributions from elsewhere should also be acknowledged (e.g. Rowley 1960; Whitehead and Langham 1965; Watson and Bell 1975).

In case clarification of terms should be needed the following definitions, as they apply to this paper, are offered:

'cereal' or 'Cerealia': pollen deriving from cultivated grasses;

'cereal-type': pollen types which morphologically correspond to those of the cultivated grasses but which could include some wild grass species;

'cereal-size': any grass pollen grain with a diameter corresponding to an accepted minimum size for many cereal pollen grains, e.g. 37μm for grains treated with sodium hydroxide (NaOH) or potassium hydroxide (KOH), followed by Erdtman's acetolysis and embedding in silicone oil (Faegri and Iversen 1975). This category would also include some grains from wild grasses.

5.2 IDENTIFICATION

In 1929, von Post had suggested that the frequencies of cereal pollen could be used to determine the expansion of tilled areas (quoted in Iversen 1941: 48). With the appearance in 1937 of the celebrated paper by Firbas, the possibilities of separating cereal pollen grains from those of wild grasses in modern and fossil samples were demonstrated. Firbas's work showed that the pollen of cereals was usually considerably larger (at around 35-50μm for KOH-treated material) than that of other grasses (typically 20-25μm), and that cultivated species had thicker pore walls and pore diameters as well as more pronounced exine sculpturing. He recognised the difficulties involved in procedures based on grain size criteria and his early use of size frequency curves indicates some overlap in size between wild and cultivated grasses (Figure 1).

Refinements to the approach taken by Firbas are seen, for example, in Troels-Smith (1955), Beug (1961), Faegri and Iversen (1964) and Whitehead and Langham (1965) where various combinations of pore, annulus and grain diameters are measured together with thickness and delimitation of the annulus (Figure 2). The simple procedure of placing a cereal-size grain in the cereal-type category is insufficient - this writer has seen Gramineae grains of about 50μm diameter in late glacial and postglacial deposits from Scotland, but in these cases the pore and annulus sizes did not meet the criteria for inclusion within the cereal-type category. The pollen size parameters are subject to the dictates of grain condition (e.g. crumpling, swelling, preservation); laboratory preparation; grain orientation beneath the cover slip (silicone oil as a mountant can minimise swelling and often enables the analyst to re-orientate the grain while viewing it through the microscope); and the quality of microscope optics and analyst's eyesight.

Paramount, however, is the inherent variability in the sizes of graminaceous pollen grains and, in fossil material, their condition. It is perhaps worth stressing that the consideration of pollen morphological characteristics in modern samples is all very well but fossil materials are frequently more difficult to assess. Pollen from lake and peat deposits is often better preserved than that from soils yet can be impossible to identify further than to 'cereal-size'. Many palynologists attempting to make determinations on grains derived from some archaeological soil contexts will recognise the frustration of doing much less well with their samples. It is these difficulties which encouraged Andersen (1979: 71) to promote the use of annulus diameter as 'the most important

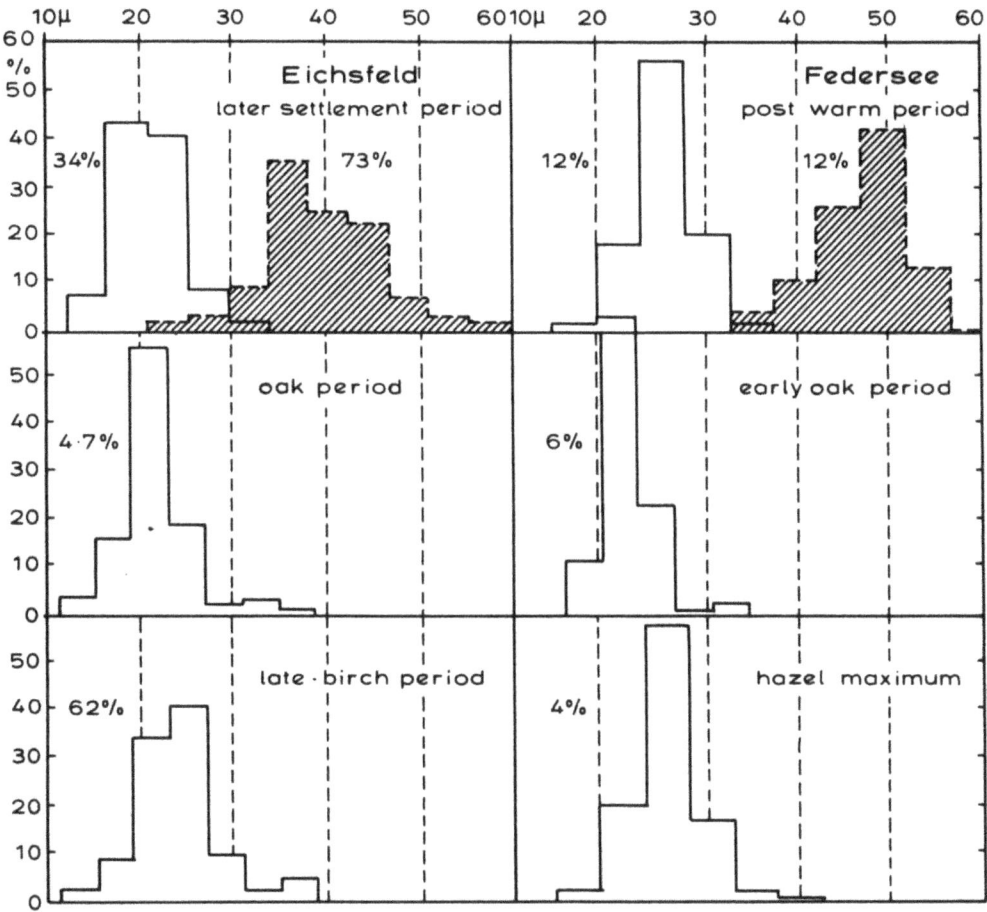

FIGURE 1 Size frequency curves of fossil grass pollen grain diameters (after Firbas 1937: Figure 2). Open curves = inferred wild grass grains; shaded curves = cereal-size grains.

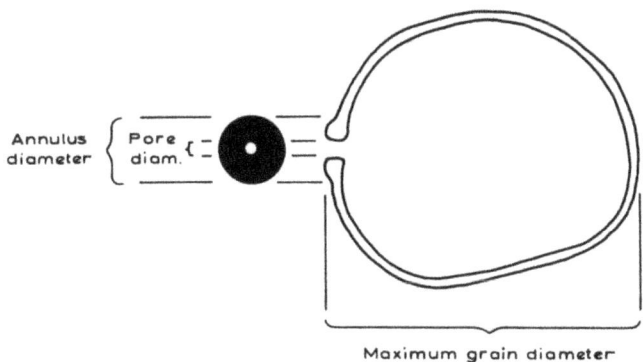

FIGURE 2 Terminology used in grass pollen grain description.

character for identification of Poaceae pollen, and size and surface sculpturing are considered of secondary importance'.

Observation of the surface sculpturing of the pollen exine is facilitated by the use of phase contrast microscopy (Grohne 1957; Beug 1961) and by electron microscope studies (Rowley 1960; Tsukada and Rowley 1964; Andersen and Bertelsen 1972; Köhler and Lange 1979). In combination with size measurements an assessment of surface pattern can place grains within sculpturing categories as an aid to identification to a lower taxonomic level (for instance, see Table 1).

Although the sophistication of scanning electron microscopy might be thought to hold out the best, though laborious, hope of identification, the work of Watson and Bell (1975) might be quoted. They note that 'notwithstanding the existence of some taxonomic order in surface patterns, we think that scanning electron microscopy offers only very limited scope for identifying grass pollens' (ibid.: 985). The research of Köhler and Lange (1979) supports that of Watson and Bell while also optimistically claiming that 'a higher degree of reliability in the identification of the pollen types may be achieved if all available LM (light microscope) and SEM (scanning electron microscope) characters are combined and if the ecological differences of those wild grass species ranging within the pollen size of cereals are taken into consideration as an exclusive factor' (ibid.: 133).

Considering the variability of surface sculpturing on grass pollen grains, it may be that a measure of the number of punctae (sculpturing elements) per grain may be fairly constant within some of the taxonomic groups identified by phase contrast and electron microscopy techniques. This observation was recorded for various strains of *Zea mays* L. pollen by Tsukada and Rowley (1964) though I am unaware of the application of the technique to other species or whether it is strain-specific.

Given the inherent morphological variability of the pollen grains of the Gramineae family, the published research to date does not seem to permit certainty over identification apart from the probability that most grains with a mean diameter (average of the sum of the largest diameter and that at right angles to it) greater than 37μm and with an annulus diameter greater than 8μm will be from cereals which may be placed in one of the sculpturing groups (Andersen 1979); and that those also with a distinctive oblong shape and eccentrically located pore may be *Secale cereale* L. (Andersen 1979: 83) although its scabrate as opposed to verrucate exine type places it within the *Hordeum* group sculpturing category (Andersen and Bertelsen 1972). Caution must, however, be exercised continuously and two recent papers serve to illustrate the problem. In deposits dated earlier than 8290±80 bp from the Yorkshire Wolds, Bush and Flenley (1987: 435) found Gramineae pollen with size characteristics typical of cereals and identified as 'cf. *Avena/Hordeum* (presumably wild varieties)'. O'Connell (1987) records *Triticum*-type pollen from Connemara which extends back to about 7570 bp but these are also presumably correctly disregarded from contention since they pre-date the main body of archaeological evidence for potential farming communities by several millennia. Pollen grains more convincingly indicative of cereals and cultivation though, have been recovered from Connemara (Molloy and O'Connell 1987).

Many palynologists have attempted to place their Cerealia grains into more specific groups than electron microscopy indicates might be warranted. One researcher at least has queried his own earlier determinations (Vorren 1986: 15). Palynologists should also bear in mind that many users of pollen-analytical data do not take on board the written (and unwritten) qualifications which surround their inferences of cereal pollen.

1 Wild grass group
   Mean annulus diameter <8µm
   Mean pollen size <37µm
   Sculpturing scabrate or verrucate

   This group contains most of the wild grasses

2 *Hordeum* group
   Mean annulus diameter 8-10µm
   Mean pollen size 32-45µm
   Sculpturing scabrate

   This group contains wild and cultivated grasses:
wild:
   Agropyron junceiforme
   Agropyron repens
   Ammophila arenaria
   Elymus arenarius
   Glyceria fluitans
   Glyceria plicata
   Hordeum murinum
cultivated:
   Hordeum vulgare
   Triticum monococcum

3 *Avena-Triticum* group
   Mean annulus diameter >10µm
   Mean pollen size >40µm
   Sculpturing verrucate

   This group contains one wild grass (*Avena fatua*) and the following cultivated species:

   Avena nuda
   Avena sativa
   Triticum aestivum
   Triticum compactum
   Triticum dicoccum
   Triticum durum
   Triticum polonicum
   Triticum spelta

4 *Secale cereale*
   Mean annulus diameter 8.9µm
   Mean pollen size 40.1µm
   Sculpturing scabrate
   Shape oblong - Andersen (1979) should be consulted for further details.

TABLE 1 Morphological groups for Cerealia pollen (after Andersen 1979).

In the interests of non-misunderstanding rather than euphony, this may require the repetitive use of 'possiblys', 'mays' and 'coulds' (something this writer may have done insufficiently for some - Edwards and Hirons 1984; Edwards 1985a and b).

5.3 STATISTICAL AND SPATIAL CONSIDERATIONS

The pollen of most cereals grown in European prehistory is produced in low quantities. This is consistent with the self-pollinating nature of *Avena, Triticum* and *Hordeum* species (though the first two may release pollen from open flowers in good weather while barley pollination occurs in closed flowers - Valle 1964, quoted in Vuorela 1972: 8). *Secale cereale* is cross-pollinated by wind and produces abundant amounts of pollen which are likely to figure more significantly in the fossil record (cf. Godwin 1975; Chambers and Jones 1984; Willerding 1986). Studies of modern pollen samples (Vuorela 1973; Caseldine 1981) show that the very large cereal-type pollen grains also travel a short distance from their host plants in comparison with other pollen types and this reduces their contribution to pollen-receiving sites (though see the special case of lake sites below). From the point of view of fossil grain detection these factors will be exacerbated for time periods covering the beginnings of agriculture because an extensive woodland cover would also lead to the filtering of non-arboreal pollen (NAP), especially under conditions of small-scale clearance activity within forested areas (Vuorela 1973, 1986). Furthermore, there is no certainty that a chosen pollen site will have been located especially close to an early farming area. Such factors as these typically result in cereal-type pollen finds of extremely low frequency - one grain in a total pollen count of 500 or 1000 terrestrial grains would not be unusual.

The short travel distance of cereal pollen evident in present-day studies is reflected in multiple profile investigations of fossil spectra. Thus Behre and Kucan (1986), amongst other examples, show a noticeable decrease in mean cereal pollen frequencies and those for other anthropogenic indicators (averaged values for the period A.D. 700-1400) over a distance of three kilometres from an inferred medieval field area in northern West Germany (Figure 3). Similarly, Lange (1986) has investigated the pollen in ditch sediments of Slavonic earthworks in central East Germany and found that the percentages of cereal and other non-arboreal pollen types decreased sharply with distance from supposed former fields.

Given the relatively low pollen productivity of cereals, the short distance pollen travels and uncertainty over the precise location of arable fields, there will be an excellent chance that the pollen-analyst will simply not come across any Cerealia grains during routine counting. It was in an effort to overcome this possibility when investigating evidence for early agriculture in Scotland that methods of optimising cereal grain detection were devised at Birmingham (Edwards *et al.* 1986; McIntosh 1986). The strategy begins with initial core site selection. Assuming that a site located close to a bog edge will be more likely to receive cereal pollen than one distant from the area of past cultivation, then a core is taken fairly close to the edge (it is always possible, of course, that a bog edge could be fringed by filtering woodland and that a core location could be optimally placed to benefit from edge and wind factors in NAP transport). Since cereal-type grains are large and easily spotted at low power, microscope slides may be scanned at say x100 magnification. This enables a number of replicate slides to be viewed and after appropriate estimates of pollen content per slide, the time taken to assess the equivalent of several thousands of grains in the search for the low frequency cereal-types becomes measured in tens of minutes rather than tens of hours. If contiguous samples are counted then this further reduces the chance of missing grains

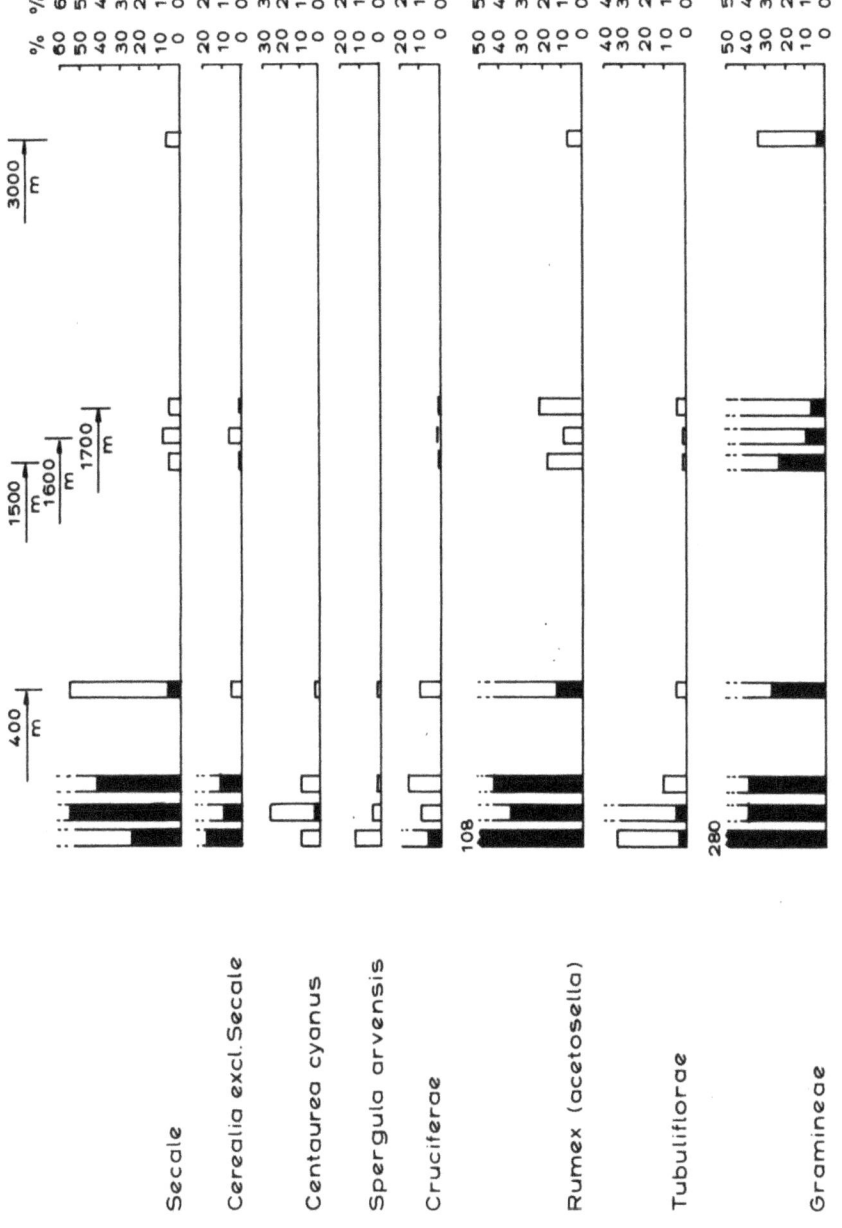

FIGURE 3 Variation in pollen representation with distance from medieval fields near Flügeln, West Germany (based on Behre and Kucan 1986, Figure 5). Black bars = % of arboreal pollen sum; white bars = 10x exaggeration.

because of the existence of a sampling interval. Employing such methods, McIntosh (1986) recorded single grains of cereal-type in the pre-elm decline levels of his Scottish sites. Estimates of such grains as a percentage of total land pollen ranged from 0.02% to 0.05%. It is possible that other means of cereal pollen concentration are possible such as selective sieving or variations in settling velocity (Edwards et al. 1986).

The examples of multiple profile research detailed above involve the study of pollen contained within a 'fixed' matrix of peat. Lake sediments present a very different situation because of the more complex set of processes involved in palynomorph transport to the sampling site and the potential for sediment and pollen redeposition (Davis et al. 1969; Edwards 1983). The fact that much of the pollen received by a lake may do so by stream transport (Bonny 1976, 1978) could, however, be a positive factor in the 'remote sensing' of early agriculture, since the pollen from distantly cultivated crops may find its way into streams adjacent to fields or crop processing areas, from where it is transported considerable distances to a lake (Edwards 1982). This mode of detection, while unsatisfactory from the viewpoint of reconstructing phases of local activity (the lake sediments are likely to contain an aggregate record from the whole catchment), could be the only means of investigating human impacts if the farmed areas contain no suitable deposits.

5.4 PATTERN AND THE CEREALIA POLLEN CURVE

In his classic 'Landnam' paper, Iversen (1941: 48-49) noted the low pollen production of the non-rye group of cereals in the supposed neolithic sections of pollen diagrams produced by Firbas, Gross and Schroder and observed that:

> 'This of course does not mean that the pollen-analytical demonstration of cereal pollen loses its value; it simply means that the cereal-pollen curve is not a definite expression of the intensity of cereal-growing...the heavy rise in the cereal-pollen curve when we get some way up into Sub-atlantic strata...need not necessarily mean that cereal cultivation has now extended so much; it may also be connected with the introduction of rye. The fluctuations in the agricultural intensity in antiquity are better reflected in the curves for weed pollen'.

This typically early appreciation of pattern in pollen diagrams is as well founded today as it was half a century ago and even in those areas or for those times in which *Secale* was less important than other cereal crops, we often see an early phase of sporadic occurrences in cereal-type pollen in the profiles followed by a later expansion (e.g. North Mains, Scotland, Hulme and Shirrifs 1985; Kirkkojärvi, Finland, Vuorela 1975; Pleszów, Poland, Wasylikowa 1986 and cf. Figure 4). The expansion in many cases may be due to a more open landscape in which the effects of pollen filtration by trees will be less marked and/or the extension of agriculture to even less suitable areas (such as closer to pollen receiving bog sites) as population expanded. If the beginnings of farming in many areas represented a speculative venture then the likely shifting nature of activity by small groups, either testing the land, taking advantage of higher soil fertility on uncultivated soils, or avoiding conflict with indigenous hunter-gatherer communities, would be unsurprising (cf. Case 1969; Coles 1976). If such enterprises were small-scale and short-lived, then their discovery must be regarded as largely fortuitous. Vuorela (1986: 57) has remarked on the difficulty of distinguishing between early agricultural phases in Finland and in the absence of Cerealia finds tantalising evidence might be provided by charcoal layers, a short-lived decline in *Picea* or a periodicity in the tree pollen succession which is difficult to explain without recourse to human interference (and it might indeed be added, without the possibilities of non-farming mesolithic impact). The difference between this and Vuorela's absolute limit in the Cerealia curve (Figure 4) is considered to be 'in many cases, only a result of sample statistics' (ibid.: 58). The empiric limit marks the start of the continuous Cerealia pollen curve and is interpreted for southern Finland as an increased number of burnt clearings yet with sufficient woodland at various stages of regeneration to hamper pollen dispersal. The rational limit is denoted by the sharp rise in Cerealia pollen following the extensive cultivation of rye in the fifteenth to nineteenth centuries. (It should be noted that palynological use of the concepts of empirical and rational limits is normally devoted to situations where only one species is involved - Godwin 1975; Smith and Pilcher 1973).

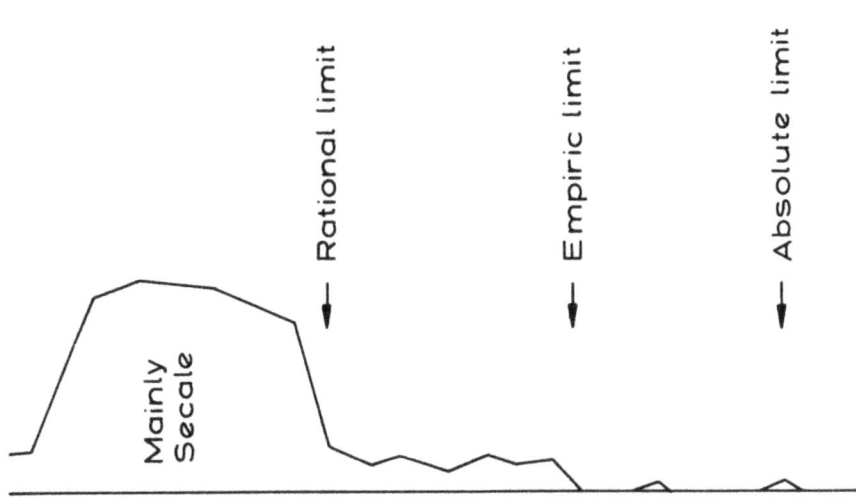

FIGURE 4 The division of the Cerealia curve (based on Vuorela 1986: Figure 6).

This patterning of the cereal pollen curve is going to be much influenced by farming practices, vegetation cover and the spatial relationship between clearing and the site of pollen deposition. The temporary reduction in *Picea* as a result of slash-and-burn agriculture is a notable phenomenon in many Finnish pollen diagrams and other parts of Europe do not appear to have an equivalent arboreal pollen indicator which could provide a clue to early agriculture, though a general fall in a number of tree pollen taxa is a common accompaniment to inferred human impact (Iversen 1941; Edwards and Hirons 1984; Aaby 1986).

5.5 SPECIES EXCLUSIVITY AND SUPPORTING PALAEOECOLOGICAL EVIDENCE

A knowledge of the ecological conditions pertaining to study sites for both the past as well as the present may enable certain grass species to be included or excluded from consideration. If, for instance, work is being carried out on an inland peat bog then it may be possible to exclude Marram Grass (*Ammophila arenaria* (L.) Link), Lyme Grass (*Elymus arenarius* L.) and Floating Sweet-grass (*Glyceria fluitans* (L.) R. Br.) from further consideration if a *Hordeum*-type grain is found. With regard to *G. fluitans* it would be necessary to be certain that the bog had not been fringed or featured standing water at some stage. Couch (*Agropyron repens* (L.) Beauv.) or Wall Barley (*Hordeum murinum* L.) may be less easily excluded if other non-arboreal taxa suggest cereal cultivation since both species are common in arable habitats (Hubbard 1984). The presence of Couch or Lyme Grass may also confuse the issue since the seed grain of *A. repens* is known to have been used in and after the eighteenth century A.D. in Finland and Sweden to eke out supplies of bread flour (Vuorela 1972), while *E. arenarius* has been used as a cereal crop in Iceland (Olafsson 1943; Tómasson 1973) and Greenland (Fredskild, forthcoming). In some areas, therefore, it may be necessary to class them as cereal crops in their own right. Palynologists will also be aware that, perversely, the pollen of einkorn (*Triticum monococcum* L.) is morphologically located within the *Hordeum* group, while other species of wheat together with some of those of oats form part of the *Avena-Triticum* group (Table 1).

If the presence of cereal-type pollen grains with all the associated uncertainties represents our most convincing single piece of evidence for possible cultivation, then a suite of appropriate weed flora in the pollen spectra (often with a reciprocal reduction in the grains of woodland taxa) will lend strong support (cf. Iversen quotation above; Groenman-van Waateringe 1983; Edwards and Hirons 1984). Extensive discussions relating to the significance of anthropogenic indicator species may be found in the literature (e.g. Turner 1964, 1986; Groenman-van Waateringe 1968, 1979a and b, 1986; Vuorela 1970, 1973, 1980, 1986; Godwin 1975; Lange 1986; Edwards 1979; Maguire 1983; Willerding 1986) and special mention must be made of the paper by Behre (1981) on the interpretation of indicators and the many relevant papers in Behre (1986) some of which are cited above. It is perhaps worth pointing out that the weed flora of the present may not accord with that at the early stages of agriculture when the woodland cover would have been much more closed; that the indicator species of value to reconstructions in, for example, northern Scandinavia or central Europe may be of less use in more southerly or western areas; and that the disturbance of weed seed banks during browsing, the use of fallow systems or simply the abandonment of clearings may produce spurious pollen data from remaining weeds and cultivars, which could have serious implications for estimates of clearance phase longevity and their related causes (Pilcher *et al.* 1971; Edwards 1979; Buckland and Edwards 1984). Even allowing for the inherent difficulties of relating present-day observations to events of the distant past, studies of the pollen rain associated with modern agriculture are regrettably few (e.g. Turner 1964; Vuorela 1973; Caseldine 1981; Groenman-van Waateringe 1986; O'Connell 1986).

Macroscopic evidence of cereal cultivation will rarely come from the same sites as the long profile pollen data since such profiles are not normally located within a settlement site or necessarily immediately adjacent to crop cultivation or processing activities. A notable exception to this has been reported from the site of Pleszów 17-20 in the area of Cracow, Poland (Wasylikowa 1986). Pollen and plant macrofossils were recovered from the palaeochannel deposits of an oxbow lake. The deposits had been accumulating perhaps one kilometre from a known Lengyel culture site (post *circa* 5830 bp) though the paper gives no information concerning the distribution of earlier *Linearbandkeramik* or later Funnel Beaker and Baden sites. The diagram of fossil taxa (Figure 5) shows a strong agreement between pollen identified as Cerealia and the macro-remains of *Triticum dicoccum* (Schrank) Schübeler and *T. monococcum* for most of the period of local Lengyel Culture occupation (5830-5380 bp). This period also corresponds to the phase of greatest non-arboreal pollen expansion in the Pleszów diagram.

There are smaller indications of Cerealia and other cultural herbaceous taxa in the pre- and post-Lengyel sections of the diagram. These could be taken to indicate smaller-scale or more distant activities on the part of agriculturalists and in view of the micro- and macrofossil evidence for the period of known local settlement, it might seem churlish to query the evidence for farming in the absence of archaeological data. A further example of the correspondence between macro- and microscopic cereal evidence is furnished by Göransson (1987a and b) from the Alvastra pile dwelling site in Sweden. Here, large amounts of barley and emmer wheat from the culture layer (*circa* 4500 bp) are also reflected in very high frequencies of Cerealia pollen.

Palynological data can take a lead in suggesting the local presence of undiscovered occupation sites as was successfully demonstrated at Crawford Lake, Ontario, Canada (Finlayson *et al*. 1973; Byrne and McAndrews 1975). Another example is provided by the inference of slash-and-burn agriculture from charcoal and pollen evidence in southern Finland from 4000 bp and since supported by macrofossil finds of *Hordeum vulgare* var. *nudum* in an archaeological context (quoted in Vuorela 1986: 54).

The analysis of microscopic charcoal has been used frequently to complement palynology in studies of early human activity (Patterson *et al*. 1987). Research in Scandinavia especially has emphasised the role of slash-and-burn cultivation in prehistoric agriculture and this is held to manifest itself in the fossil charcoal record (Iversen 1941, 1964; Tolonen 1985; Vuorela 1976, 1986; Huttunen 1980). Rowley-Conwy (1981, 1982) has queried the notion that swidden agriculture was as extensive in prehistory as some believe. In terms of early agriculture in Britain, a detailed comparison of charcoal abundance with the incidence of cereal-type pollen both at and prior to the elm decline produced equivocal results in the early deposits and a negative relationship at the elm decline (McIntosh 1986). A problem with the charcoal evidence is that a variety of causes could give rise to similar micro-charcoal patterns (Edwards, forthcoming) and peaks in charcoal may not necessarily correspond causally with episodes of agricultural activity. For the earliest phases of agriculture the picture could also be confused with continuing or residual hunter-gatherer practices in which widespread burning of vegetation may have occurred. Nevertheless, charcoal studies in palynology may well have a special value in designating those sections of our pollen profiles as areas of potential hunter-gatherer/agricultural transition which would repay further investigation through approaches involving higher sampling resolution and larger pollen sums (Green and Dolman 1988).

The cultivation of soils, especially on a prolonged basis, frequently leads to soil erosion as a result of the removal of soil particles by overland flow of surface water on inclined ground. In later periods of prehistory, studies of such characteristics as chemistry, particle-size, magnetism, accumulation rates, pollen deterioration and even the pattern of *Isoetes* microspore representation can provide an indication of agriculturally-induced erosion (Mackereth 1965; Thompson *et al*. 1975; Vuorela 1980). Palaeolimnological investigations at several lakes in County Tyrone, Northern Ireland (Hirons and Edwards 1986) suggested some small-scale erosional activity prior to the elm decline where there

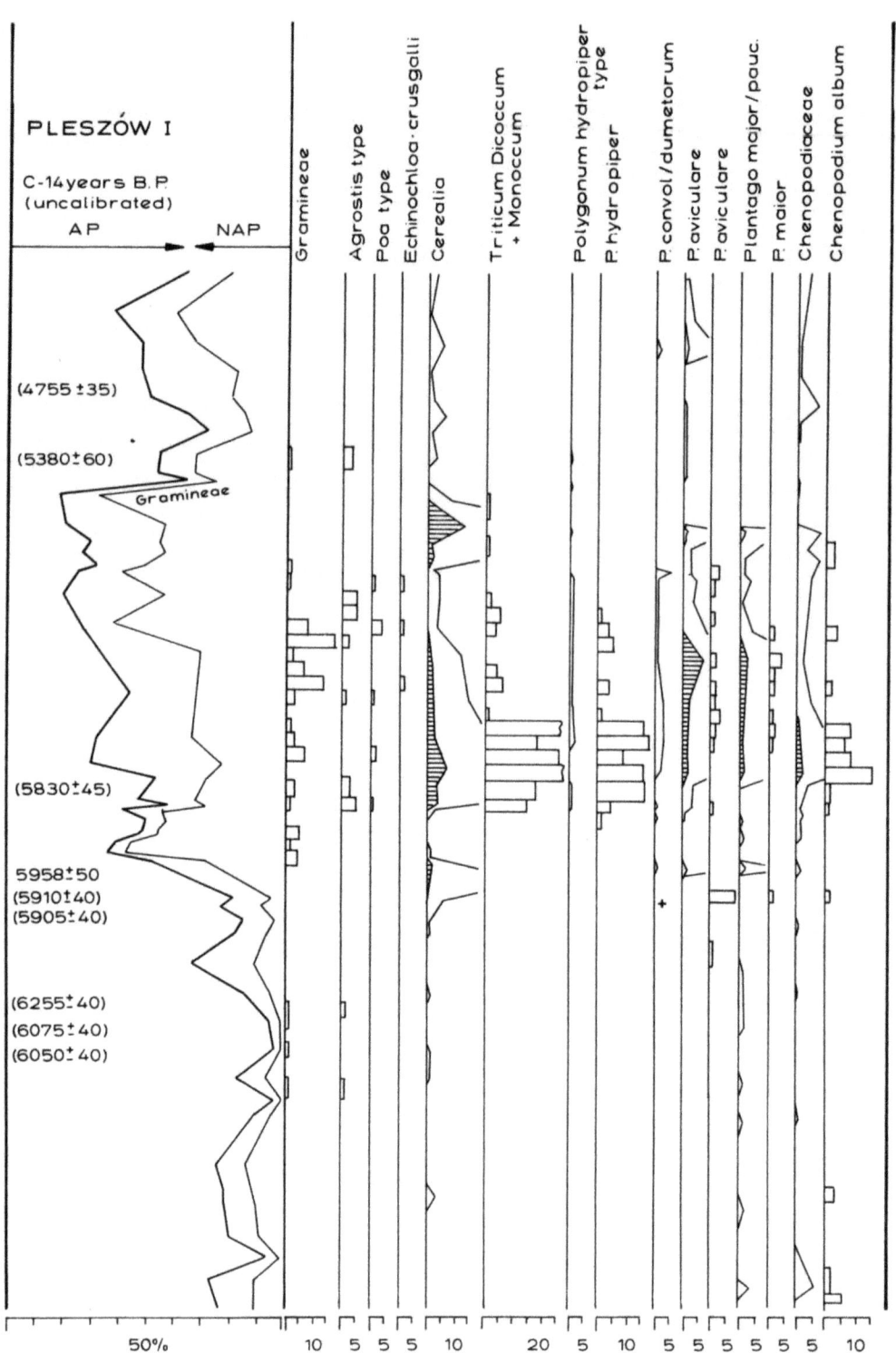

FIGURE 5 Comparison of pollen and macrofossil data from Pleszów I, Poland (based on Wasylikowa 1986: Figure 2). Curves = pollen as % total land pollen (white silhouettes at 10x exaggeration); bars = macrofossils as number of specimens in 40cc.

were also finds of cereal-type pollen. The magnitude of inferred erosional events was, however, greater at and after the first elm decline (*circa* 5200 bp). Even these later episodes are less marked than those seen in the Bronze Age levels of Braeroddach Loch, Aberdeenshire, Scotland (Edwards and Rowntree 1980). At this small and environmentally-sensitive loch, the first incidence of cereal-type pollen coincides with markedly increased levels of chemical indicators of erosion (e.g. sodium, potassium and magnesium), a rise in magnetic susceptibility values (suggesting the input of magnetite-rich material), an increase in the size of particles in the sediment and the start of a reversal in radiocarbon dates suggesting the inwash of old carbon consequent upon soil erosion. There are too few lake studies of a comprehensive nature which can provide an indication of possible erosional events for early farming periods. Given the likely small-scale environmental impacts frequently involved, then the techniques available to us may be too insensitive to detect them reliably.

5.6 STRATIGRAPHIC SIGNIFICANCE

Assuming that we are prepared to accept that grains identified to the category of cereal-type are in fact likely to be derived from the Cerealia, do we then infer cereal cultivation? Once again the response of the palynologist should be an appeal to common sense. If the grain in question comes from the analysis of a pre-Holocene interglacial core then the probability is that it is not the result of precocious agriculture! Where one places temporal limits on this line of argument is clearly difficult since, in theory, once farming has actually begun somewhere then there is always the remote possibility that the pollen grains of a cultivar could have travelled to the sampling site via, say, high level air currents. Even if the cultivation of cereals is known to have begun within a territory under investigation, Cerealia grains could have arrived at a specific sampling site by simple wind transport (O'Sullivan 1976), short-distance wind transport during harvesting or threshing (Göransson 1987a) or human conveyance (Vuorela 1973; Robinson and Hubbard 1977). Field or laboratory contamination could also compromise the stratigraphic integrity of samples and if smearing of a sample occurs (e.g. where the downward travel of the corer mixes younger material with lower older material and where the potentially contaminated outer layers are not efficiently removed) then it is equally likely to produce intrusive cultural-type herbaceous and cereal pollen grains. It is probably pushing coincidence too far to plead that the increasingly large number of sites from which pre-elm decline cereal-type pollen grains have been recovered (Edwards and Hirons 1984 and Figure 6) might be due to such contamination.

5.7 CONCLUSIONS

Much of this paper has concentrated on cereal pollen in general rather than the early phases of agriculture alone. This is inevitable since the problems of reconstruction are no less evident in many of the later phases. It is often a matter of degree and the later events, with their frequently accentuated patterns, provide a template for earlier, dimmer vestiges of the fossil pollen record.

The current uncertainty attached to the identification of cereal pollen provides us with several options and one necessity. The options must surely be positive and while including refinements in identification procedures must also include the search for confirmatory evidence such as supporting palynological and wider palaeoecological and archaeological data. Of course, the presence of a cereal-type pollen grain and a lack of

FIGURE 6 Irish and British lake and peat sites from which pre-elm decline cereal-type pollen grains have been reported:

1. Cashelkeelty (Lynch 1981)
2. Loch Namackanbeg (O'Connell 1987)
3. Dolan (Teunissen and Teunissen-van Oorschot 1980)
4. Lough Sheeauns (Molloy and O'Connell 1987)
5. Connemara National Park (O'Connell 1987)
6. Lough Doo (O'Connell *et al.* 1987)
7. Carrowkeel (Göransson 1984)
8. Ballygawley Lough (Göransson 1984)
9. Strand Hill (Göransson 1984)
10. Weir's Lough (Hirons and Edwards 1986)
11. Ballynagilly (Pilcher and Smith 1979)
12. Newferry (Smith and Collins 1971)
13. Rhoin Farm (Aros Moss) (McIntosh 1986)
14. Machrie Moor (Robinson 1981)/Moorlands (McIntosh 1986)
15. North Mains (Hulme and Shirriffs 1985)
16. North Gill 1A (Simmons and Innes 1987)
17. Willow Garth (Bush and Flenley 1987)
18. Soyland Moor (Williams 1985)
19. Martin Mere (Tooley 1985)
20. Flea Moss Wood (Simmons and Innes 1987)
21. West Heath Spa (Girling and Greig 1985)
22. Rimsmoor (Waton 1982)

WARNING! In the absence of declared identification criteria used in the determination of cereal-type grains at a number of these sites, it is important to realise that these finds may include those of cereal-size grains rather than those derived from the Cerealia. For sites 2,5,6 and 17, the early cereal-type grains are specifically excluded from contention by the authors. Not all sites have been radiocarbon-dated. (Early grains which appear in a pollen diagram from Loch Pityoulish (O'Sullivan 1976) are excluded since the associated text states that the first traces of cereal pollen are found at about 1850 bp).

THE CEREAL POLLEN RECORD AND EARLY AGRICULTURE

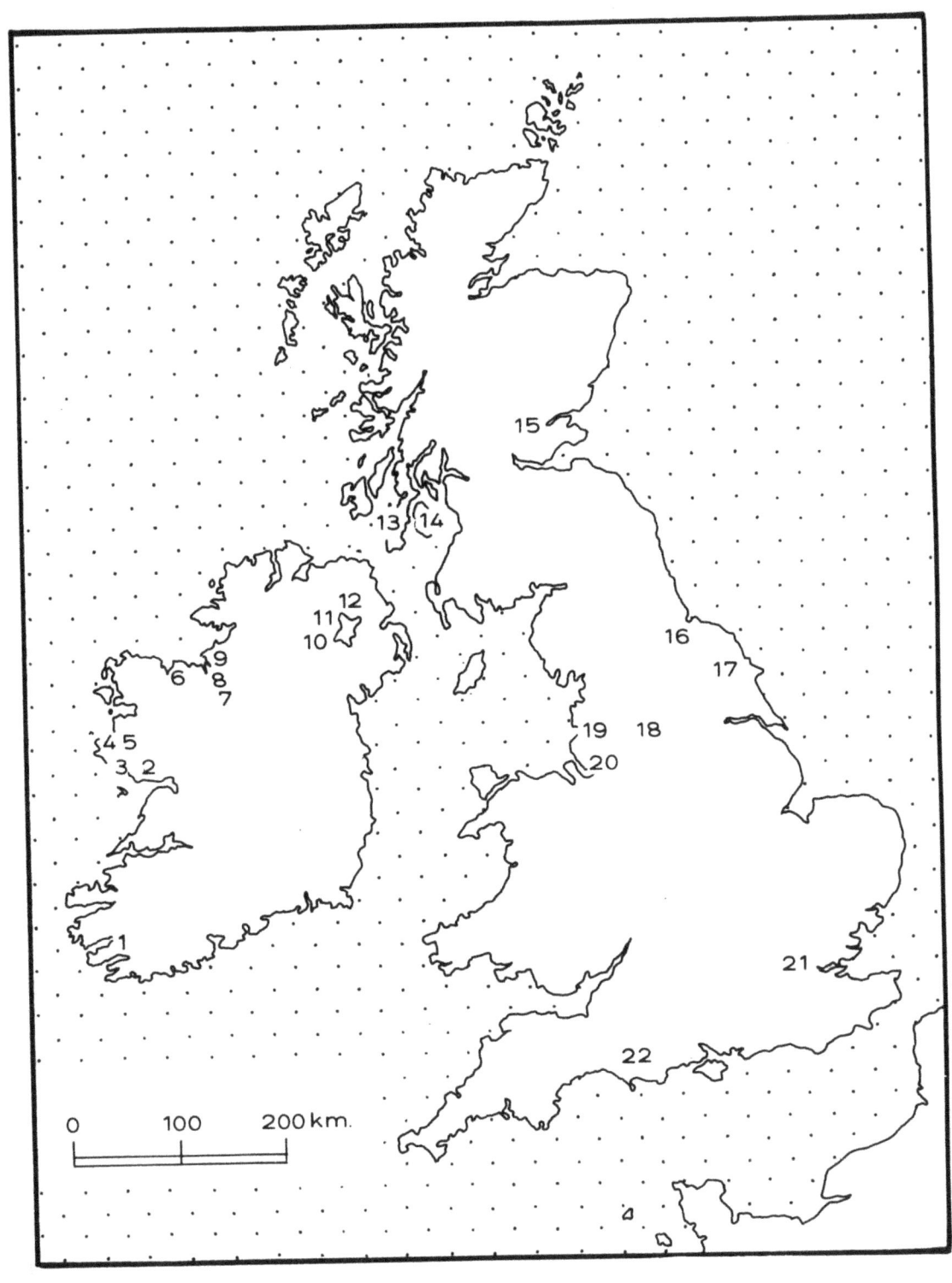

127

additional positive evidence for cultivation neither proves nor disproves a Cerealia identification. Instances of early (i.e. apparent pre-elm decline date) neolithic archaeological contexts are fairly numerous (e.g. see Figure 3 in Zvelebil and Rowley-Conwy 1986) but then so are pollen sites where the elm decline is dated earlier than a nominal 5000-5100 bp (Smith and Pilcher 1973; Edwards 1985b). The relative imprecisions of radiocarbon-dating and sampling problems serve to complicate the issue. Even the promising finds of cereal grain impressions on Ertebølle pottery from southern Sweden (Jennbert 1984) are associated with a layer dated after 5260±80 bp (Lu-1842). It might be considered, however, that the weight of evidence for early agriculture is considerably increased if a large number of palynologists are reporting cereal-type pollen grains for the thousand years or so prior to the elm decline in north west Europe (cf. Figure 6 for the British and Irish patterns and note the growing number of finds from Denmark (Kolstrup 1988) and Sweden (Göransson 1987a)) but not generally for earlier periods - that is unless there was something occurring at this stage, such as openings in woodland and the removal of screens to greater amounts of wild grass pollen, which increased the likelihood of large grain deposition and consequent detection. To these possibilities we must presumably add the increased statistical chance of finding large Gramineae grains when more palynologists are looking at more sites with closer sampling intervals and larger pollen sums. We certainly require palynologists not to hide their finds of cereal-type grains within the collective Gramineae category in order that the patterns of large grain occurrence can be properly assessed. It would also help if palynologists provided the full size characteristics and/or phase-contrast photographs of their putative cereal grains (Jóhansen 1979; McIntosh 1986; Vorren 1986; Göransson 1987a; Molloy and O'Connell 1987). The one necessity in dealing with the evidence of deposits containing very early cereal-type pollen (and how little we worry about post-elm decline finds!) would appear to be the need to qualify interpretations with suitable caution lest our words are misconstrued.

BIBLIOGRAPHY

AABY, B. (1986) Trees as anthropogenic indicators in regional pollen diagrams from eastern Denmark. In: *Anthropogenic indicators in pollen diagrams.* (ed. Behre, K.-E.) Rotterdam: A.A. Balkema pp. 73-93.

ANDERSEN, S. TH. (1979) Identification of wild grasses and cereal pollen. *Danmarks Geologiske Undersogelse Årbog 1978* pp. 69-92.

ANDERSEN, S. TH. AND BERTELSEN, F. (1972) Scanning electron microscope studies of pollen of cereals and other grasses. *Grana* 12 79-86.

BEHRE, K.-E. (1981) The interpretation of anthropogenic indicators in pollen diagrams. *Pollen et Spores* 23 225-45.

BEHRE, K.-E. (ed.) (1986) *Anthropogenic indicators in pollen diagrams.* Rotterdam: A.A. Balkema.

BEHRE, K.-E. AND KUCAN, (1986) Die Reflektion archäologisch bekannter Siedlungen in Pollendiagrammen verschiedener Entfernung - Beispiele aus der Siedlungskammer Flögeln, Nordwestdeutschland. In: *Anthropogenic indicators in pollen diagrams.* (ed. Behre, K.-E.) Rotterdam: A.A. Balkema pp. 95-114.

BEUG, H.-J. (1961) *Leitfaden der Pollenbestimmung für Mitteleuropa und angrenzende Gebiete.* Stuttgart: Gustav Fischer.

BONNY, A.P. (1976) Recruitment of pollen to the seston and sediment of some Lake District lakes. *Journal of Ecology* **64** 859-87.

BONNY, A.P. (1978) The effect of pollen recruitment processes on pollen distribution over the sediment surface of a small lake in Cumbria. *Journal of Ecology* **66** 385-416.

BUCKLAND, P.C. AND EDWARDS, K.J. (1984) The longevity of pastoral episodes of clearance activity in pollen diagrams - the rôle of post-occupation grazing. *Journal of Biogeography* **11** 243-49.

BUSH, M.B. AND FLENLEY, J.R. (1987) The age of the British chalk grassland. *Nature* **329** 434-36.

BYRNE, R. AND McANDREWS, J.H. (1975) Pre-Columban purslane (*Portulaca oleracea* L.) in the New World. *Nature* **253** 726-27.

CASE, H. (1969) Neolithic explanations. *Antiquity* **43** 176-86.

CASELDINE, C.J. (1981) Surface pollen studies across Bankhead Moss, Fife, Scotland. *Journal of Biogeography* **8** 7-25.

CHAMBERS, F.M. AND JONES, M.K. (1984) Antiquity of rye in Britain. *Antiquity* **58** 219-24.

COLES, J.M. (1976) Forest farmers: some archaeological, historical and experimental evidence. In: *Acculturation and continuity in Atlantic Europe.* (ed. De Laet, S.J.) Brugge: IV Atlantic Colloquium pp. 59-66.

DAVIS, R.B., BREWSTER, L.A. AND SUTHERLAND, J. (1969) Variation in pollen spectra within lakes (1). *Pollen et Spores* **11** 557-71.

DIMBLEBY, G. (1985) *The palynology of archaeological sites.* London: Academic Press.

EDWARDS, K.J. (1979) Palynological and temporal inference in the context of prehistory: with special reference to the evidence from lake and peat deposits. *Journal of Archaeological Science* **6** 255-70.

EDWARDS, K.J. (1982) Man, space and the woodland edge - speculations on the detection and interpretation of human impact in pollen profiles. In: *Archaeological aspects of woodland ecology.* (eds Bell, M. and Limbrey, S.) Oxford: British Archaeological Reports S146 pp. 5-22.

EDWARDS, K.J. (1983) Multiple profile studies and pollen variability. *Progress in Physical Geography* **7** 587-609.

EDWARDS, K.J. (1985a) The anthropogenic factor in vegetational history. In: *The Quaternary history of Ireland.* (eds Edwards, K.J. and Warren, W.P.) London: Academic Press pp. 187-220.

EDWARDS, K.J. (1985b) Radiocarbon dating. In: *The Quaternary history of Ireland.* (eds Edwards, K.J. and Warren, W.P.) London: Academic Press pp. 280-93.

EDWARDS, K.J. (forthcoming) Meso-neolithic vegetational impacts in Scotland and beyond: palynological considerations. In: *The Mesolithic in Europe.* (ed. Bonsall, C.) Edinburgh: John Donald.

EDWARDS, K.J. AND HIRONS, K.R. (1984) Cereal pollen grains in pre-elm decline deposits: implications for the earliest agriculture in Britain and Ireland. *Journal of Archaeological Science* **11** 71-80.

EDWARDS, K.J., McINTOSH, C.J. AND ROBERTSON, D.E. (1986) Optimising the detection of cereal-type pollen grains in pre-elm decline deposits. *Circaea* **4** 11-13.

EDWARDS, K.J. AND ROWNTREE, K.M. (1980) Radiocarbon and environmental evidence for changing rates of erosion at a Flandrian stage site in Scotland. In: *Timescales in geomorphology.* (eds Cullingford, R.A., Davidson, D.A. and Lewin, J.) Chichester: John Wiley and Sons pp. 207-23.

FAEGRI, K. AND IVERSEN, J. (1964, 2nd edition; 1975 3rd edition) *Textbook of pollen analysis.* Copenhagen: Munksgaard and Oxford: Blackwell Scientific Publications.

FINLAYSON, W.D., BYRNE, R.A. AND McANDREWS, J.A. (1973) Iroquoian settlement and subsistence patterns near Crawford Lake, Ontario. *Bulletin of the Canadian Archaeological Association* **5** 134-36.

FIRBAS, F. (1937) Der Pollenanalytische Nachweis des Getreidebaus. *Zeitschrift für Botanik* **31** 447-78.

FREDSKILD, B. (forthcoming) In: *The cultural landscape: past, present, future.* (eds Birks, H.J.B., Birks, H.H., Kaland, P.E. and Moe, D.) Cambridge: University Press.

GIRLING, M.A. AND GREIG, J. (1985) A first fossil record for *Scolytus scolytus* (F.) (elm bark beetle): its occurrence in elm decline deposits from London and the implications for Neolithic elm disease. *Journal of Archaeological Science* **12** 347-51.

GODWIN, H. (1975) *History of the British flora.* (2nd edition) Cambridge: University Press.

GÖRANSSON, H. (1984) Pollen analytical investigations in the Sligo area. In: *The archaeology of Carrowmore. Environmental archaeology and the megalithic tradition at Carrowmore, County Sligo, Ireland.* (ed. Burenhult, G.) Stockholm: Theses and Papers in North-European Archaeology 14 pp. 154-93.

GÖRANSSON, H. (1987a) *Neolithic man and the forest environment around Alvastra pile dwelling.* Stockholm: Theses and Papers in North-European Archaeology 20.

GÖRANSSON, H. (1987b) The cultural landscape during the time of the Alvastra pile dwelling. In: *Theoretical approaches to artefacts, settlement and society.* (eds Burenhult, G., Carlsson, A., Hyenstrand, Å. and Sjovold, T.) Oxford: British Archaeological Reports S366 pp. 153-63.

GREEN, D.G. AND DOLMAN, G.S. (1988) Fine resolution pollen analysis. *Journal of Biogeography* **15** 685-701.

GROENMAN-VAN WAATERINGE, W. (1968) The elm decline and the first appearance of *Plantago major*. *Vegetatio* **15** 292-96.

GROENMAN-VAN WAATERINGE, W. (1979a) Weeds. *Proceedings of the fifth Atlantic Colloquium, Dublin 1978* pp. 363-67.

GROENMAN-VAN WAATERINGE, W. (1979b) The origin of crop weed communities composed of summer annuals. *Vegetatio* **41** 57-59.

GROENMAN-VAN WAATERINGE, W. (1983) The early agricultural utilization of the Irish landscape: the last word on the elm decline? In: *Landscape archaeology in Ireland*. (eds Reeves-Smyth, T. and Hamond, F.) Oxford: British Archaeological Reports 116 pp. 217-32.

GROENMAN-VAN WAATERINGE, W. (1986) Grazing possibilities in the Neolithic of the Netherlands based on palynological data. In: *Anthropogenic indicators in pollen diagrams*. (ed. Behre, K.-E.) Rotterdam: A.A. Balkema pp. 187-202.

GROHNE, U. (1957) Die Bedeutung des Phasenkontrastverfahrens für die Pollenanalyse, dargelegt am Beispiel der Gramineepollen vom Getreide-Typ. *Photographie und Forschung* **7** 237-48.

HIRONS, K.R. AND EDWARDS, K.J. (1986) Events at and around the first and second *Ulmus* declines: palaeoecological investigations in County Tyrone, Northern Ireland. *New Phytologist* **104** 131-53.

HUBBARD, C.E. (1984) *Grasses*. (3rd edition) London: Penguin.

HULME, P.D. AND SHIRRIFFS, J. (1985) Pollen analysis of a radiocarbon-dated core from North Mains, Strathallan, Perthshire. *Proceedings of the Society of Antiquaries of Scotland* **115** 105-13.

HUTTUNEN, P. (1980) Early land use, especially the slash-and-burn cultivation in the commune of Lammi, southern Finland, interpreted mainly using pollen and charcoal analyses. *Acta Botanica Fennica* **113** 1-45.

IVERSEN, J. (1941) Land occupation in Denmark's stone age. *Danmarks Geologiske Undersogelse II* **66** 1-68.

IVERSEN, J. (1964) Retrogressive vegetational succession in the postglacial. *Journal of Ecology* **52** (Suppl.) 59-70.

JENNBERT, K. (1984) Den produktiva gåvan. Tradition och innovation i Sydskandinavien för omkring 5300 år sedan. *Acta Archaeologica Ludensia Series in 4* 16.

JOHANSEN, J. (1979) Cereal cultivation in Mykines, Faroe Islands A.D. 600. *Danmarks Geologiske Undersogelse Arbog 1978* 93-103.

KÖHLER, E. AND LANGE, E. (1979) A contribution to distinguishing cereal from wild grass pollen grains by LM and SEM. *Grana* **18** 133-40.

KOLSTRUP, E. (1988) Late Atlantic and early Sub-boreal vegetational development at Trundholm, Denmark. *Journal of Archaeological Science* **15** 503-13.

LANGE, E. (1986) Pollenanalytische Untersuchungen von Burggrabensedimenten aus der nordwestlichen Niederlausitz - ein Beitrag zu methodischen Fragen der Auswertung von Pollendiagrammen und zur slawischen Landwirtschaft. In: *Anthropogenic indicators in pollen diagrams.* (ed. Behre, K.-E.) Rotterdam: A.A. Balkema pp. 153-66.

LYNCH, A. (1981) *Man and environment in south west Ireland, 4000 B.C. - A.D. 800, a study of man's impact on the development of soil and vegetation.* Oxford: British Archaeological Reports 85.

MACKERETH, F.J.H. (1965) Some chemical observations on post-glacial lake sediments. *Philosophical Transactions of the Royal Society of London B* **250** 165-213.

MAGUIRE, D.J. (1983) The identification of agricultural activity using pollen analysis. In: *Integrating the subsistence economy.* (ed. Jones, M.) Oxford: British Archaeological Reports S181 pp. 5-18.

McINTOSH, C.J. (1986) *Palaeoecological investigations of early agriculture on the Isle of Arran and the Kintyre Peninsula.* M.Sc. thesis, University of Birmingham.

MOLLOY, K. AND O'CONNELL, M. (1987) The nature of the vegetational changes at about 5000 B.P. with particular reference to the elm decline: fresh evidence from Connemara, western Ireland. *New Phytologist* **106** 203-20.

O'CONNELL, M. (1986) Reconstruction of local landscape development in the post-Atlantic based on palaeoecological investigations at Carrownaglogh prehistoric field system, County Mayo, Ireland. *Review of Palaeobotany and Palynology* **49** 117-76.

O'CONNELL, M. (1987) Early cereal-type pollen records from Connemara, western Ireland and their possible significance. *Pollen et Spores* **29** 207-24.

O'CONNELL, M., MITCHELL, F.J.G., READMAN, P.W., DOHERTY, T.J. AND MURRAY, D.A. (1987) Palaeoecological investigations towards the reconstruction of the post-glacial environment at Lough Doo, County Mayo, Ireland. *Journal of Quaternary Science* **2** 149-64.

O'SULLIVAN, P.E. (1976) Pollen analysis and radiocarbon dating of a core from Loch Pityoulish, Eastern Highlands of Scotland. *Journal of Biogeography* **3** 293-302.

OLAFSSON, E. (1943) *Ferdabok Eggerts Olafssonar og Bjarna Palssonar um ferdir theirra a Islandi arin 1752-1757.* (2nd edition) Reykjavik.

PATTERSON, W.A. III, EDWARDS, K.J. AND MAGUIRE, D.J. (1987) Microscopic charcoal as a fossil indicator of fire. *Quaternary Science Reviews* **6** 3-23.

PILCHER, J.R. AND SMITH, A.G. (1979) Palaeoecological investigations at Ballynagilly, a neolithic and bronze age settlement in County Tyrone, Northern Ireland. *Philosophical Transactions of the Royal Society of London B* **286** 345-69.

PILCHER, J.R., SMITH, A.G., PEARSON, G.W. AND CROWDER, A. (1971) Land clearance in the Irish Neolithic: new evidence and interpretation. *Science* **172** 560-62.

ROBINSON, D.E. (1981) *The vegetational and land use history of the west of Arran, Scotland.* Ph.D. thesis, University of Glasgow.

ROBINSON, M. AND HUBBARD, R.N.L.B. (1977) The transport of pollen in the bracts of hulled cereals. *Journal of Archaeological Science* **4** 197-99.

ROWLEY, J.R. (1960) The exine structure of 'cereal' and 'wild' type grass pollen. *Grana Palynologica* **2** 9-15.

ROWLEY-CONWY, P. (1981) Slash-and-burn in the temperate European Neolithic. In: *Farming practice in British prehistory.* (ed. Mercer, R.J.) Edinburgh: University Press pp. 85-96.

ROWLEY-CONWY, P. (1982) Forest grazing and clearance in temperate Europe with special reference to Denmark: an archaeological view. In: *Archaeological aspects of woodland ecology.* (eds Bell, M. and Limbrey, S.) Oxford: British Archaeological Reports S146 pp. 199-215.

SIMMONS, I.G. AND INNES, J.B. (1987) Mid-holocene adaptations and later mesolithic forest disturbance in Northern England. *Journal of Archaeological Science* **14** 385-403.

SMITH, A.G. AND COLLINS, A.E.P. (1971) The stratigraphy, palynology and archaeology of diatomite deposits at Newferry, County Antrim, Northern Ireland. *Ulster Journal of Archaeology* **34** 3-25.

SMITH, A.G. AND PILCHER, J.R. (1973) Radiocarbon dates and vegetational history of the British Isles. *New Phytologist* **72** 903-14.

TEUNISSEN, D. AND TEUNISSEN-VAN OORSCHOT, H.G.C.M. (1980) The history of the vegetation in SW Connemara (Ireland). *Acta Botanica Neerlandica* **29** 285-306.

THOMPSON, R., BATTARBEE, R.W., O'SULLIVAN, P.E. AND OLDFIELD, F. (1975) Magnetic susceptibility of lake sediments. *Limnology and Oceanography* **20** 687-98.

TOLONEN, M. (1985) Palaeoecological record of local fire history from a peat deposit in south west Finland. *Annales Botanici Fennici* **22** 15-29.

TOMASSON, P. (1973) Meltekja á Herjólfsstödum í álftaveri. *Arbók hins inslenska Fornleifafélags* 43-61.

TOOLEY, M.J. (1985) Sea-level changes and coastal morphology in North-west England. In: *The geomorphology of North-west England.* (ed. Johnson, J.H.) Manchester: University Press pp. 94-121.

TROELS-SMITH, J. (1955) Pollenanalytische Untersuchungen zu einigen Schweizerischen Pfahlbauproblemen. In: *Das Pfahlbauproblem. Monographien zur Ur- und Frühgeschichte der Schweiz* **11** 11-58.

TSUKADA, M. AND ROWLEY, J.R. (1964) Identification of modern and fossil maize pollen. *Grana Palynologica* **5** 406-12.

TURNER, J. (1964) The anthropogenic factor in vegetational history, I. Tregaron and Whixall Mosses. *New Phytologist* **63** 73-90.

TURNER, J. (1986) Principal components analysis of pollen data with special reference to anthropogenic indicators. In: *Anthropogenic indicators in pollen diagrams.* (ed. Behre, K.-E.) Rotterdam: A.A. Balkema pp. 221-32.

VORREN, K.-D. (1986) The impact of early agriculture on the vegetation of northern Norway - a discussion of anthropogenic indicators in biostratigraphical data. In: *Anthropogenic indicators in pollen diagrams.* (ed. Behre, K.-E.) Rotterdam: A.A. Balkema pp. 1-18.

VUORELA, I. (1970) The indication of farming in pollen diagrams from southern Finland. *Acta Botanica Fennica* **87** 1-40.

VUORELA, I. (1972) Human influence on the vegetation of Katinhäntä Bog, Vihti, S. Finland. *Acta Botanica Fennica* **98** 1-21.

VUORELA, I. (1973) Relative pollen rain around cultivated fields. *Acta Botanica Fennica* **102** 1-27.

VUORELA, I. (1975) Pollen analysis as a means of tracing settlement history in south west Finland. *Acta Botanica Fennica* **104** 1-48.

VUORELA, I. (1976) An instance of slash-and-burn cultivation in southern Finland investigated by pollen analysis of a mineral soil. *Memoranda Societas Fauna Flora Fennica* **52** 29-46.

VUORELA, I. (1980) Microspores of *Isoetes* as indicators of human settlement in pollen analysis. *Memoranda Societas Fauna Flora Fennica* **56** 13-19.

VUORELA, I. (1986) Palynological and historical evidence of slash-and-burn cultivation in South Finland. In: *Anthropogenic indicators in pollen diagrams.* (ed. Behre, K.-E.) Rotterdam: A.A. Balkema pp. 53-64.

WASYLIKOWA, K. (1986) Plant macrofossils preserved in prehistoric settlements compared with anthropogenic indicators in pollen diagrams. In: *Anthropogenic indicators in pollen diagrams.* (ed. Behre, K.-E.) Rotterdam: A.A. Balkema pp. 173-85.

WATON, P.V. (1982) Man's impact on the chalklands: some new pollen evidence. In: *Archaeological aspects of woodland ecology.* (eds Bell, M. and Limbrey, S.) Oxford: British Archaeological Reports S146 pp. 75-91.

WATSON, L. AND BELL, E.M. (1975) A surface-structural survey of some taxonomically diverse grass pollens. *Australian Journal of Botany* **23** 981-90.

WHITEHEAD, D.R. AND LANGHAM, E.J. (1965) Measurement as a means of identifying fossil maize pollen. *Bulletin of the Torrey Botanical Club* **92** 7-20.

WILLERDING, U. (1986) Aussagen von Pollenanalyse und Makrorestanalyse zu Fragen der Frühen Landnutzung. In: *Anthropogenic indicators in pollen diagrams.* (ed. Behre, K.-E.) Rotterdam: A.A. Balkema pp. 135-51.

WILLIAMS, C.T. (1985) *Mesolithic exploitation patterns in the central Pennines.* Oxford: British Archaeological Reports 139.

ZVELEBIL, M. AND ROWLEY-CONWY, P. (1986) Foragers and farmers in Atlantic Europe. In: *Hunters in transition.* (ed. Zvelebil, M.) Cambridge: University Press pp. 67-93.

6

POLLEN ANALYTICAL EVIDENCE FOR THE BEGINNING OF AGRICULTURE IN SOUTH CENTRAL EUROPE

Hansjörg Küster

ABSTRACT

Palynological dates for the beginning of agriculture in central Europe are brought together. In most landscapes there is a correlation between archaeological and palynological dates, but not in the Alps or in the northern foothills of the mountains, where palynological dates are older than archaeological dates. Archaeologists are now finding early neolithic layers in the Alps which indicate that the palynological dates concerning the beginning of agriculture have been correct.

6.1 INTRODUCTION

In palaeoenvironmental research the identification of cereal pollen grains is one of the most difficult, but also one of the most interesting problems, the outcome being of significance to both pollen analysts and archaeologists. By identifying cereal pollen grains the pollen analyst can detect when the landscape was affected by agriculture for the first time. Archaeologists are provided with information about the early Neolithic in areas where it has not been possible to detect neolithic settlement layers by excavation.

6.2 IDENTIFICATION OF CEREAL POLLEN GRAINS

For more than half a century now there have been discussions on the question of whether or not one can rely on the evidence for the early Neolithic given by pollen analysts. Firbas (1937), Inge Müller (1948), Grohne (1975b), Beug (1961, 1986), Andersen (1978) and Küster (1988a) discuss the problem of the identification of cereal pollen grains. To summarise, all these authors agree that the following features are specific to cereal pollen grains:

(1) The size of the pollen grain must exceed *circa* 40µm.

(2) The diameter of the pore must exceed *circa* 4µm.

(3) The diameter of the annulus must be at least twice as big as the diameter of the pore.

(4) The outer edge of the annulus must be very distinct.

By examining all these features, it is possible to be almost completely certain that a 'cereal pollen grain' has been produced by a cereal plant: wild grass species which also have rather large pollen grains can be separated from the cereal pollen type. The pore of all *Bromus* L. species is too small; the annulus of both *Glyceria* R.Br. and *Elymus* L. is too thin. No wild grass species has a distinct outer edge to the annulus, with the exception of some wild *Hordeum* L. species; sometimes *Avena fatua* L. (which is a weed introduced to central Europe by early farmers), and *Agropyron caninum* P.B. (macrofossils of which are only very rarely recorded for the Neolithic, for example Küster 1988a). Nevertheless, it is not totally impossible that some wild grass species could have become polyploid by chance, and thus produced pollen grains which appear to be cereal pollen grains (Beug 1986).

6.3 THE EVIDENCE IN NORTH CENTRAL EUROPE

Looking at the palynological evidence for the earliest cereal cultivation, it is obvious that in most cases pollen analysis shows the same date for the beginning of the Neolithic as archaeology does (Figure 1). Palynological data support the archaeological theory that areas of loess in central Europe were settled by *Linearbandkeramik* farmers during the first half of the fifth millennium bc. Cereal pollen grains have been recorded from this period from Hungary (Járai-Komlodi 1968; Zólyomi 1971); the south west German Neckar region (Smettan 1985); the Maas and lower Rhine area (Janssen 1960; Kalis 1988; Schütrumpf 1971, 1972/73); south east Westfalia (Pott 1982; Schütrumpf 1973); the loess plains near the Harz mountains (Beug 1986; Helmut Müller 1953); Thuringia (Lange 1966); and the borders of the River Oder (Hanna Müller 1966).

The mountainous areas of central Germany and the sandy plains further to the north were settled later. Some very early records are given by Groenman-van Waateringe (1978) for the Hazendonk in the Netherlands, and by Burrichter (1969) for the Zwillbrocker Venn in the Münster area: both are dated from the end of the fifth millennium bc.

Late Atlantic evidence for cereal pollen grains is given for several places in Belgium and Luxembourg (Couteaux 1969; Damblon 1969; Dricot 1960; Munaut 1967); for the eastern parts of the Netherlands and the adjacent landscapes in Western Germany (Casparie 1972; van Geel 1978; Grahle and Müller 1967; Grohne 1957a; Kramm 1978; Pott 1984; Schwaar 1977; van Zeist 1957); for the eastern parts of Lower Saxony (Lesemann 1969; Overbeck 1952; Selle 1962); and for the eastern parts of Schleswig-Holstein (Schmitz 1952). All other areas of north west Germany show younger dates for the first cereal pollen grains.

Sub-boreal cereal pollen grains are reported from the Elbe-Weser area (Behre 1976; Körber-Grohne 1967; Schneekloth 1963); and for the western and central part of Schleswig-Holstein and Jutland (Aletsee 1959; Averdieck 1958, 1974, 1976, 1978, Averdieck *et al.* 1972; Iversen 1958; Kubitzki 1961; Menke 1969). The large islands of Rügen (Kliewe and Lange 1971) and Bornholm (Mikkelsen 1954) also seem to have been settled during the early Sub-boreal according to the palynological evidence.

There are only a few diagrams which show scattered cereal pollen grains from earlier periods. In two cases, cereal pollen grains were recorded at the transition from the Boreal to the early Atlantic: at the Heuweg in the Wendland in the extreme eastern part of Lower Saxony (Lesemann 1969), and near the Dümmersee south of Bremen (Schwaar 1979). Schwaar (1980, 1983) also reports early Atlantic cereal pollen grains from the Königsmoor between Bremen and Hamburg. An exact description of these

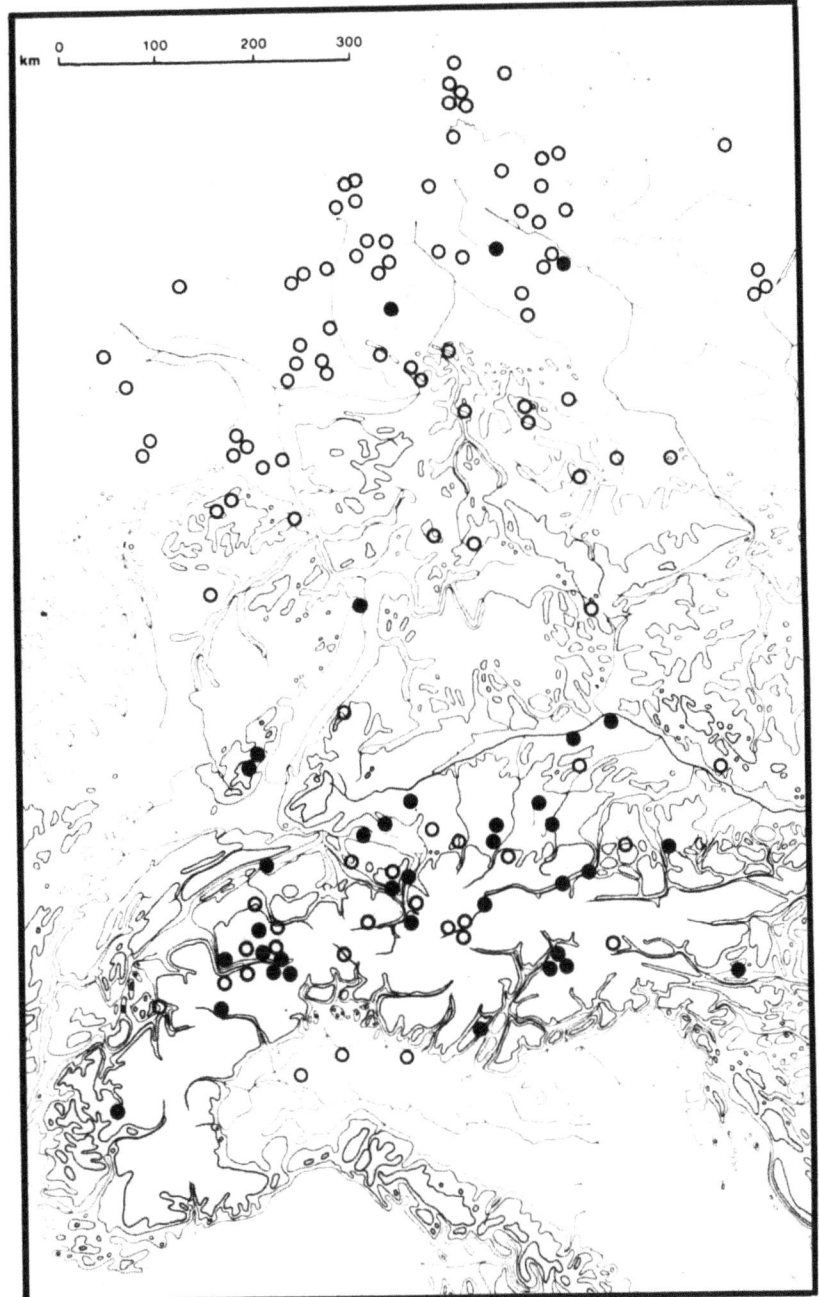

FIGURE 1 The beginnings of neolithic agriculture as shown in pollen diagrams from central Europe.

KEY:

open circles: pollen analytical evidence contemporaneous with the oldest settlements known to archaeologists

closed circles: pollen analytical evidence older than settlements known to archaeologists pollen records is not given, however, and they have not been verified. Later on, these very scattered pollen grains could have been produced by polyploid wild grasses. It can in general be concluded that archaeology and pollen analysis are in agreement with regard to the beginning of the Neolithic in north central Europe. This is also the case in Scandinavia (e.g. Berglund 1969) and in the British Isles, where Edwards and Hirons (1984) state only that the first cereal pollen grains can be observed in somewhat earlier layers than the elm decline, which may represent the early Neolithic in the British Isles. This has also been shown recently by Molloy and O'Connell (1987) in western Ireland.

However it is also in these regions that cereal pollen grains are sometimes observed in layers which are markedly earlier than the first archaeological evidence for farming (e.g. Vorren 1986; O'Connell 1987). In spite of the fact that the majority of pollen diagrams in these areas show a later date for the spread of agriculture, which does not conflict with the archaeological evidence, one cannot totally refute the impression that these very scattered cereal pollen records may reflect a very early date for small-scale agriculture even in marginal landscapes.

6.4 THE EVIDENCE IN SOUTH CENTRAL EUROPE

Turning to the more southerly parts of central Europe, another picture becomes obvious. Here too, in regions which are not covered by loess and where there is no archaeological evidence from the Neolithic, cereal pollen grains often occur in very old layers, indicating evidence of early neolithic farming. In three profiles from the Auerberg in south Bavaria, cereal pollen grains were identified in levels from the second half of the fifth millennium bc (Küster 1986, 1988a and b). In addition, profiles some fifty kilometres to the north east revealed cereal pollen grains from the early Atlantic (Bakels 1978; Kossack and Schmeidl 1974/75). The same seems to be the case in the upper Rhine area near Speyer (Lessmann 1983), although this diagram is not well dated; near Vienna (Havinga 1972); in Carinthia (Bortenschlager 1966); and at Lake Constance (Inge Müller 1948).

In northern Switzerland, early Atlantic cereal pollen grains have also been recorded several times, despite the fact that early neolithic material is not known to archaeologists in this area. Cereal pollen grains at the Nussbaumer Seen south west of Konstanz occur before 4000 bc (Rösch 1983). It is also significant that these early dates for cereal pollen grains are reported even from sites in the Alps. In the French Alps, Samuel Wegmüller (1976) found several very early cereal pollen grains near Grenoble, for example at the Col Bayard, in a layer from the middle of the sixth millennium bc. Similar dates are given for some sites in the upper Rhone Valley in Switzerland (Welten 1977, 1982); from the Simmental south of Bern (Welten 1952); from Graubünden (Burga 1976, 1980); and from the alpine Rhine Valley (Hans Peter Wegmüller 1976).

Very similar in date, or perhaps even a little bit earlier, are the oldest cereal pollen records from northern Tyrolia (Wahlmüller 1985); the area of Bressanone in southern Tyrolia (Seiwald 1980); and from the Lago di Ledro in northern Italy (Beug 1964).

In most cases this is in contrast to the archaeological evidence from these areas, and so it has often been concluded that these early cereal pollen grains reflect polyploidy amongst wild grasses rather than cereal cultivation. However these records have become so very frequent, that it is obvious that one could rely on them when one also relies on the palynological evidence for the beginning of the Neolithic in other regions. It cannot be argued that these pollen grains are just the result of occasional polyploidy in wild grass species, as there is no reason for an increase in the rate of polyploidy beginning in the early Atlantic, and not in earlier periods.

Some early cereal pollen grains from mountainous areas will only reflect early agriculture in the valleys nearby. This seems to be the case in the Vosges in eastern France (De Valk 1981; Kalis 1984), but in the Alps most of the early cereal pollen records come from lower locations in the valleys or from the foothills rather than from higher places. They cannot reflect long distance pollen transport from the Mediterranean or the Near Eastern area, because no other exotic species are present in the pollen record for the Alpine region during the Atlantic period. The dispersal of cereal pollen is very poor, except for rye, which was not cultivated at this time in central Europe.

6.5 ARCHAEOLOGICAL EVIDENCE

Therefore there seems to be some evidence for early neolithic crop husbandry in south central Europe which could be even earlier in the Alps than at the northern border of the mountains. It is interesting that archaeological evidence from this period is beginning to confirm the palynological evidence. Very recently, an early neolithic settlement was found at Sion in the Rhone valley close to the place where Welten (1977, 1982) found early neolithic cereal pollen grains (Gallay 1986; Küster 1987). The settlement is dated to about 4500 bc, and the pottery seems to have been influenced by that of the Mediterranean. The excavation report also demonstrates the difficulties of Alpine archaeology: the settlement layer had been covered by several metres of gravel and loam during the last seven millennia. Therefore it is very difficult for archaeologists to find any early settlements in this area, as the erosion and sedimentation rates are so very high. In addition, Nothdurfter (pers. comm.) reports some early neolithic settlements which have been recently detected in south Tyrolia, which confirm Seiwald's findings (1980).

As archaeologists begin to confirm the ideas held by palynologists about the beginning of the early Neolithic in central Europe, the problems of interpreting very early cereal pollen records seem to come to an end. We hope that in the future it will be possible to get a rather clearer picture of this early neolithic period from excavations in the central and northern Alps, and in southern Bavaria.

BIBLIOGRAPHY

ALETSEE, L. (1959) Zur Geschichte der Moore und Wälder des nördlichen Holsteins. *Nova Acta Leopoldina* N.F. **21** (139) 1-51.

ANDERSEN, S. TH. (1979) Identification of wild grass and cereal pollen. *Danmarks Geologiske Undersogelse Årbog 1978* pp. 69-92.

AVERDIECK, F.R. (1958) Pollenanalytische Untersuchungen zur Vegetationsgeschichte im Osten Hamburgs. *Mitteilungen der Geographischen Gesellschaft in Hamburg* **53** 161-76.

AVERDIECK, F.R. (1974) Zur Vegetations-, Siedlungs- und Seegeschichte. In: *Bosau I.* (ed. Hinz, H.) Neumünster: Offa-Bücher pp. 150-69.

AVERDIECK, F.R. (1976) Palynologische Untersuchungen zur Altersbestimmung und Vegetationsgeschichte des Alstertals. *Mitteilungen des Geologisch-Paläontologischen Instituts der Universität Hamburg.* Sonderband Alster pp. 81-89.

AVERDIECK, F.R. (1978) Palynologischer Beitrag zur Entwicklungsgeschichte des Grossen Plöner Sees und der Vegetation seiner Umgebung. *Archiv für Hydrobiologie* **83** (1) 1-46.

AVERDIECK, F.R., ERLENKEUSER, H., AND WILLKOMM, H. (1972) Altersbestimmungen an Sedimenten des Grossen Segeberger Sees. *Schriften des Naturwissenschaftlichen Vereins in Schleswig-Holstein* **42** 47-57.

BAKELS, C.C. (1978) *Four linearbandkeramik settlements and their environment: a paleoecological study of Sittard, Stein, Elsloo and Hienheim.* Leiden: University Press (also published as *Analecta Praehistorica Leidensia 11*).

BEHRE, K.-E. (1976) Pollenanalytische Untersuchungen zur Vegetationsgeschichte bei Flögeln und im Ahlenmoor (Elb-Weser-Winkel). *Probleme der Küstenforschung im südlichen Nordseegebiet* **11** 101-18.

BERGLUND, B.E. (1969) Vegetation and human influence in south Scandinavia during prehistoric time. *Oikos Suppl.* **12** 9-28.

BEUG, H.-J. (1961) *Leitfaden der Pollenbestimmung für Mitteleuropa und angrenzende Gebiete.* Stuttgart: Gustav Fischer.

BEUG, H.-J. (1964) Untersuchungen zur spät- und postglazialen Vegetationsgeschichte im Gardaseegebiet unter besonderer Berücksichtigung der mediterranen Arten. *Flora* **154** 401-44.

BEUG, H.-J. (1986) Vegetationsgeschichtliche Untersuchungen über das Frühe Neolithikum im Untereichsfeld, Landkreis Göttingen. In: *Anthropogenic indicators in pollen diagrams.* (ed. Behre, K.-E.) Rotterdam: A.A. Balkema pp. 115-24.

BORTENSCHLAGER, S. (1966) Pollenanalytische Untersuchung des Dobramoores in Kärnten. *Carinthia II* **76/156** 59-74.

BURGA, C.A. (1976) Frühe menschliche Spuren in der subalpinen Stufe des Hinterrheins. *Geographica Helvetica* **2** 93-96.

BURGA, C.A. (1980) Pollenanalytische Untersuchungen zur Vegetationsgeschichte des Schams und des San Bernhardino-Passgebietes (Graubünden, Schweiz). *Dissertationes Botanicae* **56**, Vaduz.

BURRICHTER, E. (1969) Das Zwillbrocker Venn, Westmünsterland, in moor- und vegetationskundlicher Sicht. Mit einem Beitrag zur Wald- und Siedlungsgeschichte seiner Umgebung. *Abhandlungen aus dem Landesmuseum für Naturkunde zu Münster und Westfalen* **31** (1) 1-60.

CASPARIE, W.A. (1972) *Bog development in southwestern Drenthe (The Netherlands).* 's-Gravenhage.

COUTEAUX, M. (1969) Recherches palynologiques en Gaume, au Pays d'Arlon, en Ardenne Méridionale (Luxembourg Belge) et au Gutland (Grand-Duché de Luxembourg). *Acta Geographica Lovaniensia* **8**.

DAMBLON, F. (1969) Etude palynologique comparée de deux tourbières du plateau des Hautes Fagnes de Belgique: la Fagne Wallonne et la Fagne de Clefay. *Bulletin du Jardin Botanique National de Belgique* **39** (1) 17-45.

DRICOT, E.-M. (1960) Recherches palynologiques sur le plateau des Hautes-Fagnes. *Bulletin de la Société Royale de Botanique de Belgique* **92** (1/2) 157-96.

EDWARDS, K.J. AND HIRONS, K.R. (1984) Cereal pollen grains in pre-elm decline deposits: implications for the earliest agriculture in Britain and Ireland. *Journal of Archaeological Science* **11** 71-80.

FIRBAS, F. (1937) Der pollenanalytische Nachweis des Getreidebaus. *Zeitschrift für Botanik* **31** 447-78.

GALLAY, A. (1986) Neolithikum und Frühbronzezeit im Wallis. Chronologie. Archäologische Daten der Schweiz. *Antiqua* **15** 50-61.

GEEL, B. VAN (1978) A palaeoecological study of holocene peat bog sections in Germany and the Netherlands. *Review of Palaeobotany and Palynology* **25** 1-120.

GRAHLE, H.-O. AND MÜLLER, H. (1967) Das Zwischenahner Meer. *Oldenburger Jahrbuch* **66** 83-121.

GROENMAN-VAN WAATERINGE, W. (1978) The impact of neolithic man on the landscape in the Netherlands. In: *The effect of man on the landscape: the lowland zone.* (eds Limbrey, S. and Evans, J.G.) London: Council for British Archaeology Research Report 21 pp.135-46.

GROHNE, U. (1957a) Zur Entwicklungsgeschichte des ostfriesischen Küstengebietes auf Grund botanischer Untersuchungen. *Probleme der Küstenforschung im südlichen Nordseegebiet* 6. Hildesheim.

GROHNE, U. (1957b) Die Bedeutung des Phasenkontrastverfahrens für die Pollenanalyse, dargelegt am Beispiel der Gramineepollen vom Getreide-Typ. *Photographie und Forschung* **7** 237-48.

HAVINGA, A.J. (1972) A palynological investigation in the Pannonian climatic region of Lower Austria. *Review of Palaeobotany and Palynology* **14** 319-52.

IVERSEN, J. (1958) Pollenanalytischer Nachweis des Reliktencharakters eines jütischen Linden-Mischwaldes. *Veröffentlichungen des Geobotanischen Institutes Rübel in Zürich* **33** 137-44.

JANSSEN, C.R. (1960) On the Late-glacial and Post-glacial vegetation of South Limburg (Netherlands). *Wentia* **4** 1-112.

JARAI-KOMLODI, M. (1968) The late glacial and Holocene flora of the Hungarian Great Plain. *Annales Universitatis Scientiarum Budapestinensis de Rolando Eötvös Nominatae, Sectio Biologica* **9-10** 199-225.

KALIS, A.J. (1984) *Forêt de la Bresse* (Vogezen). Diss. Utrecht.

KALIS, A.J. (1988) Zur Umwelt des frühneolithischen Menschen: Ein Beitrag der Pollenanalyse. *Forschungen und Berichte zur Vor- und Frühgeschichte in Baden-Württemberg 31* Stuttgart: Theiss pp. 125-37.

KLIEWE, H. AND LANGE, E. (1971) Korrelation zwischen pollenanalytischen und morphogenetisch-stratigraphischen Untersuchungen, dargestellt an Holozän-Ablagerungen auf Rügen. *Petermanns Geographische Mitteilungen* **115** (1) 4-8.

KÖRBER-GROHNE, U. (1967) *Geobotanische Untersuchungen auf der Feddersen Wierde*. Wiesbaden: Steiner.

KOSSACK, G. AND SCHMEIDL, H. (1974/75) Vorneolithischer Getreidebau im Bayerischen Alpenvorland. *Jahresbericht der Bayerischen Bodendenkmalpflege* **15/16** 7-23.

KRAMM, E. (1978) Pollenanalytische Hochmooruntersuchungen zur Floren- und Siedlungsgeschichte zwischen Ems und Hase. *Abhandlungen aus dem Landesmuseum für Naturkunde zu Münster in Westfalen 40* (4). Münster.

KUBITZKI, K. (1961) Zur Synchronisierung der nordwestdeutschen Pollendiagramme (mit Beiträgen zur Waldgeschichte Nordwestdeutschlands). *Flora* **150** (1) 43-72.

KÜSTER, H. (1986) Werden und Wandel der Kulturlandschaft im Alpenvorland. *Germania* **64** (2) 533-59.

KÜSTER H. (1987) (Book review). *Bayerische Vorgeschichtsblätter* **52** 284-87.

KÜSTER, H. (1988) *Vom Werden einer Kulturlandschaft*. Weinheim: Acta Humaniora.

KÜSTER H. (forthcoming) The history of the landscape around Auerberg, Southern Bavaria: a pollen analytical study. In: *Cultural landscape, past, present, and future.* (eds Birks, H.J.B., Kaland, P.E. and Moe, D.) Cambridge: University Press.

LANGE, E. (1966) Zur Vegetationsgeschichte des zentralen Thüringer Beckens. *Drudea* **5** (1) 3-58.

LESEMANN, B. (1969) Pollenanalytische Untersuchungen zur Vegetationsgeschichte des Hannoverschen Wendlandes. *Flora B* **158** 480-519.

LESSMANN, U. (1983) *Pollenanalysen an Böden im nördlichen Oberrheintal unter besonderer Berücksichtigung der Steppenböden*. Diss. Bonn.

MENKE, B. (1969) Vegetationsgeschichtliche Untersuchungen und Radiocarbon-Datierungen zur holozänen Entwicklung der schleswig-holsteinischen Westküste. *Eiszeitalter und Gegenwart* **20** 35-45.

MIKKELSEN, V.M. (1954) Studies on the sub-Atlantic history of Bornholms vegetation. *Danmarks Geologiske Undersogelse II* **80** 210-29.

MOLLOY, K. AND O'CONNELL, M. (1987) The nature of the vegetational changes at about 5000 B.P. with particular reference to the elm decline: fresh evidence from Connemara, Western Ireland. *New Phytologist* **106** 203-20.

MÜLLER, HANNA (1966) Beiträge zur Vegetationsentwicklung auf dem Mönchsheider Sander bei Chorin. *Archiv für Forstwesen* **15** (8) 857-67.

MÜLLER, HELMUT (1953) Zur spät- und nacheiszeitlichen Vegetationsgeschichte des mitteldeutschen Trockengebietes. *Nova Acta Leopoldina* **110** (16) 1-67.

MÜLLER, INGE (1948) Der pollenanalytische Nachweis der menschlichen Besiedlung im Federsee- und Bodenseegebiet. *Planta* **35** 70-87.

MUNAUT, A.V. (1967) Recherches paléo-écologiques en basse et moyenne Belgique. *Acta Geographica Lovaniensia 6.* Louvain.

O'CONNELL, M. (1987) Early cereal-type pollen records from Connemara, western Ireland, and their possible significance. *Pollen et Spores* **29** 207-24.

OVERBECK, F. (1952) Das grosse Moor bei Gifhorn im Wechsel hygrokliner und xerokliner Phasen der nordwestdeutschen Hochmoorentwicklung. *Niedersächsisches Amt für Landesplanung und Statistik, Veröffentlichungen A I: Natur, Wissenschaft, Siedlung und Planung 41.* Hannover.

POTT, R. (1982) Das Naturschutzgebiet 'Hiddeser Bent-Donoper Teich' in vegetationsgeschichtlicher und pflanzensoziologischer Sicht. *Abhandlungen aus dem Westfälischen Landesmuseum für Naturkunde in Münster* **44** (3).

POTT, R. (1984) Pollenanalytische Untersuchungen zur Vegetations- und Siedlungsgeschichte im Gebiet der Borkenberge bei Haltern in Westfalen. *Abhandlungen aus dem Westfälischen Landesmuseum für Naturkunde in Münster* **46** (2).

RÖSCH, M. (1983) Geschichte der Nussbaumer Seen (Kanton Thurgau) und ihrer Umgebung seit dem Ausgang der letzten Eiszeit aufgrund quartärbotanischer, stratigraphischer und sedimentologischer Untersuchungen. *Mitteilungen der Thurgauischen Naturforschenden Gesellschaft 45.* Frauenfeld.

SCHMITZ, H. (1952) Pollenanalytische Untersuchungen an der inneren Lübecker Bucht. *Die Küste* **1** (2) 34-44.

SCHNEEKLOTH, H. (1963) Das weisse Moor bei Kirchwalsede (Kreis Rotenburg/Hannover). *Beiheft zum Geologischen Jahrbuch* **55** 105-38.

SCHÜTRUMPF, R. (1971) Neue Profile von Köln-Merheim. Ein Beitrag zur Waldgeschichte der Kölner Bucht. *Kölner Jahrbuch für Vor- und Frühgeschichte* **12** 7-20.

SCHÜTRUMPF, R. (1972/73) Weitere Profile von Köln-Merheim und ihre Datierung. *Kölner Jahrbuch für Vor- und Frühgeschichte* **13** 23-35.

SCHÜTRUMPF, R. (1973) Pollenanalyse, Jahrringanalyse und C14-Datierung in ihrem Zusammenwirken für die urgeschichtliche Chronologie. II. Die relativ-chronologische Datierung fossiler Eichenstämme aus der Kölner Bucht und dem nördlichen Vorland des Teutoburger Waldes nach der Pollenanalyse. *Archäologisches Korrespondenzblatt* **3** 143-53.

SCHWAAR, J. (1977) Vegetationsgeschichtliche Untersuchungen im Wildenlohsmoor bei Friedrichsfehn, Krs. Oldenburg. *Abhandlungen des Naturwissenschaftlichen Vereins Bremen* **38** (19) 335-54.

SCHWAAR, J. (1979) Spät- und postglaziale Pflanzengesellschaften im Dümmer-Gebiet. *Abhandlungen des Naturwissenschaftlichen Vereins Bremen* **39** 129-52.

SCHWAAR, J. (1980) Getreidebau vor 4000 v. Chr. im niedersächsischen Tiefland? *Nachrichten aus Niedersachsens Urgeschichte* **49** 261-63.

SCHWAAR, J. (1983) Spät- und postglaziale Vegetationsstrukturen im oberen Wümmetal bei Tostedt (Landkreis Harburg). *Jahrbuch des Naturwissenschaftlichen Vereins für das Fürstentum Lüneburg* **36** 139-66.

SEIWALD, A. (1980) Beiträge zur Vegetationsgeschichte Tirols IV: Natzer Plateau-Villanderer Alm. *Berichte des Naturwissenschaftlich-Medizinischen Vereines in Innsbruck* **67** 31-72.

SELLE, W. (1962) Beitrag zur Vegetationsgeschichte des Weichselspätglazials und des Postglazials im südlichen Randgebiet der Lüneburger Heide. *Berichte der Naturhistorischen Gesellschaft Hannover* **106** 41-47.

SMETTAN, H.W. (1985) Pollenanalytische Untersuchungen zur Vegetations- und Siedlungsgeschichte der Umgebung von Sersheim, Kreis Ludwigsburg. *Fundberichte aus Baden-Württemberg* **10** 367-421.

VALK, E.J. DE (1981) *Late holocene and present vegetation of the Kastelberg (Vosges, France)*. Diss. Utrecht.

VORREN, K.-D. (1986) The impact of early agriculture on the vegetation of northern Norway - a discussion of anthropogenic indicators in biostratigraphical data. In: *Anthropogenic indicators in pollen diagrams*. (ed. Behre, K.-E.) Rotterdam: A.A. Balkema pp. 1-18.

WAHLMÜLLER, N. (1985) Beiträge zur Vegetationsgeschichte Tirols V: Nordtiroler Kalkalpen. *Berichte des Naturwissenschaftlich-Medizinischen Vereines in Innsbruck* **72** 101-44.

WEGMÜLLER, H.P. (1976) Vegetationsgeschichtliche Untersuchungen in den Thuralpen und im Faningebiet. *Botanische Jahrbücher für Systematik, Pflanzengeschichte und Pflanzengeographie* **97** (2) 226-307.

WEGMÜLLER, S. (1976) *Pollenanalytische Untersuchungen zur spät- und postglazialen Vegetationsgeschichte der französischen Alpen (Dauphiné)*. Bern: Paul Haupt.

WELTEN, M. (1952) Über die spät- und postglaziale Vegetationsgeschichte des Simmentals. *Veröffentlichungen des Geobotanischen Institutes Rübel in Zürich* **26**.

WELTEN, M. (1977) Résultats palynologiques sur le développement de la végétation et sa dégradation par l'homme a l'étage inférieur du Valais Central. *Bull. Ass. franc. etud. Quat.* **47** 303-7.

WELTEN, M. (1982) Vegetationsgeschichtliche Untersuchungen in den westlichen Schweizer Alpen: Bern - Wallis. *Denkschriften der Schweizerischen Naturforschenden Gesellschaft 95*. Basel: Birkhäuser.

ZEIST, W. VAN (1957) Some radiocarbon dates in the postglacial vegetation history of the northern Netherlands. *Veröffentlichungen des Geobotanischen Institutes Rübel in Zürich* **34** 160-65.

ZOLYOMI, B. (1971) 6000jährige Geschichte der Agrikultur in der Umgebung des Balaton-Sees aufgrund von pollenstatistischen Untersuchungen der Seesedimente. *IIIème Congrès International des Musées d'Agriculture. Résumés des Communications Présentées*. Budapest pp. 194-95.

7

BOTANICAL INVESTIGATIONS AT THE NEOLITHIC LAKE VILLAGE AT WEIER, NORTH EAST SWITZERLAND: LEAF HAY AND CEREALS AS ANIMAL FODDER

David Robinson and Peter Rasmussen

ABSTRACT

Over the years the Swiss lake village sites have furnished a wealth of information about contemporary neolithic life. The site at Weier, a settlement from the Pfyner culture, has provided information basic to the understanding of the construction of lake villages, revealed evidence of the earliest repeatedly cultivated and manured arable fields in western Europe and formed the basis of Troels-Smith's theories on human involvement in the elm decline. In this paper the preliminary results of two recent pieces of work on material from Weier are discussed. In the first, material from the surface of the arable fields, which had been washed down and incorporated into the lake sediments, was analysed. It was found to be largely composed of animal dung used in manuring the fields and there is evidence that the animals' diet was supplemented with cereals. In the second, samples from a byre in the settlement, which contained a mixture of dung and leaf hay residues, were analysed and the findings interpreted in the light of a recent study involving the feeding of leaf hay and cereals to three cows, a goat and a sheep.

7.1 INTRODUCTION

The neolithic lake village at Weier in Cantone Schaffhausen was discovered in the early part of this century. Investigations by the Danish National Museum and the local Schaffhausen Museum have been going on at the site since the early 1950s and the work summarised here is in continuation of this project.

The Weier valley lies at *circa* 460m above sea-level and in the Atlantic period a lake approximately 100x500m covered the valley floor. The village was built on a dried out gyttja islet at a time of climatically-induced low water level (Troels-Smith 1955). A gently sloping terrace borders the valley floor, from which hills rise moderately sharply to between 40m and 80m.

Excavation of the village area revealed the remains of houses, byres and barns set within a wooden palisade (Figures 1 and 2). A wooden causeway linked the village to the lake shore. Houses were identified by the presence of hearths. In byres there was midden material, composed of dung and fragmented plant material, permeated by the remains of puparia of *Musca domestica* L. (house fly) (Guyan 1981). Such partially-decayed midden material, rather than fresh dung, is the preferred egg-laying site for this species (West 1951). Microscopic examination of the material also revealed ova of the parasitic worms *Trichuris* L. (whipworm) and *Fasciola hepatica* L. (liver fluke) (Nansen unpublished). In the structures identified as barns there were leaf hay residues comprising small branches, twigs and leaf fragments of *Betula, Quercus, Tilia, Acer, Fraxinus* and *Ulmus* (Guyan 1954). Radiocarbon and dendrochronological dating of wood from the village revealed three separate occupations, beginning at a level dated to *circa* 3100 bc (average value from 12 radiocarbon dates) and spanning three centuries in all (Troels-Smith 1981).

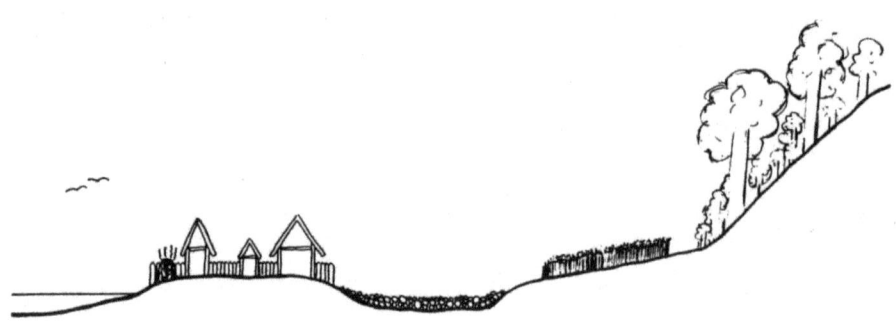

FIGURE 2 Sketch of the neolithic village and arable plots on the terrace linked by the wooden causeway.

In addition to the main excavations, a transect was examined approximately 12m to the east of the village. This ran from the terrace down into the lake sediments and peat deposits which now cover the valley floor. In the terrace deposits a horizontal grey stripe containing charcoal, carbonised grain and small pottery fragments was located. This was interpreted as a cultivation layer and is contemporary with the occupations although it is not clear how many of the occupation periods it corresponds to. It could be traced down into the lake sediments where it became thicker, more organic and richer in well-preserved macrofossils. Analysis of the layer where it lay in the lake sediments revealed considerable quantities of charcoal, carbonised grain (mostly *Triticum aestivum s.l.*, but also *T. dicoccum, T. monococcum* and both naked and hulled forms of *Hordeum vulgare*), seeds of *Linum usitatissimum* and *Papaver somniferum*, and seeds and fruits of wild food plants and annual arable weed seeds. In addition there were puparia of *Musca domestica* L. and quantities of small stones, sand, silt and clay (Jørgensen 1975; Fredskild 1978; Troels-Smith 1981, 1984). Although less diverse in terms of economic species, this plant macrofossil assemblage has much in common with that described from the village (Winiger 1971) and also with those from several other Swiss lake village sites (Heer 1865; Villaret-von Rochow 1967 *inter alia*). It is clear that the material had been washed down from the terrace during heavy rain or the spring thaw. It apparently represents a mixture of manure and domestic refuse from the village and seeds and fruits of plants growing *in situ* on the terrace. The water level of the lake at this time was such that direct deposition of refuse from the village into the lake and its subsequent incorporation into the lake sediments was not possible. Collectively, the evidence was interpreted as showing the presence of repeatedly

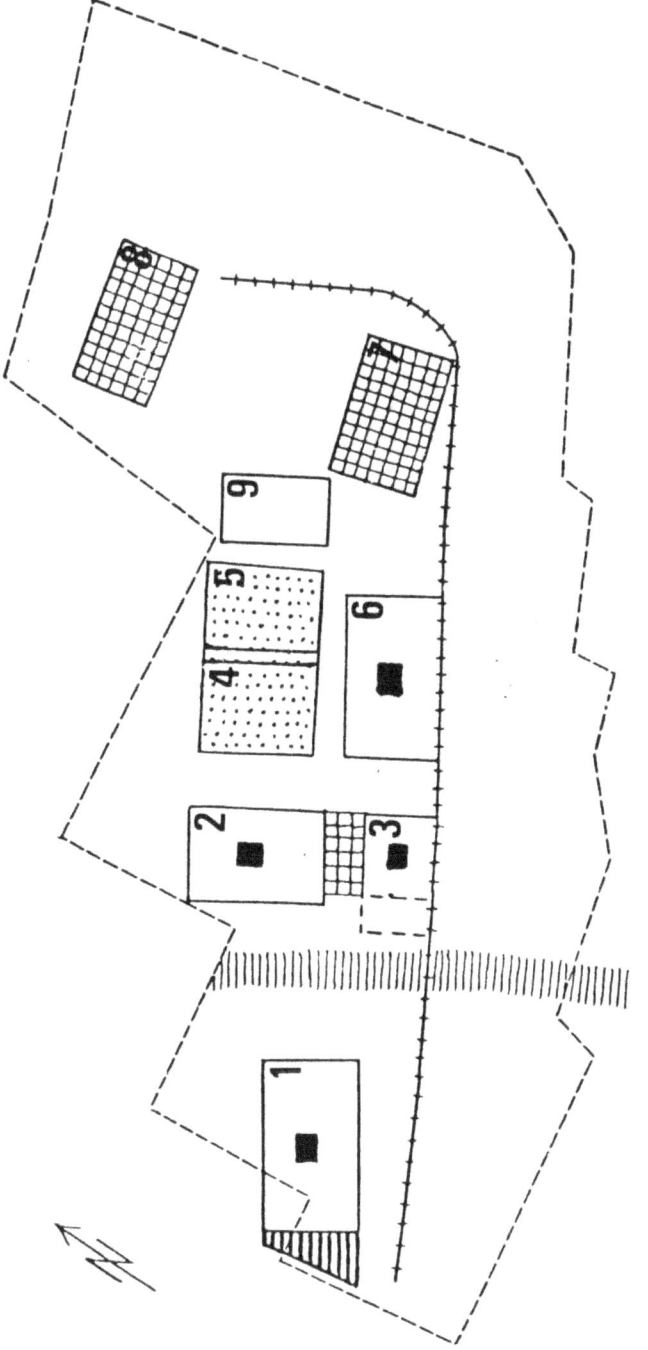

FIGURE 1 Excavation of the village at Weier (after Guyan 1981). 1, 2, 3, 6 and 9 are houses, 4 and 5 are barns for leaf hay and 7 and 8 plus the area between houses 2 and 3 are byres. The western part of house 1 is a goat byre. The wooden causeway, which linked the village to the terrace, runs between house 1 and houses 2 and 3. (Scale 1:250).

cultivated and manured arable plots on the terrace, and as such is the earliest European example of this practice (Troels-Smith 1981, 1984).

It was evident from changes in the plant macrofossil assemblages that the lake shore and terrace vegetation suffered considerable disruption during the period of occupation. Tall herb communities, reed swamp and *Betula/Alnus* carr vanished completely to be replaced some time later by a community largely dominated by *Ranunculus sceleratus* on areas of bare mud. Later, when human pressure on the lake shore was reduced, the original lake shore vegetation recovered somewhat.

7.2 RECENT WORK

The latest set of investigations concentrated on the analysis of material collected during the re-excavation of the terrace-lake shore transect in 1981 and on material from a byre in the village. The main aims of the re-excavation of the transect were to provide material which could be used to clarify a radiocarbon dating anomaly, whereby carbonised grain from the washed-in culture layer gave a significantly later date than the natural deposits which overlay it (Troels-Smith 1981), and to examine the culture layer itself in greater detail. To the first end, almost 450 carbonised cereal grains, which were individually collected and measured in on the profile during the excavation, were identified and measured (Tables 1 and 2). The species present, their relative abundance and the size of the grain were broadly in agreement with those published from the dated sample (Jørgensen 1975) and as such will only be commented upon briefly here. Higher proportions of both *Hordeum vulgare* and *Triticum monococcum/dicoccum* grains and a lower proportion of unidentified *Triticum* grains were recorded in the present study. This probably results from the routine examination of the surface cell patterns (cf. Körber-Grohne and Piening 1980) which made possible the identification of many broken and deformed grains which otherwise would have been recorded as unidentified. The material has not been dated as yet.

In the investigation of the samples from the culture layer, analyses were performed at closer vertical intervals than was the case in Fredskild's work (1978). Particular attention was also paid to very small plant fragments. The picture which emerged however was broadly similar to that described by Fredskild (Table 3).

A mixture of food, economic and arable weed species predominates and there is clear evidence for the disruption of the natural lake-shore vegetation although this re-establishes at a later stage. Some new information was forthcoming however! In the earliest washed-in layers, remains of *Linum usitatissimum* dominate the assemblage and arable weed seeds are poorly represented. This early abundance of *Linum* was not evident in Fredskild's analyses. In later layers arable weed seeds are more in evidence and cereal remains, together with seeds of *Papaver somniferum*, largely replace those of *Linum*. Whether this represents some kind of crop rotation is impossible to say without knowing the timescale over which the individual layers accumulated (Figure 3).

Undoubtedly the most fascinating new evidence was the discovery, in discrete bands within the washed-in layer, of high concentrations of compacted cereal testa fragments and large, almost complete, spikelets referable to *Triticum aestivum s.l.* all of which were uncarbonised. These were associated with similarly aggregated dicotyledonous leaf fragments. It is clear that this material too became incorporated into the sediments by way of being washed down from the cultivated terrace. Possible sources of the cereal debris include seed grain sown as spikelets, rather than threshed grain or even

	Jørgensen 1975		Robinson present study	
	Number	%	Number	%
Hordeum vulgare	28	1,4	11	1,4
H. vulgare (hulled)	24	1,2	27	6,1
H. vulgare (naked)	10	0,5	27	6,1
Triticum sp.	555	27,9	34	7,6
T. aestivum	802	42,2	264	59,3
T. cf. aestivum			8	1,8
T. dicoccum	16	0,8	14	3,1
T. cf. dicoccum			18	4,0
T. monococcum			4	0,9
T. mono/dicoccum	2	0,1	8	1,8
Unidentified	562	28,2	32	7,2
Total	1999		445	

TABLE 1 The carbonised grain from Weier.

	Number of grains	Length mm			Breadth mm			Thickness mm			Length X 100 Breadth			Thickness X 100 Breadth		
		min.	max.	aver.	min.	max.	aver.	min.	max.	aver.	min.	max.	aver.	min.	max.	aver.
Weier (present study)	150	3,0	7,1	4,8	2,3	4,7	3,6	2,1	4,6	3,0	104	174	134	51	110	83
Weier (Jørgensen 1975)	87	2,8	6,3	4,7	2,5	4,7	3,6	2,0	4,0	2,9	107	155	131	65	100	83
Burgäschisee-Süd (Villaret-von Rochow 1967)	16	3,6	5,9	4,7	2,2	4,3	3,4	2,2	3,8	2,9	115	166	139	72	100	85
Robenhausen (Flaksberger 1930)	30	3,1	6,4	4,3	2,5	4,7	3,2	2,0	3,5	2,7	110	183	135	67	100	83

TABLE 2 Carbonised grains of *Triticum aestivum* from Weier and two other neolithic Swiss sites (after Jørgensen 1975).

Layer		1	2	3	4	5A	6	5B	7	8	byre
Volume of sample (ml)		50	40	40	40	20	40	40	40	40	20
Crop and Food Plants											
Corylus avellana	nut shell							0,6			
Hordeum vulgare	rachis internode			8			2,5				
Linum usitatissimum	seed	98	185,6	54	3	5		5	3		
	seed*	2	2,5								
	capsule segment	0,6	110	195	8	5		25	8		40
Papaver somniferum	seed		7,5	53,5	25	83	15	10	3		
P. cf. somniferum	seed			3	13						
Malus domestica	seed				3						
Rubus idaeus/fruticosus	achene			3	3	55	23	10	0,6	0,6	
Sambucus sp.	seed	0,4		1,6	0,6					1	
Triticum sp.	caryopsis*		2,5			2					
	testa fragment		rare	ab	vab	ab	rare	ab			occ
T. cf. aestivum	caryopsis*			2,5			2,5	1			
	rachis internode*	2	15	30	50	10	8	12			10
	rachis internode			vab	ab	ab	occ	occ			
	spikelet			vab	ab	ab	occ				
Unidentified	straw fragment			ab					occ		
Arable Weeds and Wasteland Plants											
Arenaria serpyllifolia	seed			3		5					
Brassica rapa	seed*					5					
Campanula sp.	seed									3	
Cerastium sp.	seed					5					
Chenopodium sp.	seed					5	3				
C. album	seed							10	5	0,6	
Cirsium sp.	achene								3		
Fragaria vesca	achene					60					
Galeopsis sp.	nutlet				3						
Hieracium sp.	achene		3								
Hypericum sp.	seed							3			10
H. perforatum	seed					10					
Lapsana communis	achene			8	5	25	1	3			10
Silene alba	seed			3		5	3				10
Bilderdykia convolvulus	fruit		3	3		5		3			20
Potentilla sp.	seed					1					
P. erecta	seed	4			8	5	13	15	3		
Ranunculus cf. repens	achene	2									
Solanum nigrum	seed	2									
Sonchus asper	achene					5					
Urtica dioica	seed			5					3		
U. urens	seed								3		
Caryophyllaceae	seed							5			
Compositae	achene			3					3	10	
Cruciferae	seed	2		3							
Umbelliferae	fruit	0,4									
Aquatic Plants											
Characeae	oospores	4									
Najas flexilis	fruit	7,2		3							
Potamogeton obtusifolius	fruit stone	2									
P. cf obtusifolius	fruit stone		0,6								

TABLE 3 (Continued on opposite page).

Layer		1	2	3	4	5A	6	5B	7	8	byre
Wetland Plants											
Caltha palustris	seed	2									
Carex sp.	nutlet	18	10			10	5		18	410	
Filipendula ulmaria	achene		3		3						
Lycopus europeus	nutlet	2	3			5			13	108	
Ranunculus sceleratus	achene			3		5			200	2000+	
Solanum dulcamara	seed	2								5	
Myoson aquaticum	seed*	2			5					5	
	seed				3						
Viola sp.	seed				3						
Trees and Shrubs											
Alnus glutinosa	fruit	11								13	
Betula sp.	fruit	38	3			20		15	10	35	
B. pendula	cone scale*	2									
	cone scale	14	3						13	3	10
Tilia sp.	fruit	2			3						
Viburnum opulus	fruit			3							
Unidentified	buds	ab	ab	ab	vab	occ	ab	occ	occ	occ	
	leaf fragments	occ	ab	occ	ab	occ			occ		
	twigs		ab	ab	ab	occ	ab	occ	ab		
	wood fragments	occ	ab	ab		occ	vab	occ	ab	ab	
	charcoal	occ	ab	ab	ab	ab	ab	ab	ab	occ	ab
Bryophytes											
Overall abundance		vab	occ	rare	rare	occ	rare	rare		rare	rare
Brachytheciaceae		rare	rare	rare		rare					
Ceratodon purpureus		ab	rare			rare		rare			
Homalothecium sp.					rare	rare					
Hypnum cupressiforme		vab	rare		rare						rare
Neckera complanata						rare	rare				
Pleurozium schreberi		occ									
Rhizomnium punctatum		rare				rare			rare		
Thamnobryum alopecurum									rare		
Thuidium sp.						rare					
T. delicatulum			rare								
Ulota crispa			rare								
Unidentified			rare	rare	rare		rare			rare	
Animal remains											
Daphnia	ephippia	14			3						
Diptera (fly)	puparia	8	35	75	6	35	10	13	35	13	5
Fish	scales	4									
Insect	fragments	rare	rare	occ	occ	occ	occ	rare	rare	rare	occ
Trichuris sp.	ova				+						
cf. Taenia sp.	ova				+						
Miscellaneous											
Cenococcum geophilum	fruiting bodies	24		13		20			3	3	
Fungus (unidentified)	fruiting bodies	2	5	3					1		

TABLE 3 Macrofossils from Weier: the analysis of the 1981 samples. The values are expressed as number of macrofossils per 100ml sediment or as estimated abundance. Where fragments of seeds were recovered, the total number was divided by five to give the corresponding number of whole seeds. Abbreviations: occ = occasional; ab = abundant; vab = very abundant; + = present but not quantified; * = carbonised.

PLANT GROUPS

Layers	Open water	Lake shore	Trees/shrubs	Apophytes	Arable weeds	Economic plants	Bryophytes
Byre		+			+	+	+
8		++	++++	+	+	+	+
7		++	++	+++	+	+	
5B		+	+	+	++	+	+
6		+		+	+	++	+
5A		+	+	++	+++	++	+
4		+		++	+	++	
3	+	++	+	+	+	+++	+
2	+	++	+	+	+	+++	+
1	++	+++	+++	+	+	++	+++

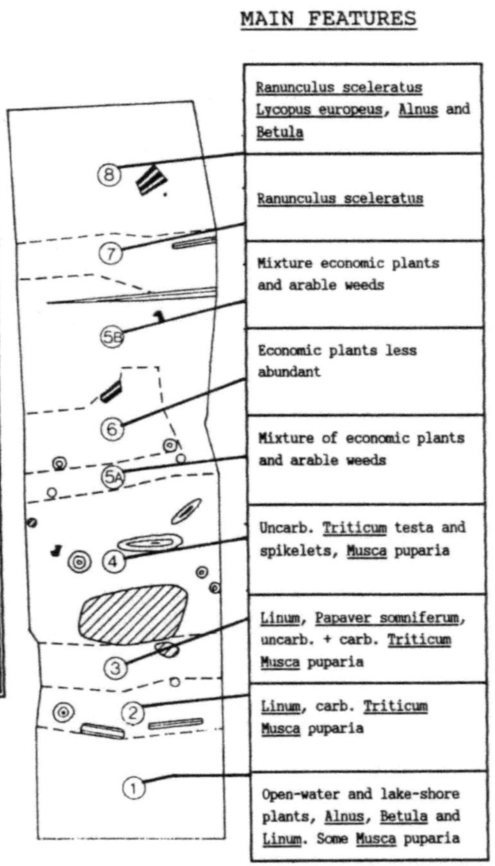

MAIN FEATURES

⑧ Ranunculus sceleratus, Lycopus europeus, Alnus and Betula

⑦ Ranunculus sceleratus

Mixture economic plants and arable weeds

⑥ Economic plants less abundant

Mixture of economic plants and arable weeds

Uncarb. Triticum testa and spikelets, Musca puparia

④ Linum, Papaver somniferum, uncarb. + carb. Triticum, Musca puparia

③ Linum, carb. Triticum, Musca puparia

① Open-water and lake-shore plants, Alnus, Betula and Linum. Some Musca puparia

2 cm

FIGURE 3 Detailed stratigraphy of the analysed sediments and summary of the plant macrofossil analyses presented in Table 3. Abundance is recorded on a scale from + (present) to +++ (very abundant).

grain lost during the harvest. However the nature of the fragments and the way in which they are concreted together strongly suggest that it is faecal material which is represented. Additional evidence for this is present in the form of eggs of *Trichuris* L. (whipworm) and possibly *Taenia* L. (tapeworm). The former were occasionally found adhering to cereal testa fragments. The *Trichuris* eggs are in the size range for *T. trichiura* and *T. suis* which affect man and pig respectively (Thienpont *et al.* 1979). It is notoriously difficult, because of the overlapping size ranges, to identify the ova to species level, and it has not been possible to determine whether human or animal faeces are represented. A sample from the byre lying between houses 2 and 3 (Figure 1) was examined, and it was also found to contain cereal testa fragments albeit in lower concentrations. Unfortunately this does little to identify the source of the material and it is very likely that both human and animal faeces are present. Remains of *Rubus*, *Malus* and *Sambucus* were recovered and they are common in human faeces and were present in intact human coprolites found during the excavations (Fredskild 1978). It seems very unlikely that human and animal faeces were kept segregated within the village and it is not unreasonable to suspect that human faeces were deposited directly in the byre. There is however strong circumstantial evidence for believing that a large part of the cereal debris has animal dung as its immediate source. Firstly, the material is very coarse. Testa fragments representing half and quarter grains are relatively common and are directly associated with large numbers of rachis internodes and virtually intact spikelets. These are not the products of flour milling and the human population would have found this mixture most unpalatable. Secondly, the cereal debris is mixed with compacted bodies of fine dicotyledonous leaf fragments which probably had leaf hay or fresh leaves (both animal fodder) as their source. If it is the case that animals were being fed on wheat spikelets then this begs the question - what kind of management practices allowed the wheat crop to be used in this way? Post-harvest stubble grazing is one possibility but it seems unlikely that this would result in such concentrations of faeces being present in the lake sediments. Furthermore puparia of *Musca domestica* L. are present in quantity, and this species never reproduces in fresh faeces (West 1951). It very specifically favours larger accumulations of animal faeces and rotting vegetable matter, such as middens, which both generate heat and provide a large stable food supply (West 1951; Sacca 1964; Overgaard-Nielsen unpublished). The evidence is therefore more consistent with the source of the cereal debris being manure from stall-fed animals.

It was partly with this thought in mind that further detailed analyses of samples from the same byre (between houses 2 and 3 on Figure 1) were carried out. This work is still in progress and will only be outlined here. It will be published in full in Rasmussen (forthcoming). In the byre there were three successive floor layers, presumably corresponding to the three periods of occupation of the site. The floors were made of split trunks or branches and were separated by midden material as identified by its general appearance and its content of Muscid puparia and intestinal parasite eggs (Troels-Smith 1984; Nansen unpublished). On analysis, the material was found to contain a mixture of small twigs, leaf and cereal bran fragments in addition to a wide variety of seeds and fruits, mostly uncarbonised. Individual goat faeces were also recovered and these contained leaf, wood and cereal bran fragments (Rasmussen forthcoming). In an attempt to find out which feeding practices could produce such assemblages in a byre, a series of experiments was carried out at the Lejre Historical-Archaeological Research Centre. Three cows, a goat and a sheep were fed either on leaf hay (ash - *Fraxinus* and elm - *Ulmus*) or a mixture of leaf hay and whole wheat (*Triticum*) grains (Plates I and II).

PLATES I AND II The leaf hay and cereal feeding experiments (photographs H. Rasmussen).

In the case of the cows, the branches of the dried leaf hay were pulled intact through the bars of the feeding rack. The cows ate the leaves and the smaller branches, and the residue fell to the floor where it was trampled underfoot, fragmented and became mixed with dung, which was a mixture of leaf and fine wood fragments. Most of the whole wheat grains appeared to pass intact and undigested through the cows and they littered the area to the rear of the animals. This is in sharp contrast to the fossil material at Weier where few intact whole grains were seen and where the bran was clearly fragmented. It is therefore likely that if grain was fed to cattle at Weier then it was first pounded or soaked to make it more digestible, or mixed with chaff so that the animals were encouraged to chew it. All of these practices are known from more recent times and one or other of them would have been necessary if cows were to derive any benefit from feeding with grain. The concentrations of bran and chaff which were present in the 'washed-in layer' at Weier would be consistent with both the mixing of the grain with chaff or the feeding of pounded wheat spikelets.

In contrast to the cows, the sheep and goat were very able to chew and digest the whole wheat grains, and the mixture of leaf, wood and cereal bran fragments which appeared in the faeces was very similar to that seen in the fossil goat faeces from Weier.

The differences seen in these experiments between the ability of cattle on the one hand and sheep and goats on the other, to digest whole grain are not new. The agricultural literature advises against the feeding of whole grain to cattle because the digestibility is so low (MacLeod *et al.* 1972). The differences can apparently be attributed to two things: firstly cattle chew the grain much less thoroughly than sheep and goats and secondly cattle have a much larger opening from the rumen to the psalterium which allows a greater number of whole grains to pass through the rumen undigested (Ørskov and Ranvig 1979). It is very unlikely that this differing ability to digest whole grain did not also exist in the Neolithic.

To return to the fossil material, nearly 1600 twig fragments from the manure layers in the byre were picked out, identified and had their annual rings counted. The results are presented in Tables 4 and 5. From Table 4 it can be seen that ash (*Fraxinus*), lime (*Tilia*) and willow (*Salix*) are by far the most frequently represented. Elm (*Ulmus*) in contrast, is very poorly represented, which is surprising in view of elm's known high palatability and nutritional content. On the other hand ivy (*Hedera*) and clematis (*Clematis*) are present in sufficient amounts to suggest that they are not incidental but were an important part of the collected fodder, probably in the depths of winter when little else was available. Mistletoe (*Viscum*) was probably only used as fodder in times of the severest shortage.

With regard to the age of the twig fragments (Table 5), the majority fall in the 1-4 year category, with most of the remainder being under 10 years of age. Fragments with up to as many as 19 annual rings were encountered, but these were of a diameter similar to those in the 1-10 year categories. The age and size distributions of the twigs suggest very strongly that there has been selective harvesting of leaf hay with two aims in mind: firstly to produce fodder with the greatest possible leaf content combined with the least possible wood, and secondly to ensure that the woody material which was present could be chewed by the animals.

Age in years	Number	Weight g	%	%
1	264	48,8	13,7	↓
2	197	38,4	10,8	↓
3	216	76,1	21,4	↓
4	177	52,9	14,9	↓
5	105	26,9	7,6	↓
				68,4
6	87	42,7	12,0	↓
7	59	32,6	9,2	↓
8	32	15,7	4,4	↓
9	19	6,4	1,8	↓
10	15	5,1	1,4	↓
				28,8
11	2	1,3	0,4	↓
12	5	0,5	0,1	↓
13	3	2,0	0,6	↓
14	4	0,7	0,2	↓
15	4	2,0	0,6	↓
				1,9
16	-	-	-	↓
17	2	1,0	0,3	↓
18	2	0,5	0,1	↓
19	1	1,7	0,5	↓
				0,9
Σ	1194	355,3		

TABLE 5 Age distribution of twig fragments in byre samples as a percentage of total weight.

		Number	Weight (g)	%
Ash	(Fraxinus)	339	114,3	27,1
Lime	(Tilia)	285	81,5	19,3
Willow	(Salix)	307	71,7	17,0
Alder	(Alnus)	215	39,7	9,4
Ivy	(Hedera)	186	28,7	6,8
Clematis	(Clematis)	69	28,6	6,8
Hazel	(Corylus)	60	27,2	6,4
Oak	(Quercus)	68	15,1	3,6
Elm	(Ulmus)	45	14,5	3,4
Birch/Alder	(Betula/Alnus)	15	0,6	0,1
Misteltoe	(Viscum)	5	0,3	0,07
Total		1594	422,2	

TABLE 4 Species distribution of twig fragments in byre samples as a percentage of total weight.

7.3 CONCLUSIONS

Much has been written on the subject of feeding animals on leaf hay in prehistory and the feeding of grain has also been suggested (Jarman *et al.* 1982). Much of this has been rather theoretical and has been in connection with describing various phenomena seen in pollen diagrams or in describing what probably happened. These analyses from Weier provide the first strong archaeological evidence for these practices and furthermore provide detailed information about the species involved. As to why these particular foodstuffs were used, it is clear that a store of winter fodder would have been crucial if the community and its animals were to survive. In a largely wooded landscape leaf hay would have been more readily collected than grass and herb hay for example, but it would still have been difficult to collect and store the relatively vast quantities which were required if all the animals were to survive. Analysis of the Weier bone material shows that a large proportion of the animal stocks were slaughtered before the onset of winter, which presumably reflects this difficulty (Soergel 1969). However the capacity for overwintering stock apparently increased with the successive occupations at Weier, as the bone analyses show that the proportion slaughtered in the autumn was lower in each successive occupation period. It may well be that one of the reasons for this increased winter carrying capacity was the feeding of cereals, probably in the form of pounded ears or spikelets, as a supplement to leaf hay. In which case cereal cultivation to produce animal food would have been as important as that for human consumption.

NOMENCLATURE

In the text and in the tables the nomenclature of the flowering plants follows *Flora Europaea* (1964-1980), that of bryophytes, Watson (1981) and that of algae, Round (1973). Authorities for non-plant species are given in the text.

ACKNOWLEDGEMENTS

Thanks are due to dr. phil. J. Troels-Smith who collected this material and made it available for study. The samples were collected in 1956 and in 1981 as part of a research project supported by Gemeinde Thayngen, the Carlsberg Foundation and the Danish Natural Sciences Research Council. The present work on the byre samples is also supported by the Carlsberg Foundation.

BIBLIOGRAPHY

FREDSKILD, B. (1978) Seeds and fruits from the neolithic settlement Weier, Switzerland. *Botanisk Tidsskrift* **72** (4) 189-201.

GUYAN, W.U. (1954) Das jungsteinzeitliche Moordorf von Thayngen-Weier. In: *Das Pfahlbauproblem. Herausgegeben zum Jubiläum des 100-jährigen Bestehens der Schweizerischen Pfahlbauforschung 1854-1954.* Schaffhausen.

GUYAN, W.U. (1981) Zur Viehhaltung im Steinzeitdorf Thayngen-Weier II. *Archäologie der Schweiz* **4** (3) 112-19.

HEER, O. (1865) *Die Pflanzen der Pfahlbauten.* Neujahrsblatt der Naturforschenden Gesellschaft aus dem Jahr 1866, Zürich.

JARMAN, M.R., BAILEY, G.N. AND JARMAN, H.N. (eds) (1982) *Early European agriculture.* Cambridge: University Press.

JØRGENSEN, G. (1975) *Triticum aestivum* s.l. from the neolithic site of Weier in Switzerland. *Folia Quaternaria* **46** 7-21.

KÖRBER-GROHNE, U. AND PIENING, U. (1980) Microstructure of the surfaces of carbonised and non-carbonised grains of cereals as observed in scanning electron and light microscopes as an additional aid in determining prehistoric findings. *Flora* **170** 189-228.

MACLEOD, N.A., MACDEARMID, A. AND KAY, M. (1972) A note on the use of field beans (*Vicia faba*) for growing cattle. *Animal Production* **14** 111-13.

NANSEN, P. (unpublished) *Anvendelse af parasitologiske analyser i den arkaeologiske forskning.* Institut for veterinaer mikrobiologi og hygiejne, Den kongelige Veterinaer- og Landbohøjskole, København 1984.

ØRSKOV, E.R. AND RANVIG, H. (1979) *Fodring af får og lam.* København: Landbrugets Informationskontor.

OVERGAARD-NIELSEN, B. (unpublished) *Fortolkning af forekomsten af fluepuparier i udvaskningslag fra neolitisk ager ved bopladsen Thayngen-Weier, Schweiz.* Århus Universitet 1983.

RASMUSSEN, P. (forthcoming) Feeding animals with leaf hay in the Neolithic: archaeobotanical evidence from Weier, Switzerland. *Journal of Danish Archaeology.*

ROUND, F.E. (1973) *The biology of the algae.* (2nd edition) London: Edward Arnold.

SACCA, G. (1964) Comparative bionomics in the genus *Musca*. *Annual Review of Entomology* 9 341-58.

SOERGEL, E. (1969) Stratigraphische Untersuchungen am Tierknochen - Material von Thayngen Weier. *Archäologie und Biologie, Forschungsberichte, Deutsche Forschungsgemeinschaft.* pp. 157-71.

THIENPONT, D., ROCHETTE, F. AND VANPARIJS, O.F.J. (1979) *Diagnosing helminthiasis through coprological examination.* Beerse, Belgium: Janssen Research Foundation.

TROELS-SMITH, J. (1955) Pollenanalytische Untersuchungen zu einigen Schweizerischen Pfahlbauproblemen. In: *Das Pfahlbauproblem. Monographien zur Ur- und Frühgeschichte der Schweiz* 11 11-58.

TROELS-SMITH, J. (1981) Naturwissenschaftliche Beiträge zur Pfahlbauforschung. *Archäologie der Schweiz* 4 (3) 98-111.

TROELS-SMITH, J. (1984) Stall-feeding and field-manuring in Switzerland about 6000 years ago. *Tools and Tillage* 5 (1) 13-25.

TUTIN, T.G., HEYWOOD, V.H., BURGES, N.A., VALENTINE, D.H., WALTERS, S.M. AND WEBB, D.A. (1964-1980) *Flora Europaea 1-5.* Cambridge: University Press.

VILLARET-VON ROCHOW, M. (1967) Frucht- und Samenreste aus der neolitischen Station Seeberg, Burgäschisee-Süd. *Acta Bernensia* 2 (4) 21-64.

WATSON, E.V. (1981) *British mosses and liverworts.* (3rd edition) Cambridge: University Press.

WEST, L.S. (1951) *The housefly.* New York.

WINIGER, J. (1971) Das Fundmaterial von Thayngen-Weier im Rahmen der Pfyner Kultur. *Monographien zur Ur- und Frühgeschichte der Schweiz 18* pp. 1-173.

8

THE EVIDENCE FOR EARLY RYE CULTIVATION IN NORTH WEST EUROPE

F.M. Chambers

ABSTRACT

Although previously regarded as a late Iron Age or Roman crop introduction to Britain and the near continent, recent finds of the pollen and seeds of rye from these areas now suggest an earlier introduction. Possible reasons for rye's adoption as a crop, and the uses to which it may have been put are discussed. However, rye's status in prehistory, as crop or weed, remains in some doubt.

8.1 INTRODUCTION

This paper considers the evidence for early rye cultivation in north west Europe. By 'early', I mean prehistoric, for until very recently, the generally accepted view was that rye, as a crop plant, was an Iron Age or, more likely, a Roman introduction to Britain; and a similar view prevailed with regard to its introduction to the near continent. This view derived principally from the work of Helbaek (1964, 1971), who maintained that rye was introduced to these areas in Roman times. Expansion into the rest of north west Europe apparently occurred by the end of the first millennium A.D.

New evidence has since come to light which suggests a far greater antiquity for rye in Britain, and which, by extension, might prompt re-interpretation of occasional rye records in pre-Roman levels on the near continent.

The British evidence consists of pollen grains of *Secale* from peat and lake horizons dating from the second millennium bc, plus a small number of *Secale* grains (seeds) from late Bronze Age contexts on some British sites.

These records raise a number of questions, some of which were addressed in a short note in *Antiquity* (Chambers and Jones 1984), and which are now re-examined. There are additional questions which were not then addressed but which are now considered here. In particular, this paper considers whether there is convincing evidence for rye being present in north west Europe in the pre-Roman Iron Age, in the Bronze Age, or in earlier times, and if so, what such presence indicates in terms of the status of rye, either as a weed, or as a fodder, food-grain or other type of crop.

8.2 LINES OF EVIDENCE

In tracing the spread of rye in prehistory, there are various lines of evidence which can be considered. These include:

(1) pollen grains;
(2) macro-remains (waterlogged and carbonised grains, rachis and chaff fragments and grain impressions on pottery);
(3) early historical writings;
(4) linguistic evidence - recorded names for rye;
(5) historically recorded or inferred uses for rye;
(6) prevailing crop husbandry practices and hence the likelihood of adopting rye as a crop.

These will be considered, and an assessment reached as to rye's likely status in prehistory.

8.2.1 Pollen evidence for rye

The *Secale* genus includes annual, short- and long-lived perennial, wild, 'weedy' and cultivated taxa (Sencer and Hawkes 1980). In considering the significance of pollen records of rye, it is first necessary to establish whether the *Secale* pollen curve in published pollen diagrams, or the *Secale* column in tables of pollen data, unambiguously refer to *Secale cereale* L. emend. Sencer (cultivated rye).

Many experienced palynologists feel confident of separating *Secale* pollen both from other cereal types and from wild grasses. However, critical examination of type-slides of *Secale* shows that confidence to be somewhat misplaced, as examples of *Secale* pollen can be found which on grounds of size, shape and internal pore diameter might normally be assigned either to the general category of wild grasses (Gramineae or Poaceae, $<40\mu m$), or to the undifferentiated Cerealia category, $>40\mu m$. This is because the diagnostic features of *Secale* pollen, such as a distinctive prolate shape, eccentrically positioned pore and the apparent puckering of the distal end of the grain are not always evident. Hence the *Secale* curve in a pollen diagram produced by an experienced palynologist will always tend to underestimate the presence of the genus, though perhaps not significantly. It is less likely that other Cerealia or Gramineae pollen would be mistaken for *Secale*. Enthusiastic amateurs could make such errors, because the proximal surface pattern is not dissimilar to those of some other grains.

So far as Britain and north west Europe are concerned, the only *Secale* species likely to be encountered is *S. cereale*. *S. montanum* might be encountered in Turkey, but its distribution becomes progressively sparser westward, and although it is locally common in mountain regions and reaches as far west as Spain, it is unlikely to contribute to the *Secale* curve in the area considered in this paper. Zohary (1971) considered it the ancestor of *S. cereale*, so its believed gene centre in Turkey may have some relevance to the origin and spread of cultivated rye. Hillman (1978a: 164), however, cited Stutz' (1972) analysis of the distribution of potential ancestors of cultivated rye, including *S. anatolicum* and *S. vavilovii*, and firmly pointed 'to East Anatolia and Transcaucasia as the home of the immediate ancestors of *Secale cereale*, as well as the primary gene centre for the genus as a whole'. None of these other species is native to north west Europe.

Pollen evidence for rye in published pollen diagrams from north west Europe in horizons pre-dating the Roman period appears sparse. In part this may be due to an unfortunate presumption concerning the date of rye's introduction: in the absence of radiocarbon dates, first *Secale* pollen records in Britain have been assumed to be from the Roman period (cf. Moore and Chater 1969), whereas the 'rational limit' (*sensu* Smith and Pilcher (1973) for pollen taxa), at which records increase, has been assumed to be Anglo-Saxon or Viking in age (cf. Godwin 1975: 414-15). Nevertheless, in recent years an increasing number of British researchers have recorded pollen from horizons apparently dating from the Bronze Age (Chambers 1980; Maguire *et al.* 1983; Waton 1983; Jones *et al.* 1985). Details of some of these records were provided in Chambers and Jones (1984).

On the continent, pollen evidence from the Netherlands suggests the presence of *Secale* in the Iron Age during the sixth century B.C., and of regular occurrence from the second century B.C., which led to the conclusion that rye was grown outside the Roman-occupied territory at that time (van Geel 1978). Pollen of *Secale* and *Centaurea cyanus* (cornflower) in a diagram produced by van Geel *et al.* (1983) from the northern Netherlands indicated that rye cultivation had developed in that locality by 1680±40 bp (GrN-9068). Further east and north, in southern Finland, the introduction of rye has been dated to the period A.D. 450-1000 using pollen analysis (Tolonen *et al.* 1979; Donner 1984).

Very early *Secale* records are therefore not the norm, but as several sites in southern and western Britain have now produced *Secale* records from the Bronze Age, it seems necessary to reappraise the contention that rye is an Iron Age or Roman crop introduction from the continent. In this reappraisal, it must be remembered that *Secale* produces an abundance of wind-dispersed pollen, and hence it tends to be over-represented relative to the other old world cereals, which are self-pollinating (Godwin 1975).

8.2.2 Macro-fossil remains of rye

If pollen records of rye from prehistoric horizons in north west Europe can be considered sparse, then macro-remains of rye from prehistoric contexts in this area are, if anything, sparser. They contrast with records further east in Europe where rye grains have been found in Austria, southern Germany and Poland from neolithic sites (Wernick 1961; Klichowska 1975; Piening 1982) and evidence for rye cultivation claimed from Bronze Age sites in Czechoslovakia (Tempír 1966, 1968). The early dates in central Europe clearly pre-date evidence from sites further west. Kreuz (pers. comm.), for example, reports a single grain from a neolithic *bandkeramik* site in central Germany in horizons dating from *circa* 5500 B.C. This is regarded as a weed element. It is the much later *Volkerwanderungszeit* (early migration period) which typically might be expected to have more abundant rye records, indicating rye cultivation (Behre *et al.* 1982, pers. comm.)

In the Netherlands, van Zeist (1976) claimed two 'early' records, but early in this context referred to the second and first centuries B.C. and the first century A.D. Van Zeist argued that these records provided evidence of rye cultivation rather than weed elements.

In Denmark, Osterbolle (Jutland) yielded rye seeds from the first century A.D., which was regarded by Helbaek (1971) as the introductory period there, though the rye seed grain records were described as of a pathetic quality. 'Better' records now exist from the site of Praestestien, Esbjerg in the Roman Iron Age from the fourth century A.D. (Robinson pers. comm.).

In Britain, pre-Roman Iron Age sites with rye seed records are few (Helbaek 1952; Green 1981), whilst only very recently has it been possible to extend British records back to the 'later Bronze Age' with some seed grains from a site in Scotland (Barclay and Fairweather 1984). The latter were regarded by the authors as a weed element.

There are various possible reasons for this paucity of prehistoric rye macro-remains, other than the obvious; namely, that rye was either a rare or absent element in the range of prehistoric crop plants in north west Europe. For example, several factors militate against pre-Iron Age *Secale* records, amongst which are:

(1) Sparseness of macro-remains evidence from archaeological sites for all cereals, due to failure on the part of excavators to search systematically for such evidence. Only in recent years have large-scale sampling techniques, such as flotation, been refined and used to any extent in north west Europe.

(2) Reluctance to flout conventional wisdom on the introductory period for rye, which might be expressed as caution on the part of archaeobotanists to attribute 'early' grains to rye, due, for example, to possible confusion with *Triticum* (wheat). Even experienced analysts exhibited this tendency: Helbaek's observations on the seed grains from the Roman fortress of *Isca* (Caerleon, Gwent) led him to conclude that some were 'tantalisingly suggestive of rye (*Secale cereale* L.) without any proper proof being available' (Helbaek 1964: 159). Charring of wheat and rye grains, for example, can cause distortion such that the grains may resemble one another.

(3) A failure to search for macro-fossil evidence other than seeds. The rachis segments of rye are characteristic and should provide additional evidence for *Secale*. However, 'bran' fragments, found in waterlogged deposits, are distinguishable from those of *Triticum* only in conditions of very good preservation when the 'transverse cells of the pericarp are preserved' (Dickson 1987: 100). Nevertheless, if they are preserved and are recognisable, Dickson (1987) warns that this may lead to a false impression of the importance of rye, as its bran fragments seem to preserve better than those of *Triticum*. Unfortunately, waterlogged cereal remains from clearly stratified prehistoric contexts are not common.

(4) The likelihood, or otherwise, of *Secale* grain being found in a charred condition. Hulled prehistoric cereal crops, including non free-threshing varieties of *Hordeum* and *Triticum*, require crop processing sequences in wet climates involving drying and then parching to separate the grain from the chaff (Hillman 1981); if these processes were not carefully controlled, the chances of charred grain being produced as a result are greater than charring events in the crop processing of free-threshing cereals such as *Secale*. Hence charred rye grains on archaeological sites may tend to be under-represented relative to those of hulled cereals; indeed, the hulled cereals might be expected to predominate over free-threshing cereals in prehistoric contexts (Hillman 1981: 124, Figure 1).

Dennell (1983) has pointed out that the abundance of remains of a plant species is not a reliable indicator of whether or not that plant species was cultivated, and this observation seemed to Chambers and Jones (1984) to be particularly pertinent in the case of rye, when attempting to assess whether sparse rye seed records indicate rye's status as crop or weed.

A summary of those factors leading to under- or over-estimates of the importance of rye is presented in Table 1.

(A) Factors influencing the relative importance of rye in prehistory from north west European contexts.

Tendency to under-estimate rye's representation due to:

    (i) caution on the part of archaeobotanists (and palynologists);

    (ii) fewer charred macro-remains relative to those of hulled cereals;

    (iii) poor preservation of waterlogged bran fragments and hence difficulty in separate identification;

Tendency to over-estimate rye's representation due to:

    (i) relative abundance of wind-blown pollen of allogamous rye (cf. other cleistogamous cereals with low representations);

    (ii) better preservation of waterlogged bran fragments than other cereals.

(B) Factors consciously or unconsciously influencing researchers in the estimation of rye as weed or crop

    (1) weed: sparse (e.g. 0.2% Total Land Pollen) pollen representation; sparse (*vis-à-vis* barley or wheat) macro-remains, especially seeds; 'poor quality' seed grains; expectations as to status based on age of material, e.g. age is pre-Roman.

    (2) crop: more abundant (e.g. >1% Total Land Pollen) pollen; relatively abundant macro-fossil remains; cleaned seeds or 'quality' seed grain; macro-remains consistent with crop processing of rye; expected age Roman or post-Roman.

TABLE 1 Factors influencing the estimation of the importance of rye.

8.2.3 Early historical writings

Some indication of the antiquity, prevalence and status of rye in prehistory might be gained from early historical writings. There are problems here with the reliability that can be placed upon authors and their sources. For example, Tacitus claimed to have written about Britain on the basis of ascertained fact, which he contrasted with his predecessors, whom he suggested had to rely on guesswork (Tacitus A.D. 98, in translation 1970: 60); but scholars now consider Tacitus to have overstated his case (ibid.: 11). In any case, his surveys of both Britain and Germany fail to incorporate any detailed information on crop types, other than to exclude the vine and olive from native cultivation in Britain (ibid.: 63) and to exclude fruit trees (unspecified) from Germany (ibid.: 104). A more illuminating reference comes in the writings of his contemporary, Pliny: '[*Secale*]...is a very poor food and only serves to avert starvation; its stalk carries a large head but is a thin straw; it is of a dark sombre colour, and exceptionally heavy. Wheat is mixed in with this to mitigate its bitter taste, and all the same it is very unacceptable to the stomach even so. It grows in any sort of soil with a hundred-fold yield, and serves of itself to enrich the land' (Pliny, translated by Rackham 1950: 279). The dismissive tone of Pliny's piece, plus the brevity of it, emphasise the marginality of rye cultivation to the Roman world by the first century A.D.

8.2.4 Linguistic evidence

Linguistic evidence for the origins of rye has been discussed by Sencer and Hawkes (1980). German, Scandinavian and Anglo-Saxon names for rye of *roggo, rugr, ryge* and *rig* are considered to have originated somewhere in central to south western Asia, but it is not clear whether this attribution is based in etymology or on the expected source area for rye. They drew attention to the central Asian names *chou-dar, jou-dar*, and *gandam-dar*, meaning 'a weed growing amongst wheat and barley', and later concluded that, on a continental scale, rye spread 'as a weed in wheat and barley and became a crop in its own right in several places independently' (Sencer and Hawkes 1980: 300). Despite their claimed geographical origin, one cannot place the same interpretation on the meanings of the non-Celtic north west European names for rye. In this context, Chambers and Jones (1984) pondered a connection between the verb 'to rogue' (when used in the context of cleaning of cereal crops) and *roggen*, but there is no evidence for this - the derivation of 'rogue' remains unclear.

Sencer and Hawkes felt able to conclude that: 'Linguistic evidence points to early acquaintance with rye of the people living in the Caucasus and the north east Black Sea region. The name of rye in the Caucasian languages was retained by the Greeks who traded it and Celts who borrowed it and spread it during their migrations. These are revealed by the facts (*sic*) that the Greek, Celtic and Latin names of rye are derivatives of its name in the Caucasian languages' (Sencer and Hawkes 1980: 311).

8.3 USES FOR RYE CROPS

In the historic period in north west Europe, rye has traditionally been associated with areas of marginal cereal agriculture. It is less exacting than other cereals, particularly wheat, in terms of climatic requirements (Jones 1981), and is reasonably tolerant of acid, light-textured, nutrient-deficient soils. Chambers and Jones (1984) pointed to its suitability for light, sandy or loessic soils or for upland marginal arable areas. It can be sown as a mixed crop (maslin) to extend the range of other cereals, usually wheat (Hillman 1978a). Its uses include:

    (1)   bread crop;
    (2)   other food-grain use;
    (3)   maslin, especially to extend range of wheat;

(4) fodder crop;
(5) grain for beverage;
(6) straw for thatching.

Of these, Green (1981) suggested that in early historical times, rye may have been specifically grown as a fodder crop, or its long straw used for thatching. The deliberate cultivation of rye grain for brewing was suggested by Helbaek (1964) in his description of sprouted rye seeds from the Roman fortress of *Isca* (Caerleon), Gwent, but this interpretation is not widely held. Nevertheless, although Roman rye 'beer' may be controversial, a variety of beverages from rye whisky to rye coffee are well attested in more recent times; indeed the Cossack practice of sowing rye for vodka production in otherwise unfruitful soils was graphically described by Nobel prize-winning author, Mikhail Sholokhov (quoted in Chambers and Jones 1984: 222).

In assessing evidence for rye from prehistoric times, it is important to recognise that rye may not have experienced the crop-processing sequences for a normal bread crop, as its uses may have been different, and therefore one cannot necessarily expect to find the sort of macroscopic evidence one would normally associate with staple bread crops. Further, as rye may have been a weed in other cereals, and as its seed grain has not changed significantly upon domestication (cf. Dennell 1983), it is not possible to ascertain convincingly from rye seed-grain shapes whether rye was then crop or weed, although poor quality grain has suggested a weed status to some authors.

8.4 CROP HUSBANDRY PRACTICES

Although agricultural societies in historical times seem governed by traditional practices, the diffusion of a particular crop innovation can be surprisingly rapid (cf. Overton 1985). The suggestion that rye may have spread as a weed in wheat and barley until being adopted as a crop in its own right, therefore bears closer examination.

Assessment of suitable cereal crops for an area depends on a range of factors, both practical and intuitive. In general, wheat is least tolerant of poorer soils, which might nevertheless grow satisfactory crops of barley, or failing them, oats. There are, of course, significant differences between varieties of each of these cereals, not least in whether the crops are autumn-sown or spring-sown. Given this considerable variety, it is necessary to consider why any farmer, village or culture would wish to adopt rye.

To societies used to dealing with hulled grains, the adoption of rye would not merely have meant a crop innovation, but also a different sequence of crop processing, as rye is a free-threshing cereal. However, this presumes that rye would have been grown as a grain crop: if its main purpose were to produce straw for thatching, then its free-threshing properties might have been a bonus. What is perhaps more significant is the adoption of rye into a mixed crop or maslin. It is not likely that rye would have been introduced in association with a hulled cereal such as spelt (*Triticum spelta*), a wheat which was also regarded as an Iron Age introduction to Britain (e.g. Frere 1967), but which has controversial records from the Neolithic (Helbaek 1952; Field *et al.* 1964) and confirmed records from the later Bronze Age (Hinton, in Drewett 1982). Sowing with naked wheats would be more likely, but this could not guarantee successful crop harvesting and processing. There would be difficulties of differences in ripening time in a rye/barley mix (Hillman pers. comm.), and even the expected rye/bread-wheat mix implied by Hillman (1978a) might lead to differences in time of ripening (Greig pers. comm.). These differences might vary in significance from place to place, or vary in the same location but between growing seasons. Hence, the particular mix would not be expected to be the same across north west Europe. Legge (pers. comm.) has suggested that a rye/oats mix would be more likely to have been sown and been successful in parts of Scotland, and presumably this would have involved naked rather than hulled varieties of oats.

A major problem with rye is its susceptibility to the fungal infestation of ergot. This disease is prevalent in damp, poor growing seasons, is very difficult to avoid, and given rye's marginalisation in arable economies, might be expected to be a recurrent problem with rye cultivation. Indeed, the later Bronze Age records from Myrehead, Central Scotland include evidence for ergot (Barclay and Fairweather 1984). Grain infested with ergot is unsuitable for human consumption as it can give rise to ergotism - a particularly unpleasant condition from which a range of complications, from gangrene to dementia, may develop. It is even possible that ergot-infested rye was deliberately cultivated for its hallucinogenic effects.

Sencer and Hawkes' (1980) suggestion that rye spread as a weed in wheat and barley until being adopted as a crop in its own right in several places independently, may imply that rye was adopted as a crop in circumstances where its success as a weed was obvious, and therefore was deliberately cultivated. Hence, the adoption of rye might be regarded as a move 'down market', which neighbouring farmers, villagers or cultures may have viewed as an admission of previous poor crop husbandry.

8.5 CONCLUSIONS

Physical evidence of rye in the form of pollen and macro-fossil remains from prehistoric and early historical contexts in north west Europe is sparse compared with later records. Early historical writings do not point to rye being a major crop in the Roman world. Linguistic evidence in the Middle East points to rye being a weed in crops of wheat and barley before being adopted as a crop in its own right, but that sequence cannot necessarily be assumed for north west Europe. The adoption of rye by any community would probably have been strongly influenced by the prevailing crop husbandry practices, by the perception or otherwise of rye's attendant disease problems, by rye's probable low status, and by failures of other, more highly regarded or more demanding cereal crops, especially on marginal land or after a sequence of poor seasons.

If the sparse seed grain records of rye from pre-Roman Iron Age contexts are accepted as evidence for cultivation of rye, then Bronze Age pollen and seed grain records ought to be accepted as possibly confirming cultivation in those periods too. If, however, the Bronze Age records are regarded merely as weeds, then similar caution should apply to the Iron Age, and some of the Roman (cf. Hillman 1978b) records. The presence of sparse pollen records of rye in prehistory in north west Europe can be regarded as probably indicating cereal cultivation, for, if not a crop, rye can be regarded as a segetal species. Rye probably did not become a major crop until the early first millennium A.D. in areas such as Denmark and the Low Countries, and may not have been a major mono-crop in Britain until Anglo-Saxon or medieval times. However, the accumulating evidence does suggest an earlier antiquity for rye; in such early times it may be envisaged as a minor component, confined to acid, sandy or loessic soils; these perhaps already rather degraded through poor crop and soil husbandry.

In cultivation, rye's uses were not confined to that of a bread crop, but probably also included beverage, fodder and straw production. A major function of rye may have been to extend the climatic and edaphic ranges of other cereals, notably wheat, in the form of maslin crops.

ACKNOWLEDGEMENTS

Thanks are due to David Robinson for information on rye seed records in Denmark; to Dr K.-E. Behre and Angela Kreuz for information on sites in Germany; to Richard Head for information on the diffusion of recent crop innovations, and to Gordon Hillman for clarifying various aspects of rye crop ecology.

BIBLIOGRAPHY

BARCLAY, G.J. AND FAIRWEATHER, A.D. (1984) Rye and ergot in the Scottish later Bronze Age. *Antiquity* **58** 126.

BEHRE, K.-E., BRANDT, K., KUCAN, D., SCHMID, P. AND ZIMMERMAN, W.H. (1982) *Mit dem Spaten in die Vergangenheit: 5000 Jahre Siedlung und Wirtschaft im Elbe-Weser-Dreieck.* Wilhelmshaven: Landesinstitut für Marschen- und Wurtenforschung.

CHAMBERS, F.M. (1980) *Aspects of vegetational history and blanket peat initiation in upland South Wales.* Ph.D. thesis, University of Wales.

CHAMBERS, F.M. AND JONES, M.K. (1984) Antiquity of rye in Britain. *Antiquity* **58** 219-24.

DENNELL, R.W. (1983) *European economic prehistory: a new approach.* London: Academic Press.

DICKSON, C. (1987) The identification of cereals from ancient bran fragments. *Circaea* **4** 95-102.

DONNER, J. (1984) Some comments on the pollen-analytical records of cereals and their dating in southern Finland. *Fennoscandia archaeologica* **1** 13-17.

DREWETT, P. (1982) Later Bronze Age downland economy and excavations at Black Patch, East Sussex. *Proceedings of the Prehistoric Society* **48** 321-400.

FIELD, N.H., MATTHEWS, C.L. AND SMITH, I.F. (1964) New neolithic sites in Dorset and Bedfordshire, with a note on the distribution of neolithic storage pits in Britain. *Proceedings of the Prehistoric Society* **30** 352-81.

FRERE, S. (1967) *Britannia - a history of Roman Britain.* London: Routledge and Kegan Paul.

GEEL, B. VAN (1978) A palaeoecological study of Holocene peatbog sections in Germany and the Netherlands based on the analysis of pollen, spores and macro- and microscopic remains of fungi, algae, cormophytes and animals. *Review of Palaeobotany and Palynology* **25** 1-120.

GEEL, B. VAN, HALLEWAS, D.P. AND PALS, J.P. (1983) A late Holocene deposit under the Westfriese Zeedijk near Enkhuizen (Prov. of Noord-Holland, The Netherlands): palaeoecological and archaeological aspects. *Review of Palaeobotany and Palynology* **38** 269-335.

GODWIN, H. (1975) *History of the British Flora.* (2nd edition) Cambridge: University Press.

GREEN, F.J. (1981) Iron Age, Roman and Saxon crops: the archaeological evidence from Wessex. In: *The environment of man: the Iron Age to the Anglo-Saxon period.* (eds Jones, M. and Dimbleby, G.) Oxford: British Archaeological Reports 87 pp. 129-53.

HELBAEK, H. (1952) Early crops in southern England. *Proceedings of the Prehistoric Society* **18** 194-233.

HELBAEK, H. (1964) The Isca grain: a Roman plant introduction to Britain. *New Phytologist* **63** 158-64.

HELBAEK, H. (1971) The origin and migration of rye, *Secale cereale* L.; a palaeoethnobotanical study. In: *Plant life of South-West Asia.* (eds Davis, P.H., Harper, P.C. and Hedge, I.C.) Edinburgh: University Press pp. 265-78.

HILLMAN. G.C. (1978a) On the origins of domestic rye - *Secale cereale*: the finds from aceramic Can Hasan III in Turkey. *Anatolian Studies* **28** 157-74.

HILLMAN, G.C. (1978b) Remains of crops and other plants from Carmarthen (Church Street). *Cambrian Monographs and Collections* **1** 107-12.

HILLMAN, G.C. (1981) Reconstructing crop husbandry practices from charred remains of crops. In: *Farming practice in British prehistory.* (ed. Mercer, R.J.) Edinburgh: University Press pp. 123-62.

JONES, M. (1981) The development of crop husbandry. In: *The environment of man: the Iron age to the Anglo-Saxon period.* (eds Jones, M. and Dimbleby, G.) Oxford: British Archaeological Reports 87 pp. 95-127.

JONES, R., BENSON-EVANS, K. AND CHAMBERS, F.M. (1985) Human influence upon sedimentation in Llangorse Lake, Wales. *Earth surface processes and landforms* **10** 227-35.

KLICHOWSKA, M. (1975) Najstarze zboza z wykopalisck Polskich (Die altesten Getreidearten aus den polnischen Ausgrabungen). *Archaeologia Polski* **20** 83-143.

MAGUIRE, D., RALPH, N. AND FLEMING, A. (1983) Early land use on Dartmoor: palaeobotanical and pedological investigations on Holne Moor. In: *Integrating the subsistence economy.* (ed. Jones, M.) Oxford: British Archaeological Reports S181 pp. 57-105.

MOORE, P.D. AND CHATER, E.H. (1969) The changing vegetation of west-central Wales in the light of human history. *Journal of Ecology* **57** 361-79.

OVERTON, M. (1985) The diffusion of agricultural innovations in early modern England: turnips and clover in Norfolk and Suffolk, 1580-1740. *Transactions of the Institute of British Geographers* (N.S.) **10** 205-21.

PIENING, U. (1982) Botanische Untersuchungen an verkohlten Pflanzenresten aus Nordwürttemburg. *Fundberichte aus Baden-Württemburg* **7** 239-71.

RACKHAM, H. (1950) *Pliny's Natural History - English translation Volume V Libri XVII-XIX*. London: Heinemann.

SENCER, H.A. AND HAWKES, J.G. (1980) On the origin of cultivated rye. *Biological Journal of the Linnaean Society* **13** 299-313.

SMITH, A.G. AND PILCHER, J.R. (1973) Radiocarbon dates and vegetational history of the British Isles. *New Phytologist* **72** 903-14.

STUTZ, H.C. (1972) On the origin of cultivated rye. *American Journal of Botany* **59** 59-70.

TACITUS, C. (A.D. 98, in translation 1970) *The Agricola and the Germania*. Harmondsworth: Penguin Books.

TEMPIR, Z. (1966) Vysledky palaeoethnobotanického studia pestováni zemedelskych rostlin na území CSSR (Results of palaeoethnobotanical studies on the cultivation of agricultural plants in the CSSR). *Vedecké Práce Ceskoslovenského Zemedelského Muzea* 27-144.

TEMPIR, Z. (1968) Archeologické nálezy zemedelskych rostlin a plevelú v Cechách a na Morave (Archaeological finds of agricultural plants and their weeds in Bohemia and Moravia). *Vedecké Práce Ceskoslovenského Zemedelského Muzea* 15-88.

TOLONEN, K., SIIRIANEN, A. AND HIRVILUOTO, A.-L. (1979) *Iron Age cultivation in SW Finland*. Finskt Museum 1976 5-66.

WATON, P.V. (1983) *A palynological study of the impact of man on the chalkland landscape of central southern England*. Ph.D. thesis, University of Southampton.

WERNECK, H.L. (1961) Ur- und frühgeschichtliche sowie mittelalterliche Kulturpflanzen und Hölzer aus den Ostalpen und dem südlichen Böhmerwald. *Archaeologia Austriaca* **30** 69-117.

ZEIST, W. VAN (1976) Two early rye finds from the Netherlands. *Acta Botanica Neerlandica* **25** 71-79.

ZOHARY, D. (1971) Origin of South-West Asiatic cereals: wheats, barley, oats and rye. In: *Plant life of South-West Asia*. (eds Davis, P.H., Harper, P.C. and Hedge, I.C.) Edinburgh: University Press pp. 235-60.

IV

THE ADOPTION OF AGRICULTURE IN THE BRITISH ISLES: THEORY AND EVIDENCE

9

CATTLE AND SHEEP IN BRITAIN AND NORTHERN EUROPE UP TO THE ATLANTIC PERIOD: A PERSONAL VIEWPOINT

Barbara Noddle

ABSTRACT

The remains of the aurochs, *Bos primigenius*, in Britain are discussed in the light of recent discoveries in the region of the Severn Estuary and the Bristol Channel. The accepted theory that this species became smaller during the Holocene is not borne out, but animals from central and eastern Europe are smaller than those from northern Europe.

The author's views on the process of domestication are presented. Neolithic and Bronze Age cattle and sheep are discussed. There is evidence for at least two types of cattle and sheep which probably originated from different parts of continental Europe.

9.1 INTRODUCTION

During the years 1986 and 1987, three aurochs were discovered in the region of the Severn Estuary and the Bristol Channel. The most complete specimen was found at Uskmouth, and was excavated by S. Parry for Newport Museum, Gwent. The others were a limbless head and torso from Rhymney, Cardiff, and a large skull found in the sand dunes at Rhossili, Gower by the National Trust. Besides these, a second specimen has been recovered from Charterhouse Warren Farm Swallet, Mendip (Levitan *et al.* 1988). All these specimens are Atlantic in date and this suggests that the population of *Bos primigenius* in this area might have been quite high, so a search of the literature and visits to a number of local museums were carried out. This proved only too successful and the results discussed below were limited by the time available rather than the material.

Agriculture, including domestic livestock, was introduced into the British Isles during the Atlantic period. This involved the transport of already domesticated livestock from the European mainland, and possible local domestication of wild animals already present. These processes are discussed, as well as the forms of neolithic cattle and sheep, using published material as well as unpublished faunal assemblages from Skara Brae, Orkney (1974 excavations) and Polnagallum, Ulster with which the author has been associated.

9.2 WILD CATTLE

Fossil bones, especially skulls of the wild ox, *Bos primigenius*, are frequently collected in the British Isles. Ungulate horns, in particular those of *Bos*, seem to have an intrinsic appeal to the human race, and have formed a large part of the subject matter of entire books, for example Conrad (1959). In 1839 the skull of the Melksham (Wiltshire) aurochs was sold at auction for the then enormous sum of £25, and it still adorns the Town Hall. Though there has been some confusion with *Bison*, the description of the *Urus* by Julius Caesar (*De Bello Gallica* Book 6, Chapter 9), written in about 65 B.C., was sufficiently accurate for skulls and horn cores to be recognised instantly by educated persons, although Whitehead (1953) believed that it was based on hearsay. One nineteenth century name for the species was *Bos urus* (Caesar) (Owen 1846). There are also persistent folk myths of giant wild cattle, such as the Dun Cow of Warwick, roaming the countryside even up to the medieval period. However, these also occur in Ireland, where aurochs remains have never been found (Ganz 1981).

In mainland Britain, aurochs bones have been found dating from at least the Hoxnian Interglacial onwards; a possible specimen from the previous interglacial, found at Boxgrove, Suffolk, has not yet been confirmed. Though not numerous, they were widespread, and findspots include Clacton, Essex (Oakley and Leakey 1937), Mendip, Somerset (Savage and Richards 1980), and Renfrewshire (Ritchie 1920). Far more were found dating from the Ipswichian Interglacial, sometimes in large groups such as those recovered from brickearth at Ilford, Essex (Brady 1874) and Barrington, Cambridgeshire (Fisher 1879), but more often in smaller numbers as are recorded frequently in geological journals of the nineteenth century. They have also been dated, as at Brundon, Essex, to interstadials of the Wolstonian Glaciation (Szabo and Collins 1975) and to the late Devensian interstadial in the Mendips, which has been dated to 12,000 B.C. (Currant 1986). These, and numerous specimens dating from the Holocene, have usually been discovered as the result of agricultural activities such as draining and marl digging, or during excavations for industrial purposes. Most of the thirty-eight specimens described by Smith (1872) came from Scottish marl pits, and he even observed a pair of skulls decorating a farm gateway. Eleven skulls, four of which still remain in the local museum, were recovered during the excavation of Preston Docks, and others were found in similar circumstances at Cardiff (Parry pers. comm.). However, aurochs was not often recorded amongst the numerous bones found during the construction of the Sharpness and Tewkesbury docks on the east bank of the Severn, although Pleistocene specimens are common enough in the gravel deposits only a few miles to the north east (Noddle 1987b). Evidence of *Bos primigenius* is also lacking from the reclaimed polders of Holland (Clason 1967).

Though aurochs bones have been found in caves such as those of Derbyshire and Mendip, arriving both by accident and human intervention (Campbell 1977; Currant 1986), the animal seems to have favoured estuarine areas. The majority of Scottish specimens have been found near the Tay and Tweed (Smith 1872) and no less than three specimens from the same area south of the River Trent were exhibited at a meeting of the Midland Scientific Society in 1864 (Anon. 1864-5). These may well have been overwhelmed by sediment-laden floods resulting from soil erosion brought about by early agricultural activities (Shotton 1978). This would seem to be the case with the two specimens that have been excavated recently from the alluvial flats west of the Severn which instigated the writing of this paper. Both date from the Bronze Age and no *Bos primigenius* has yet been found in Britain later than this period (Clutton-Brock 1986), although the recent specimen from Charterhouse Warren Farm Swallet, Mendip is 300 years later than the dates she has given (Levitan *et al.* 1988). Long immersion in flood water is suggested by the limbless condition of the recently discovered animal from Rhymney, and another limbless skeleton reported from Potsdam (Teichart 1987), seemingly also a flood victim, dates from the same period.

There must be a reason for this mass demise, although aurochs did not of course become extinct at this period on the European continent. Was it that the population of this species had become so large that it was forced to graze in dangerous situations? Grigson (1978) has suggested that these animals, which are nearly all males, were frustrated bulls, and the behaviour of Chillingham cattle, a more or less feral herd, gives some support to this (Hall 1979). Few aurochs females have been recovered and the majority of these are from the well-researched group found in Danish bogs (Degerbøl and Fredskild 1970). There is considerable sexual dimorphism in the species, and the smaller female is not so readily distinguished from domestic stock as is the male. Three skulls dating from the neolithic period at Maiden Castle, which were considered to be a specific neolithic type by Jackson (1943a), have been shown to be *Bos primigenius* females by the painstaking measurements of Grigson (1973). It is possible that aurochs females were taken into domestic herds (see below) and that the intractable males, which probably attempted to disrupt these, were disposed of by driving them into bogs; none of these bog drowned specimens show any butchery marks.

Remains of aurochs calves are also rare, so it is unfortunate that the recent report on such a specimen from Denmark contained little information on the bones themselves (Richter 1982). One of the Severn bank specimens referred to above (Rhymney) was associated with some calf bones (as well as a large specimen of *Cervus elephas*). A metacarpal bone was more mature but rather smaller than that of a three month old Friesian heifer in my collection, but the shaft of the bone was very narrow, suggesting malnutrition, rather than slow growth in *Bos primigenius*.

The aurochs skull and horn core have been extensively studied, in particular by Grigson, whose thesis (1973) covered all previous studies. It was intended that the present paper would concentrate on post cranial bone, but during visits to various west of England museums some skulls and horn cores were observed that have not been described or studied since the nineteenth century. The usual shape of the *Bos primigenius* horn core is well known from numerous drawings of skulls and from the upper Pleistocene cave paintings of France and Spain. Grigson (1973) refers to this shape as the 'primigenius spiral', but all the skulls she worked on were of postglacial date and Boyd Dawkins (1866) considered that Pleistocene horn cores were variable in shape. There are a number of *Bos primigenius* bones in Devizes museum collected by M.E. Cunnington from the Westbury (Wiltshire) drift (museum code 1818). These include a large horn core with a smooth surface, whose curvature follows the arc of a circle, with no hint of a spiral; the ratio of the basal diameters is 1.59. Lydekker (1912) illustrated one of the Pleistocene skulls from Ilford, which appears to have horn cores of similar shape. This circular form has been observed in two other aurochs skulls of unknown date. The first of these is also at Devizes and comes from Cherhill; it was described in considerable detail by Woods (1839), who found it so different from the Melksham aurochs that he considered it domestic. Cunnington (1906) re-examined the circumstances of its discovery and demonstrated that it did not, as Woods supposed, originate from a Bronze Age barrow, but from a glacial deposit. The other skull is exhibited in Bridgwater Museum; it was donated at about the turn of the century, and was stated to have come from 'a cave in a quarry in Cannington Park', this site being a Carboniferous Limestone outlier of the Mendip hills. Both Pleistocene and Holocene aurochs have been recovered from such caves, but the best efforts of Mrs. Langdon at the museum on my behalf failed to elicit further detail. The above evidence, though tenuous, suggests that the Pleistocene aurochs of south west England may have lacked the primigenius spiral. Holocene specimens conformed to the norm, as can be seen from the Melksham specimen and another from a peat bog near Newbury, Berkshire (Adams 1853).

Both metrical and non-metrical data are used when describing long bones, but the former normally predominates. Measurements can be used to delineate both size and shape, the latter being most easily described by means of ratios. The absolute size of a mature bone is determined not only by its genetic potential, but also by environmental factors, primarily nutrition during the growth period. Shape is also affected by nutrition, but genetic factors are probably more important; non-metrical variation is usually genetic in origin (Noddle 1978, 1983b). There can be little doubt that the recent Uskmouth specimen is the most complete aurochs skeleton yet found in the British Isles, lacking only the calvarium. It was meticulously excavated and measured by S. Parry, who kindly supplied me with his findings. There are a number of differences in shape between this and a modern specimen, with which it is most immediately comparable because of the similar size. In several respects the aurochs is similar to the red deer, the most obvious instance being the terminal phalanx, in which the dorsal profile presents an absolutely straight edge without any upward curvature towards the joint surface. The anterior and posterior surfaces of the calcaneum process appear absolutely parallel, but when attempts were made to demonstrate this difference metrically, it was found that this parallel structure is common to all early stock and only modern animals taper proximally. The scapula of the aurochs has a relatively longer blade than the modern animal, with less proximal flare; in this respect it is similar to the Soay sheep when compared with more advanced animals (Noddle 1978). However, this again is difficult to quantify because this bone continues growing during much of the life of the animal.

Metrical studies in the present paper concentrate on the metapodial bones, because of the numerous references to these bones in the literature. Figures 1a and b present the data available for *Bos primigenius*, and the measurements are presented in several groups. Pleistocene aurochs are considered separately, and only British specimens are included. Holocene specimens are grouped under nationality. It has long been assumed that *Bos primigenius*, as well as many other species, became smaller with the passage of time (Kurtén 1968), therefore the difference in size between the Barrington and other Pleistocene aurochs was unexpected. The Barrington measurements are taken from Browne (1983); a large number of bison bones were also obtained from this site, but Browne claims to have found valid distinctions between the two species (*Bos* is normally larger than *Bison*). However, the precise date of this and the Ilford deposit has not been obtained, so that although the two groups occur in the same interglacial, climatic conditions may have been different. The Holocene aurochs show a distinct reduction in size going from north to south, with the Scandinavian specimens being the largest, and the Swiss, despite their male sex, the smallest. Portuguese cattle bone from the neolithic period could not be separated into wild and domestic forms (von den Driesch and Boessneck 1975). Jarman (1969) attempted to demonstrate a decrease in size of *Bos primigenius* with time in southern Europe, but his data were scanty. The majority of the large British specimens date from the Bronze Age, so there is little sign of size reduction here.

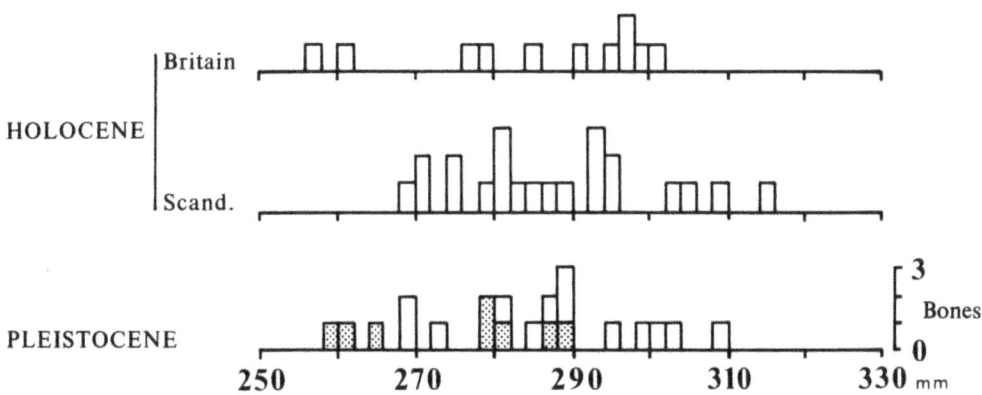

FIGURE 1a Length of metatarsal of aurochs (*Bos primigenius*).

Sources: for unpublished measurements made by the author, the locations of the bones are given.

PLEISTOCENE   Badger Hole (Wells Museum)
                   Barrington (Browne, 1983) - stippled
                   Clacton (Browne, 1983; Oakley and Leakey 1937)
                   Pinhole (Campbell 1977)
                   Trafalgar Square (Browne 1983)

HOLOCENE   Britain:
                   Brundon (Hopwood 1939)
                   Lowes Farm (Shawcross and Higgs 1961)
                   Charterhouse (Everton 1975)
                   Edinburgh (Royal Scottish Museum)
                   Shenstone (Birmingham University Museum of Geology)
                   Swaffham (Cambridge University Zoological Museum)
                   Uskmouth (Parry, unpublished)
                   Woodhenge (Jackson 1929)

                   Scandinavia:
                   Denmark (Degerbøl and Fredskild 1970)

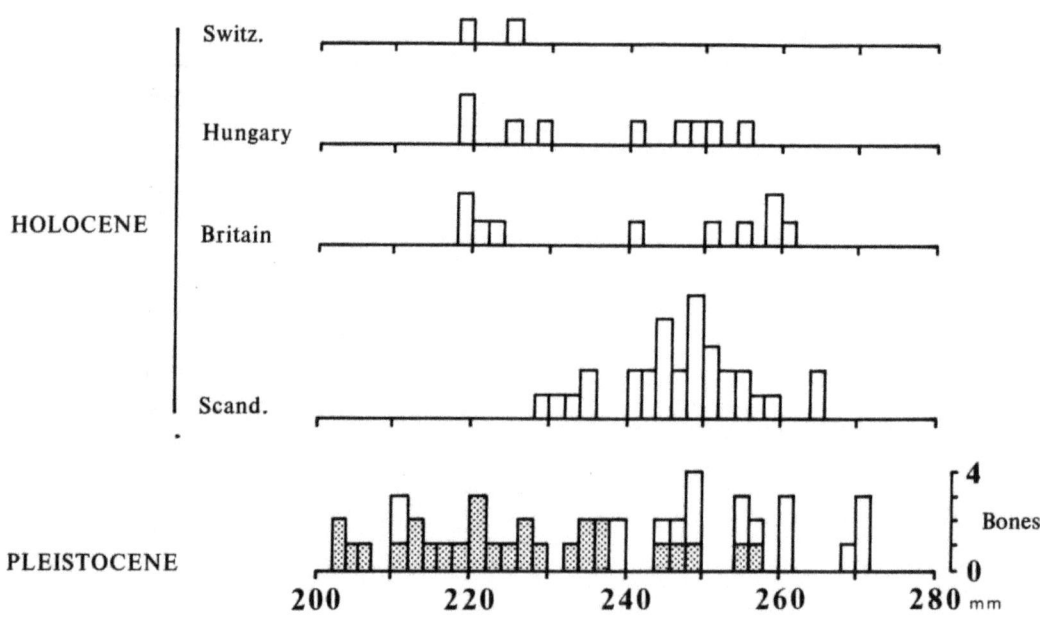

FIGURE 1b Length of metacarpal of aurochs (*Bos primigenius*).

Sources: For unpublished measurements made by the author, the locations of the bones are given.

PLEISTOCENE Avon (Gloucester Museum)
Barrington (Browne 1983) - stippled
Clacton (Browne 1983; Oakley and Leakey 1937)
Derbyshire (Royal Scottish Museum)
Greys (Browne 1983)
Ilford (Browne 1983)
Swaffham (Cambridge University Zoological Museum)
Trafalgar Square (Browne 1983)
Westbury (Devizes Museum)

HOLOCENE Switzerland:
Etival (Campy *et al.* 1983)
Farges (Chaix and Valton 1984)
Hungary: (Bökönyi 1962)
Britain:
Charterhouse (Everton 1975)
East Ham (Banks 1961)
Lowes Farm (Shawcross and Higgs 1961)
Marden (Harcourt 1971b)
North Marden (Browne 1986)
Star Carr (Fraser and King 1954)
Thatcham (King 1962)
Uskmouth (Parry, unpublished)
Scandinavia:
Denmark (Degerbøl and Fredskild 1970)
Sweden (During 1986)

9.3 DOMESTICATION

The *Pocket Oxford Dictionary* defines the term domestic as meaning 'of the household', and with reference to animals 'kept by or living with man'. These definitions are much too wide to be used in the present context, since they include animals as diverse as grain weevils, pet monkeys and dairy cows, that is to say individuals which are parasitic, captive or tamed, as well as domesticated. Zeuner (1963) and Higgs and Jarman (1969) distinguish between tame and domestic animals, but they do not follow up the resulting implications. The process of taming takes place within the lifespan of an individual, and probably involves considerable close human contact. The animal may be captive, although wild animals can be tamed by the judicious supply of food, including salt; Geist (1971) gives a graphic account of the behaviour of big horn sheep when supplied with this commodity. The process of domestication as I understand it involves controlling and exploiting the natural behaviour of an animal, including its breeding and feeding activities over a number of generations. Both physique and behaviour may become genetically altered in the course of events, the animal having come under selection pressure exerted by the stress of human contact, restriction of diet and choice of mate (Clutton-Brock 1981). Since natural behaviour in the wild forbears of a domestic animal is peculiar to that species, it can be argued that domestication has not followed the same pathway in all cases. This invalidates much discussion on the subject which has sought a single cause and explanation. The species currently under discussion, cattle and sheep, associate with each other in groups and I believe that it was these groups which were domesticated and not their individual members, a view supported by Herre and Rohrs (1977) and Baskin (1974).

One explanation for the creation of the first domesticate is the arrival of a young animal in a hunters' camp following its newly killed mother. Such an animal, in the unlikely event of its survival, would become tame, but not domesticated. If female, it might continue to live in the same area as the people, departing only to mate, and this has been the case with a number of hand reared antelope in Africa. To form a domesticated herd from the progeny of this one animal would be a very slow process indeed, suggesting deliberate planning, and in any case such offspring tend to be less tame than their mother.

The domestication of herd animals was probably a slow process and a consequence of concurrent events rather than a deliberate act. There can be no doubt that hunter-gatherers were very familiar with the behaviour of their prey species and well able to manipulate it. However, such manipulation frequently resulted in the death rather than the domestication of the animal. An example of this is the technique of the American High Plains Indians described as buffalo jumping. Buffalo herds were induced to locate themselves in a particular small area, by gentle guidance or by rousing the animals' natural curiousity to follow a dancing figure. The animals were then stampeded along a prepared route over a natural hazard such as a cliff face, so that they were either killed outright, or could be slaughtered easily. Such fatal stampedes were also organised by upper palaeolithic people in the Old World (Kowalski 1967) and on occasion they, like the Indians (Frison 1973, 1974), are believed to have used enclosures (Kehoe 1973).

The archaeological record leaves little doubt that the first ungulates domesticated were sheep, or possibly goats (Herre and Rohrs 1977). By the time that cattle were subjected to the domestication process it could therefore have been deliberate, but I think it unlikely that utilitarian motives or the selection of 'options' so beloved by present-day scholars played much part. It is presumed that the first domestication took place when agriculture was sufficiently advanced for human groups to have become settled, and that there was competition between them and herbivorous animals both for cultivated crops and natural concentrations of food plants. However, these herbivores were also a source of food, possibly of a valued ritual nature, as well as skins, bone, sinew and so on. Following the adoption of a settled way of life, the human population was increasing and larger areas were required for agriculture (Cohen 1977). In these circumstances it seems likely that the groups of herbivores came under increasing supervision, and that their numbers were reduced. Individual human groups might have come to regard individual ungulate groups as their property, to be conserved for their particular use. The use of enclosures on a regular overnight basis, as opposed to an occasional killing ground, might have come into being, leading to selective slaughter and the taming of individual animals for purposes other than slaughter. Seasonal migration could also be prevented.

This scenario does not necessarily apply to the domestication of cattle, which is believed to have taken place in different areas of the Near East from that of the caprines. In northern Europe herding may well have been more important than agriculture (Greenfield 1988 provides a recent summary of these views). The present-day herding of reindeer has been cited as a model by Zeuner (1963), but the long seasonal migrations undertaken by Scandinavian reindeer makes the Lappish system of husbandry less applicable. The more sedentary Siberian race of reindeer is more suitable, and its association with the Eskimo Chukchi is well described by Leeds (1965). This involves (as does the Lappish system) the taming of individual animals for traction, dairy production etc. Leeds also describes the continual interchange of animals between the herded and fully wild animals in the area, the attraction and retention of female animals being particularly desirable. I believe that this process played a part in the husbandry of the small numbers of domestic cattle originally introduced to the British Isles.

9.4 TRANSPORT OF LIVESTOCK ACROSS WATER

Because wild sheep did not exist in the British Isles (see below) they must have been imported. The morphology of neolithic cattle, in England at least (Grigson 1973), indicates that they too were bred mainly from imported domesticates. The means whereby this took place require consideration.

The transport of sheep and goats, and particularly cattle, across considerable stretches of open sea in primitive boats presents formidable problems. The matter was discussed at some length by Case (1969), who argued that the boats used must have been something like the Irish skin-covered curragh. This led him to suppose that the animals were trussed up for transport, so that their hooves could not tear the boat fabric, but bearing in mind the physiology of the artiodactyl rumen, this seems unlikely. The rumen produces methane gas, and if this cannot be dispersed the organ becomes distended and the animal rapidly dies. Eructation is impossible if the animal is in lateral recumbancy and is impaired in sternal recumbancy. Also an immobile animal is likely to suffer from pressure on the nerves and other tissues. The transport of loose sheep or goats on board a curragh may not have been as difficult as Case thought, since the boat could have been protected by, for example, large strips of bark. Case believed that the animals could not be persuaded to swim 'even if towed by ropes', but I believe that the method used by the inhabitants of the smaller Aran Islands, and observed as late as the 1970s, is feasible. The animals were driven into the water, but once out of their depth they became quite passive, allowing the head to be held by a man sitting in the stern of the curragh, which was then rowed out to a waiting ship. In this position the animal is free to move its limbs and to eructate, and it can also be loosed when near its destination, thus eliminating the problem of unloading. During the Atlantic period the seas were warmer than at present, and the animals would probably have been able to withstand immersion for about twelve hours. The animals do not have to be very tame; the same method was employed recently to rescue animals trapped by the rising waters behind the Kariba Dam (Nigeria).

Cattle and deer are capable of swimming considerable distances unaided. Haldane (1973) has studied the history of cattle droving; in the case of Hebridean Island stock this frequently started with one or more enforced swims, although losses were sometimes high. There is strong circumstantial evidence that a red deer stag could cross from Jura to Mull, a distance of at least twenty kilometres through strong currents, a feat considered unremarkable by the head keeper who gave me the information. It seems likely that the aurochs could swim, but whether it could cross the Pentland Firth from Caithness to Orkney is a matter for conjecture. Although this crossing is notorious nowadays for its roughness, conditions might have been different in the Atlantic period.

9.5 DOMESTIC CATTLE

Knowledge of many aspects of the neolithic period has increased in a spectactular manner during the last decade or so. For example, painstaking fieldwork by Fleming (e.g. 1979) has indicated that livestock ranching may have been an important agricultural activity. The study of crops and their accompanying weeds has flourished in the hands of Martin Jones, James Greig and their colleagues. (For a recent summary see Moffett *et al.* this volume). The considerable extent of neolithic occupation has been demonstrated on the one hand by fieldwork such as that of Hunt (1987) in northern Scotland and by the dedicated excavation work of Pryor and his colleagues in the East Anglian fens (e.g. Pryor 1986).

Unfortunately, the study of wild and domestic artiodactyls has advanced little in this time, although it is hoped that current work on material from Maiden Castle, Flag Fen and Potterne will remedy this situation. The full publication of reports on Hambledon Hill and the aptly named Snail Down are still awaited, as are the results of the 1974 excavations at Skara Brae. In consequence, the origin of British domestic cattle remains a matter for speculation, but a number of possibilities exist. Aurochs may have been domesticated locally at several sites. Domestic cattle from Europe may have been imported on a number of occasions. Both events might have taken place simultaneously and the results interbred, which the author considers the most likely possibility. This would almost certainly have resulted in a number of regional types, and it is possible to demonstrate this from the literature. Again, size must be an important criterion for lack of other information, although descriptions of the skull and horn cores are very valuable when available.

9.5.1 The South and East

The best known neolithic cattle of the British Isles come from Wessex. The majority of Jackson's reports, listed in the bibliography, are on this material, as are the large collections studied by Harcourt (Durrington Walls 1971; Marden 1976; Mount Pleasant 1979). The largest assemblage is, however, from Windmill Hill, and it has been studied in the greatest detail by Grigson (1973, 1982; Jope and Grigson 1965). These cattle were intermediate in size between the aurochs and the succeeding Celtic ox of the later

Bronze and Iron Ages. The animals had long, fairly thin horn cores, more or less circular in cross section, in contrast with the Celtic short elliptical horn (Armitage and Clutton-Brock 1976).

At present insufficient material exists from other English regions to demonstrate different races. A number of bone reports for East Anglian sites have been published, including Hockwold on the fen edge (Cram 1967) and two of the Fengate sites (Harmon 1978; Biddick 1980), but these are not large assemblages and there is little cranial material. A superficial examination by the author of some of the Flag Fen bones indicates small, slight animals, but the assemblage from this site will eventually be large and important. Legge (1981) has given a partial account of the cattle from Grimes Graves.

9.5.2 The Midlands

An intriguing situation exists in the Midlands, if certain mid-nineteenth century reports are taken at their face value. The site now covered by the Shustoke reservoir, near Birmingham, was first investigated by Crosslove, who reported in 1866 that he had found a number of bones, including those of *Bos primigenius*, in a postglacial peat bed. He deposited these bones in the museum which is now the Geological Museum of Birmingham University, where they remain to be examined. *Bos primigenius* is undoubtedly present, but so also is a very small circular horn core. Subsequent work by Kelly and Osborne (1963-64) demonstrated that this peat pre-dated the neolithic elm decline, but not by very long, and they also found an early medieval site some hundreds of metres away. It is quite possible that *Bos primigenius* on occasion bore small horns as a result of trauma in early life, and the polling gene, which Hammond *et al.* (1971) estimated occurred once in about 60,000 individuals, presumably also occurred in the wild species. A drawing, published by Duckworth (1911-12), of a shorthorned *Bos longifrons* skull was said to have been found with *Bos primigenius* in the Trent area. Another Midland specimen is a complete skeleton excavated from a Bronze Age ditch at North Eynham, near Oxford. The animal appeared to have died naturally shortly after giving birth, the calf's skeleton lying as if it had been attempting to suckle. This cow seems to be of the Celtic Shorthorn variety, but the Bronze Age date is not proved (Grey and Noddle unpublished). Another specimen which might be included in a Midland group is the skeleton from Giant's Grave, Skendleby, Lincolnshire (Jackson 1935a). Definitive evidence of Midland neolithic cattle comes from Grendon Quarry, Northampton (Gouldwell 1985). He described two horn cores with basal ratios (greatest basal diameter divided by least basal diameter) of 1.40 and 1.24, but confusingly there was also a larger specimen dating from the Iron Age with a ratio of 1.36. Celtic Shorthorns, typical of the Iron Age, have more elliptical horn cores; Armitage and Clutton-Brock (1976) gave measurements averaging 1.57. The recent excavation of a barrow at Irthlingborough, Northampton, which contained bones and particularly skulls of no less than a hundred *Bos primigenius* and domestic cattle (Selkirk 1987) should provide a great deal more information. At the moment there seems to be no record of a long-horned cattle type in the area.

9.5.3 The Atlantic group

Another possible group of neolithic and early Bronze Age cattle might be termed Atlantic, the majority of the specimens being Irish.

9.5.3.1 Ireland

Bone is not well preserved in the peaty soils which cover much of Ireland, and archaeologists have not often taken much trouble with animal bone even when it has been excavated. A major exception to this is the great tomb of Newgrange; the bones here were reported by Wijngaarden-Bakker (1974), who included a survey of the relevant literature then available in her report. McCormick (1985-6) has subsequently made a survey of all the bones reported in other prehistoric tombs, but this includes only a few unmeasured fragments. A recent paper publishing material not included in the excavation's original report on the neolithic site of Lough Gur, County Limerick, describes a few very fragmentary cattle bones (Butler 1987) though these could not be measured. Caulfield (1983) states that they came from large animals. To this can be added the unpublished material from Polnagallum, Ulster, which dates from at least the third millenium bc, and has been briefly examined by myself. This material is of considerable interest because it includes horn cores, which were lacking at Newgrange. Three of these were nearly circular across the section (ratio of basal diameters 1.22-1.27) but the fourth was more elliptical (1.56). All were short, and the curvature was reminiscent of the present day Jersey breed. A similar skull from peat in the Severn Estuary said to be Bronze Age is in the Welsh National Museum.

9.5.3.2 The Western Isles

The remains from the Western Isles are somewhat disappointing, because although numerous they are exceedingly fragmented, and identification, let alone measurement, is very difficult; indeed, the majority of the fragments seem to be waste from bone working (Simpson 1976). However, there are enough sufficiently large pieces at Rhosmor (Sergeantson unpublished) and Dalmore (Lewis) on which the author is currently working, to indicate a small animal. A skeleton from Bryn Celli Ddu, Anglesey (Jackson 1930) may also be included in this group. Apart from describing this as a 'small ox' Jackson did not study it further, as he considered the remains to be modern, but Lynch (1970) thought otherwise.

9.5.4 The Northern Isles

The final group to be described, from the Northern Isles, Orkney and Shetland, is distinctive (see below), with numerous and well preserved bones. The first description of these animals was that of Watson (1931), on the material excavated by Childe at Skara Brae. At that time the antiquity of this settlement was not realised (Childe's book bore the title *A Pictish Village*) and Watson, an experienced palaeontologist, found himself puzzled by the cattle bones in particular, even thinking that they might be a small form of island aurochs, but finally deciding that they came from a massive domesticate. Watson's report included photographs of horn cores, which showed a marked sexual dimorphism and would be classified as longhorns. They were not sufficiently massive to be *Bos primigenius*, neither did they bear the characteristic spiral; they are roughly similar to the present day North Devon breed. Watson's record is fortunate, since the only skull which was discovered during Clarke's unpublished 1974 excavations could not be lifted. It certainly resembled a miniature aurochs (complete length 520mm compared with 720mm for the recently excavated Rhymney specimen) but it was right at the bottom of the excavated area, and might have pre-dated the settlement. Some hair was recovered from it, which was described by Ryder (1980), and he took the animal to be an aurochs.

Similar cattle, dated to about 2500 B.C., were examined from Knap of Howar, on the much smaller island of Papa Westray (Ritchie 1983). The first report was made by Platt (1937a) and she does not seem to have connected them with the Skara Brae material. Indeed, the only information from Watson's report which seems to have caught the attention of archaeologists was the early age at death of so many of the animals, which seems to have given rise to the myth of autumn killing, that is to say the preservation of only a small number of breeding stock through their first winter. However, Platt did recognise the individuality of the Orcadian stock, and subsequent excavation (Ritchie 1983) produced similar material which was reported upon by Noddle (1983). On the strength of a single metatarsal from Scord of Brouster (Noddle 1986) it is suggested that the same type of cattle were present in Shetland.

These neolithic Orcadian cattle are quite unlike their English counterparts, being much larger and probably very much heavier, since the bones are very robust. Indeed, the bones are more massive than any other domestic cattle (other than modern) examined by the author and are only exceeded in size by Scandinavian material, with which they seem to have some affinity (Degerbøl and Fredskild 1970; Hatting 1978). The type does not seem to have been very long lasting. The dating of those Orcadian chambered tombs excavated in the 1930s is not certain, but they are nevertheless all classified as neolithic, and more recently excavated examples cover a period of about six hundred years, centring on 3000 B.C. (Barker 1983). Platt (1933, 1934, 1937a, b and c) examined the bone from most of the early excavations, and apart from finding one longhorn core from Mid Howe, Rousay, she does not comment on any similarity to the Knap of Howar material. She also published a separate report on a partial bovine skull which was definitely short-horned (Platt 1933), as did Ritchie (1934). Why this desirable (to modern eyes) type was allowed to die out is of course a matter for conjecture. One may speculate about infertility or biennial breeding, or even that the animals had an insatiable appetite that a heavily populated Orkney could not find the grazing to appease. Certain non-pathological lesions which, amongst archaeological animals, have only been found on the terminal phalanx of these Orcadian cattle, (they are universally present in modern stock) suggests that these animals might have been capable of rapid growth to match the brevity of the northern summer.

The above remarks about size, based on bone width data not readily available in the literature, are belied by Figures 2a and b, which present only the lengths of metapodial bones. Degerbøl and Fredskild (1970) and Grigson (1969) are of the opinion that width of bone is a more reliable indicator of weight. The cattle from the Northern Isles were probably thick-set, powerful and heavy; Midland and East Anglian stock were considerably shorter and lighter than all the others presented in Figure 2. These animals were somewhat later in date, all being designated Bronze Age. It has been demonstrated on a number of occasions, and in different parts of the world, that there is a decrease in bovine size as time passes (Jewell 1962, 1963; Bökönyi 1974). The horn core evidence, limited though it is, suggests that a long-horned type had been replaced (if it was ever present) by a short-horned one. This might be reinforced by environmental factors; the East Anglian Breckland is not likely to have been an area supporting rich grazing.

9.5.5 Conclusion

In summary, there would appear to have been three types of early domesticated cattle present in the British Isles. The Wessex and the Northern Isles animals are reasonably well characterised, whereas the Atlantic, Midland and East Anglian ones are far less so

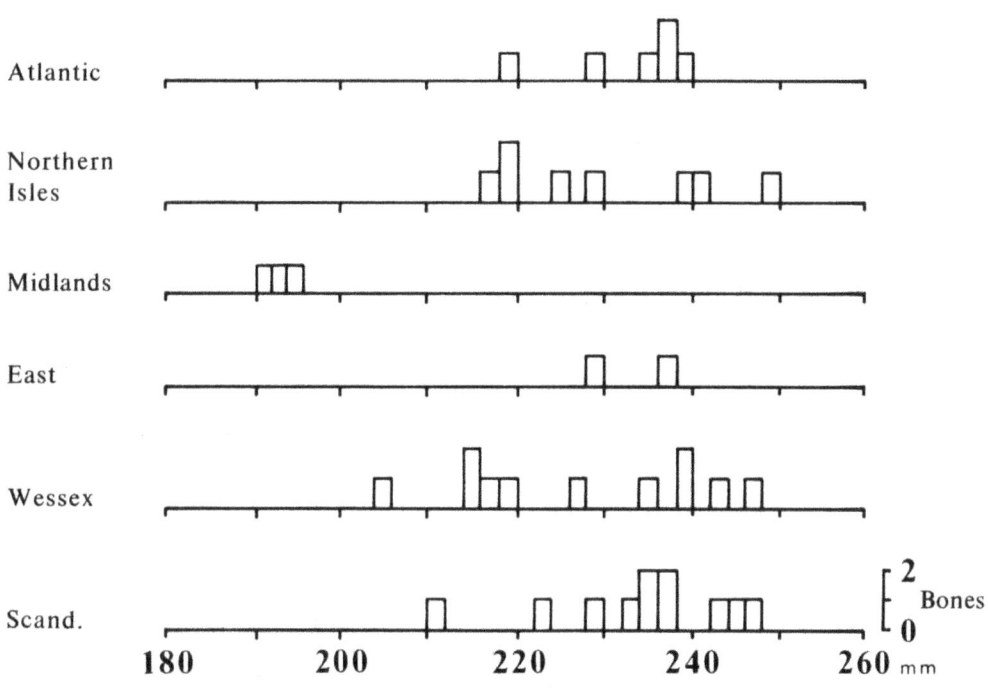

FIGURE 2a Length of metatarsal of neolithic and Bronze Age domestic cattle.

Sources:
Atlantic   New Grange (Wijngaarden-Bakker 1974)
                 Polnagallum (the author, unpublished)
Northern
Isles   Knap of Howar (Noddle 1983a)
                 Knowe of Rousay (Platt 1937b)
                 Scord of Brouster (Noddle 1987a)
                 Skara Brae (Watson 1931; the author, unpublished)

Midlands   New Wintles (the author, unpublished)
                 Skendleby (Jackson 1935a)

East   Hockwold (Cram 1967)
                 Minnis Bay (Jackson 1943b)

Wessex   Durrington Walls (Harcourt 1971a)
                 Fussells Lodge (Grigson 1966)
                 Gorsey Bigbury (Wijngaarden-Bakker 1967)
                 Stonehenge (Jackson 1935b)
                 Windmill Hill (Jope and Grigson 1965)

Scandinavia   Denmark (Degerbøl and Fredskild 1970)

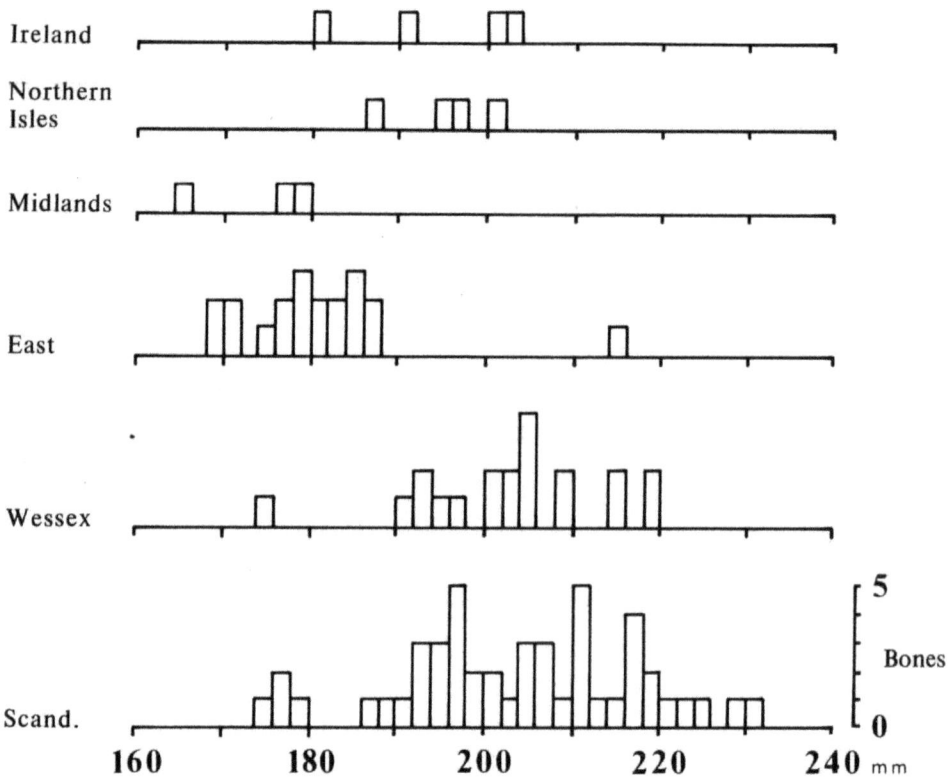

FIGURE 2b Length of metacarpal of neolithic and Bronze Age domestic cattle.

Sources:
Ireland   Newgrange (Wijngaarden-Bakker 1974)
             Polnagallum (the author, unpublished)

Northern
Isles   Skara Brae (Watson 1931; the author, unpublished)

Midlands   New Wintles (the author, unpublished)
             Skendleby (Jackson 1935a)

East   Fengate (Biddick 1980)
             Grimes Graves (Legge 1981)
             Hockwold (Cram 1967)
             Minnis Bay (Jackson 1943b)

Wessex   Cherhill (Grigson 1983)
             Durrington Walls (Harcourt 1971a)
             Isle of Wight (Shackley 1976)
             Mount Pleasant (Harcourt 1979)
             Windmill Hill (Jope and Grigson 1965)

Scandinavia Denmark (Degerbøl and Fredskild 1970)
             Sweden (During 1986)
             Troldebjerg (Higham and Message 1969)

and may not be distinct from each other. Their separation from the succeeding Celtic variety is based on scanty horn core evidence, and the two forms may have co-existed. It can only be stated categorically that the Polnagallum animals did not have Celtic horns. There is, however, an interesting Dutch neolithic calvarium, with almost circular horn cores (Clason and Brinkhuizen 1978). As to the origin of these types, it is suggested that they are related to those on the nearest part of the European continent. The importation of cattle from Scandinavia at such an early date seems unlikely, but prevailing weather conditions may have made it much easier than anticipated. In a comparison of the size of these animals, environmental factors cannot be ignored. In general both British and Scandinavian cattle are larger than those from eastern and southern Europe, and the same was true of their wild forbears.

9.6 WILD AND DOMESTIC SHEEP

There is rather less to say about sheep than cattle in Britain during the Atlantic period, since the areas in which wild sheep were indigenous do not include the British Isles. Kurtén (1968) refers to an '*Ovis antiqus*' which dates from the Cromerian; whilst this specimen, a horn core, certainly looks superficially like that of a sheep, the internal sinus structure is much more like that of an antelope (Gentry, pers. comm.), perhaps either a form of gazelle, which was present during the lower and mid Pleistocene (Hooijer 1945), or *Soereglia*, which has been confirmed from Westbury-sub-Mendip (Bishop 1982). A number of authors have identified sheep bones in France dating to the mesolithic period and earlier, but all came from cave sites which have been used to house domestic sheep, and which have not always been well excavated. The wild sheep of Corsica are said to be feral (Poplin 1979), but there is no doubt that all British sheep were originally imported by man.

It is generally agreed that sheep were first domesticated in the Near East; suitable wild specimens were available in the Zagros mountains and also perhaps further south (Clark 1964; Harrison 1968). They seem to have spread westward in two basic groups, one around the shores of the Mediterranean, the other northwards, and then west into Scandinavia (Waterbolk 1968). The long-tailed woolly sheep were probably developed in the Mediterranean area, whilst the northern sheep remained short-tailed with little wool. Woolly sheep do not seem to have been present in north west Europe during the Atlantic period. Indeed, sheep were few in number in this area compared with southern Italy, where large flocks appear to have been kept at a very early period (Whitehouse 1968, 1986). Sheep from Wessex and other English areas were few in number and small in size, being generally compared to the Soay (Jope and Grigson 1965) but the characteristic elliptical horn core of the ewe of that type does not seem to have been described before the Iron Age.

Sheep were more numerous in the more northerly areas; though they were few at Newgrange (Wijngaarden-Bakker 1974), there were more at Polnagallum and in the Northern Isles they form a substantial part of the domestic animal remains. These neolithic Orcadian sheep were substantially taller than those from further south. No elliptical horn cores have been found, and the male horn core is exactly the same shape and has the same relationship with the frontal bone as does the present day Orkney native sheep, though the latter is smaller. The shape of the scapula indicates only a primitive animal, but the position of the nutrient foramen of the femur is characteristic. The animal was more primitive than the present day Soay, in that sexual dimorphism, exemplified by the length of the metapodials, is greater than in the present day Soay (Noddle 1978, 1983). It is interesting that the sheep from Polnagallum seemed to be more like those from Orkney than those from Wessex, but there were too few to be certain.

As with the cattle, there is limited evidence that British sheep originated from two sources: the area bordering the English Channel, and Scandinavia. However, even at this early date, there may be some confusion resulting from the presence of goat, at least in England; many neolithic bone reports identify goat, and at Windmill Hill there was a complete skeleton. However, goat is less likely in the Northern Isles at this time, as it is very sensitive to change in day length, and consequently comes into oestrus in July, even as far south as Arran (McTaggart 1970). This results in kids being born in January, a situation which the feral goats of the Hebrides can survive, but which nevertheless must place them at a disadvantage.

9.7 CONCLUDING REMARKS

The early domesticates of the British Isles do not seem to foster very much interest amongst archaeologists, who can apparently refer to neolithic peoples as 'the first farmers' without mentioning their agricultural practices. It is unfortuate that later agricultural activities have removed much of the available evidence in areas of good soil, and that poor soils are frequently too acid to allow preservation of bone. However, the fenland and estuarine sites of East Anglia and, more recently, the Severn Estuary seem to have great potential in this respect, and it is to be hoped that excavators in these areas will make proper provision for archaeozoological investigations.

ACKNOWLEDGEMENTS

I am most grateful to the staff of the museums mentioned for their assistance in the examination of the specimens.

BIBLIOGRAPHY

ADAMS, J. (1853) A geological sketch of the valley of the Kennet. *Wiltshire Archaeological and Natural History Magazine* **11** 261-86.

ANON. (1864-5) Catalogue of an exhibition held at Ashby-de-la-Zouch, July 1863. *Transactions of the Midland Scientific Society* (for 1864-5) **40**.

ARMITAGE, P.L. AND CLUTTON-BROCK, J. (1976) A system for classification and description of the horn cores of cattle from archaeological sites. *Journal of Archaeological Science* **3** 329-48.

BANKS, C. (1961) Report on the recently discovered remains of the wild ox (*Bos primigenius* Boj.) from East Ham. *London Naturalist* **41** 54-58.

BARKER, G. (1983) The animal bones. In: *Isbister: a chambered tomb in Orkney.* (Hedges, J.W.) Oxford: British Archaeological Reports 114 pp. 133-50.

BASKIN, L.M. (1974) Management of ungulate herds in relation to domestication. In: *Behaviour of ungulates Volume 2.* (eds Giest, V. and Walker, F.) Switzerland: U.I.C.N.

BIDDICK, K. (1980) Animal bones from the second millennium ditches, Newark Road subsite, Fengate. In: *Excavations at Fengate, the third report.* (Pryor, F.) Northampton Archaeological Monograph 1 pp. 217-32.

BISHOP, M.J. (1982) *The mammal fauna of the early middle Pleistocene cavern infill site of Westbury-sub-Mendip, Somerset.* Special Paper on Palaeontology 28 London: Palaeontology Association.

BÖKÖNYI, S. (1962) Zur Naturgeschichte des Ures in Ungarn und das Problem der Domestikation des Hausrindes. *Acta Archaeologica Hungarica* 14 175-214.

BÖKÖNYI, S. (1974) *History of domestic mammals in central and eastern Europe.* Budapest: Akadémiai Kiadó.

BOYD DAWKINS, W. (1866) On the fossil British oxen, Part 1: *Bos urus* (Caesar). *Quarterly Journal of the Geological Society of London* 22 391-401.

BRADY, SIR A. (1874) *Catalogue of the Pleistocene vertebrae from the neighbourhood of Ilford, Essex.* Hertford: privately printed.

BROWNE, S. (1983) Investigations into the evidence for postcranial variation in *Bos primigenius* (Bojanus) in England and the problem of its differentiation from *Bison priscus*. *Bulletin of the Institute of Archaeology, University of London* 20 1-42.

BROWNE, S. (1986) The animal bone. In: The excavation of a neolithic oval barrow at North Marden, West Sussex, 1982. (Drewett, P.) *Proceedings of the Prehistoric Society* 52 40-41.

BUTLER, V.C. (1987) Identification of bone material from circles J,K,L and Site 10. In: Lough Gur excavations by Sean P. O'Riordain: further neolithic and Beaker habitations at Knockadoon. (Grogan, E. and Eogan, G.) *Proceedings of the Royal Irish Academy* 87C 501-2.

CAMPBELL, J.B. (1977) *The Upper Palaeolithic of Britain: a study of man and nature in the Late Ice Age.* Oxford: University Press.

CAMPY, M., CHAIX, L., EICHER, U., RICHARD, H., AND URLACER, J-P. (1983) L'aurochs (*Bos primigenius* Boj.) d'Etival (Jurs, France): La séquence tardi et postglaciaire sur les plateaux Jurassiens. *Revue de Palaeobiologie* 2 61-85.

CASE, H. (1969) Neolithic explanations. *Antiquity* 43 176-86.

CAULFIELD, S. (1983) The neolithic settlement of North Connaught. In: *Landscape archaeology in Ireland.* (eds Reeves-Smyth, T. and Hamond, F.) Oxford: British Archaeological Reports 116 pp. 195-215.

CHAIX, L. AND VALTON, B. (1984) Note sur un aurochs (*Bos primigenius* Boj.) subatlantique de Jurs gesien (Ain, France). *Revue de Palaeobiologie* 3 185-90.

CLARK, J.L. (1964) *The great ark of the wild sheep.* University of Oklahoma Press.

CLASON, A.T. (1967) *Animal and man in Holland's past.* Groningen: A. and B. Wolters.

CLASON, A.T. AND BRINKHUIZEN D.C. (1978) Swifterbant; mammals, birds and fishes: a preliminary report. *Helenium* **18** 69-82.

CLUTTON-BROCK, J. (1981) *Domesticated animals from early times*. London: Heinemann and British Museum (Natural History).

CLUTTON-BROCK, J. (1986) *New dates for old animals: reindeer and the aurochs in prehistoric Britain*. Archaeozoologica Mélanges. Grenoble: La Pensée Sauvage.

COHEN, M.N. (1977) *The food crisis in prehistory*. New Haven: Yale University Press.

CONRAD, J.R. (1959) *The horn and the sword*. London: MacGibbon and Kee.

CRAM, C.L. (1967) Report on the animal bones from Hockwold. In: Excavations at Hockwold-cum-Wilton, Norfolk 1961-2. (Salway, P.) *Proceedings of the Cambridge Antiquarian Society* **60** 75-80.

CROSSLOVE J. (1866) Geology of the Birmingham area. In: *Birmingham and its environs*. (ed. Crosslove, J.) British Association for the Advancement of Science pp. 63-88.

CUNNINGTON, M.E. (1906) Notes on the Bronze Age barrow at Manton near Marlborough. *Wiltshire Archaeological and Natural History Magazine* 110-17.

CURRANT, A.P. (1986) The late glacial mammal fauna of Gough's Cave, Cheddar, Somerset. *Proceedings of the University of Bristol Speleological Society* **17** 286-304.

DEGERBØL, M. AND FREDSKILD, J. (1970) The urus (*Bos primigenius* Bojanus) and neolithic domesticated cattle (*Bos taurus domesticus* Linn) in Denmark. *Kongelige Danske Videnskabernes Selskabo Biologica Skripta* **17** 1-177.

DRIESCH, A. VON DEN AND BOESSNECK, J. (1975) Zur grobe des ures, *Bos primigenius*, auf der Iberischen Halbinsel. *Suggetierkundliche Mitteilungen* **24** 66-77.

DUCKWORTH, W.L.H. (1911-12) The Sudbury calvarium, a revised and extended description. *Journal of anatomy and physiology* (3rd series) **7** 328-49.

DURING, E. (1986) The fauna of Alvasta: an osteological analysis of animal bones from a neolithic pile dwelling. *Ossa* **12** 1-209.

EVERTON, R.F. (1975) A *Bos primigenius* from Charterhouse Warren Farm, Blagdon, Mendip. *Proceedings of the University of Bristol Speleological Society* **14** 75-82.

FISHER, O. (1879) On a mammaliferous deposit at Barrington near Cambridge. *Quarterly Journal of the Geological Society* **35** 670-77.

FLEMING, A. (1979) The Dartmoor reaves: boundary patterns and behavior patterns in the second millenium B.C. In: *Prehistoric Dartmoor in its context*. (ed. Maxfield, V.A.) Exeter: Devon Archaeological Society.

FRASER, F.C. AND KING, J.E. (1954) The faunal remains. In: *Excavations at Star Carr*. (Clark, J.G.D.) Cambridge: University Press pp. 70-95.

FRISON, G.C. (1973) *Prehistoric hunters of the High Plains.* New York: Academic Press.

FRISON, G.C. (1974) Archaeology of the Caspar Site. In: *The Caspar Site: a Hell Gap bison kill on the High Plains.* (ed. Frison, G.C.) New York: Academic Press pp.1-112.

GANZ, J. (1981) *Early Irish myths and sagas.* U.S.A.: Dorset Press.

GEIST, V. (1971) *Mountain sheep.* Chicago: University Press.

GOULDWELL, A.J. (1985) The animal bones. In: Grendon Quarry excavations. (Gibson, A.M. and McCormick, A.) *Northampton Archaeology* **20** 49-51.

GREENFIELD, R.J. (1988) The origins of milk and wool production in the Old World: a zooarchaeological perspective from the central Balkans. *Current Anthropology* **29** 553-87.

GRIGSON, C. (1966) The animal remains from Fussell's Lodge long barrow. In: The Fussell's Lodge long barrow excavations 1957. (Ashbee, P.) *Archaeologia* **100** 63-73.

GRIGSON, C. (1969) The uses and limitations of differences in absolute size in the distinction between the bones of aurochs (*Bos primigenius*) and domestic cattle (*Bos taurus*). In: *The domestication and exploitation of plants and animals.* (eds Ucko, P.J. and Dimbleby, G.W.) London: Duckworth pp. 277-94.

GRIGSON, C. (1973) *The comparative craniology of Bos taurus Linn., Bos primigenius Boj. and Bos namadicus Falc.* Ph.D. thesis, University of London.

GRIGSON, C. (1978) Late glacial and early Flandrian ungulates of England and Wales: an interim review. In: *The effect of man on the landscape: the lowland zone.* (eds Limbrey, S. and Evans, J.G.) Council for British Archaeology Research Report 21 pp 46-56.

GRIGSON, C. (1982) Sexing neolithic domestic cattle skulls and horn cores. In: *Ageing and sexing animal bones from archaeological sites.* (eds Wilson, B., Grigson, C. and Payne, S.) Oxford: British Archaeological Reports 109 pp. 25-35.

GRIGSON, C. (1983) Mesolithic and neolithic animal bones. In: Excavations at Cherhill, north Wiltshire 1967. (Evans, J.G. and Smith, I.) *Proceedings of the Prehistoric Society* **49** 64-72.

HALDANE, A.R.B. (1973) *The drove roads of Scotland.* Newton Abbot: David and Charles.

HALL, S.J.G. (1979) Studying the Chillingham wild cattle. *The Ark* **6** 72-79.

HAMMOND, J., MASON, I.L. AND ROBERTSON, T.J. (1971) *Hammond's farm animals.* London: Edward Arnold.

HARCOURT, R.A. (1971) The animal bones from Durrington Walls. In: *Durrington Walls: Excavations 1966-68.* (Wainwright, G.J. and Longworth, I.H.) Reports of the Research Committee of the Society of Antiquaries of London 29 pp. 338-50.

HARCOURT, R.A. (1979) The animal bones. In: *Mount Pleasant, Dorset: excavations 1970-1971.* (Wainwright, J.G.) Reports of the Research Committee of the Society of Antiquaries of London 37 pp. 214-33.

HARMON, M. (1978) The animal bones. In: *Excavations at Fengate, Peterborough, England. The second report.* (Pryor, F.) The Royal Ontario Museum Archaeological Monograph 5 pp.177-88.

HARRISON, D.L. (1968) On three mammals new to the fauna of Oman, Arabia, with a description of a new species of bat. *Mammalia* 32 317-25.

HATTING, T. (1978) Lidso: zoological remains from a neolithic settlement. *Archaeol. Studia* 5 193-207.

HERRE, N. AND ROHRS, W. (1977) Zoological considerations in the origins of farming and domestication. In: *Origins of Agriculture.* (ed. Reed, C.A.) The Hague: Mouton Publications.

HIGGS, E.S. AND JARMAN, M.R. (1969) The origins of agriculture: a reconsideration. *Antiquity* 43 31-41.

HIGHAM, C. AND MESSAGE, M. (1969) An assessment of a prehistoric technique of bovine husbandry. In: *Science and archaeology.* (eds Brothwell, D. and Higgs, E.) London: Thames and Hudson pp. 315-30.

HOOIJER, D.A. (1945) A fossil gazelle (*Gazella schrendarae* nov. sp.) from the Netherlands. *Zoologische Mededeelingen* 25 55-64.

HOPWOOD, A.T. (1939) Fossil mammals. In: Excavations at Brundon, Suffolk 1935-37. (Moir, J.R. and Hopwood, A.T.) *Proceedings of the Prehistoric Society* 5 13-29.

HUNT, D. (1987) *Early farming communities in Scotland: aspects of economy and settlement 4500-1250 B.C.* Oxford: British Archaeological Reports 159, volume 1.

JACKSON, J.W. (1929) Report on the animal remains found at Woodhenge, Durrington, Wiltshire. In: *Woodhenge.* (Cunnington, M.E.) Devizes: Simpson pp. 61-69.

JACKSON, J.W. (1930) The ox bones. In: The chambered cairn of Bryn Celli Ddu. (Hemp, W.J.) *Archaeologia* 80 213.

JACKSON, J.W. (1935a) On the skeleton of a small ox of Beaker age from Giant's Hills. In: The excavation of the Giant's Hills long barrow, Skendleby, Lincolnshire. (Phillips, C.W.) *Archaeologia* 85 96-98.

JACKSON, J.W. (1935b) The animal remains from the Stonehenge excavations of 1920-26. *The Antiquaries Journal* 15 434-40.

JACKSON, J.W. (1943a) Animal bones. In: *Maiden Castle, Dorset.* (Wheeler, R.E.M.) Reports of the Research Committee of the Society of Antiquaries of London 12 pp. 360-71.

JACKSON, J.W. (1943b) Appendix 2. In: A report on the late Bronze Age site excavated at Minnis Bay, Birchington, Kent, 1938-40. (Worsfold, F.H.) *Proceedings of the Prehistoric Society* **9** 41-44.

JARMAN, M.R. (1969) The prehistory of upper Pleistocene and recent cattle. Part 1: East Mediterranean, with reference to north west Europe. *Proceedings of the Prehistoric Society* **35** 236-66.

JEWELL, P.A. (1962) Changes in size and type of cattle from prehistoric to medieval times in Britain. *Zeitschrift für Tierzüchtung und Züchtungsbiologie* **77** 159-67.

JEWELL, P.A. (1963) Cattle from British archaeological sites. In: *Man and cattle.* (eds Mourant, A.E. and Zeuner, F.E.) Royal Anthropological Institute Occasional Paper 18 pp. 80-91.

JOPE, M. AND GRIGSON, C. (1965) Faunal remains. In: *Windmill Hill and Avebury: excavations by Alexander Keiller 1925-1939.* (Smith, I.) Oxford: Clarendon Press pp. 141-67.

KEHOE, T.F. (1973) *The Gull Lake site: a prehistoric bison drive site in south western Saskatchewan.* Milwaukee: Public Museum.

KELLY, M. AND OSBORNE, P.J. (1963-4) Two faunas and floras from the alluvium at Shustoke, Warwickshire. *Journal of the Linnean Society* **176** 37-65.

KING, J.E. (1962) Report on the animal bones. In: Excavations at the Maglemosian sites at Thatcham, Berkshire, England. (Wymer, J.) *Proceedings of the Prehistoric Society* **28** 355-61.

KOWALSKI, K. (1967) The Pleistocene extinction of mammals in Europe. In: *Pleistocene extinctions.* (eds Martin, P.S. and Wright, H.E.) New Haven: Yale University Press pp. 349-65.

KURTÉN, B. (1968) *Pleistocene mammals of Europe.* London: Weidenfeld and Nicolson.

LEEDS, A. (1965) Reindeer herding and Chuckchi social institutions. In: *Man, culture and animals.* (eds Leeds, A. and Vayden, A.P.) Washington: American Association for the Advancement of Science pp. 87-127.

LEGGE, A.J. (1981) The agricultural economy. In: *Grimes Graves, Norfolk: excavations 1971-72, volume 1.* (Mercer, R.J.) Department of the Environment Research Report 11 London: Her Majesty's Stationery Office pp. 79-103.

LEVITAN, B.M., AUDSLEY, A., HAWKES, C.J., MOODY, A., MOODY, P., SMART, P.L. AND THOMAS, J.S. (1988) Charterhouse Warren Farm Swallet, Mendip, Somerset: exploration, geomorphology, taphonomy and archaeology. *Proceedings of the University of Bristol Speleological Society* **18** 171-239.

LYDEKKER, R. (1912) *The ox and its kindred*. London: Methuen.

LYNCH, F. (1970) *Prehistoric Anglesey: the archaeology of the island to the Roman conquest*. Llamgefni: Anglesey Antiquarian Society.

McCORMICK, F. (1985-6) Faunal remains from prehistoric Irish burials. *The Journal of Irish Archaeology* 3 37-48.

McTAGGART, H.S. (1970) Observations on the behaviour of an island community of feral goats. *British Veterinary Journal* 127 399-400.

NODDLE, B.A. (1978) Some minor skeletal differences in sheep. In: *Research problems in zooarchaeology*. (eds Brothwell, D.R., Thomas, K.D. and Clutton-Brock, J.) London: Institute of Archaeology Occasional Papers 3 pp. 133-41.

NODDLE, B.A. (1983a) Animal bone from Knap of Howar. In: Excavation of a neolithic farmstead at Knap of Howar, Papa Westray, Orkney. (Ritchie, A.) *Proceedings of the Society of Antiquaries of Scotland* 113 92-100.

NODDLE, B.A. (1983b) Size and shape, time and place: skeletal variation in cattle and sheep. In: *Integrating the subsistence economy*. (ed. Jones, M.) Oxford: British Archaeological Reports S181 pp. 211-38.

NODDLE, B.A. (1987a) Animal bones. In: *Scord of Brouster: an early agricultural settlement on Shetland, excavations 1977-1979*. (Whittle, A.) Oxford: University Committee for Archaeology Monographs 9 p. 132.

NODDLE, B.A. (1987b) Mammalian remains from the Cotswolds region. In: *Studies in palaeoeconomy and environment in south west England*. (eds Balaam, N.D., Levitan, B. and Straker, V.) Oxford: British Archaeological Reports 181 pp. 31-50.

OAKLEY, K.P. AND LEAKEY, M. (1937) Report on excavations at Jaywick Sands, Essex (1934), with some observations on the Clactonian Industry, and on the fauna and geological significance of the Clacton Channel. *Proceedings of the Prehistoric Society* 3 217-60.

OWEN, R. (1846) *A history of British fossil mammals and birds*. London: John van Voorst.

PLATT, M.I. (1933) Notes on the skull of an ancient ox from Rousay, Orkney. *Scottish Naturalist* 199 17-24.

PLATT, M.I. (1934) Report on the animal bones. In: A long stalled chambered cairn or mausoleum (Rousay type) near Midhowe, Rousay, Orkney. (Callender, J.G. and Grant, W.G.) *Proceedings of the Society of Antiquaries of Scotland* 68 348-50.

PLATT, M.I. (1937a) The animal bones. In: Howar, a prehistoric structure on Papa Westray, Orkney. (Traill, W. and Kirkness, W.) *Proceedings of the Society of Antiquaries of Scotland* 71 309-21.

PLATT, M.I. (1937b) Report on the animal bones. In: A long stalled cairn at Blackhammer, Rousay, Orkney. (Callender, J.G. and Grant, W.G.) *Proceedings of the Society of Antiquaries of Scotland* 71 306-8.

PLATT, M.I. (1937c) Report on the animal bones. In: A neolithic double-chambered cairn of the stalled type and later structures on the Calf of Eday, Orkney. (Calder, C.S.T.) *Proceedings of the Society of Antiquaries of Scotland* **71** 152-54.

POPLIN, F. (1979) Origine du mouflon de Corse dans une nouvelle perspective paléontologique: par marronage. *Annales de Génétique et de Sélection animale* **11** 133-43.

PRYOR, F. (1986) Fenland project. *History and Archaeology Review* **1** 5-20.

RICHTER, J. (1982) Adult and juvenile aurochs, *Bos primigenius* Boj., from the Maglemosian site of Ulkestrup Lyng Ost, Denmark. *Journal of Archaeological Science* **9** 247-59.

RITCHIE, A. (1983) Excavation of a neolithic farmstead at Knap of Howar, Papa Westray, Orkney. *Proceedings of the Society of Antiquaries of Scotland* **113** 40-121.

RITCHIE, J. (1920) *The influence of man on animal life in Scotland: a study in faunal evolution.* Cambridge: University Press.

RITCHIE, J. (1934) Report on bones of ox. In: The broch of Midhowe, Rousay, Orkney. (Callender, J.G. and Grant, W.G.) *Proceedings of the Society of Antiquaries of Scotland* **68** 515-16.

RYDER, M.L. (1980) Hair remains throw light on early British prehistoric cattle. *Journal of Archaeological Science* **7** 389-92.

SAVAGE, R.J.G. AND RICHARDS, C. (1980) Merck's Rhinoceros from Worlebury Hill, Weston-Super-Mare, Avon. *Proceedings of the University of Bristol Speleological Society* **15** 219-26.

SELKIRK, A. (1987) Irthlingborough. *Current Archaeology* **9** 331-33.

SHACKLEY, M.L. (1976) Palaeoenvironmental evidence from a late third millenium B.C. peat bed at New Slede Bridge, Isle of Wight. *Journal of Archaeological Science* **3** 385-89.

SHAWCROSS, F.W. AND HIGGS, E.S. (1961) The excavation of a *Bos primigenius* at Lowe's Farm, Littleport. *Proceedings of the Cambridge Antiquarian Society* **54** 3-16.

SHOTTON, F.W. (1978) Archaeological inferences from the study of alluvium in the lower Severn and Avon valleys. In: *The effect of man on the landscape: the lowland zone.* (eds Limbrey, S. and Evans, J.G.) Council for British Archaeology Research Report 21 pp. 27-31.

SIMPSON, D.D.A. (1976) The later neolithic and Beaker settlement site at Northton, Isle of Harris. In: *Settlement and economy in the third and second millenium B.C.* (eds Burgess, C.B. and Miket, R.) Oxford: British Archaeological Reports 33 pp. 221-31.

SMITH, J.A. (1872) Notes on the ancient cattle of Scotland. *Proceedings of the Society of Antiquaries of Scotland* **9** 587-674.

SZABO, B.J. AND COLLINS, D. (1975) Ages of fossil bones from British interglacial sites. *Nature* **254** 680-81.

TEICHART, L. (1987) Knochenfunde vom ur (*Bos primigenius* Bojanus 1827) am Schlaatz bei Potsdam. *Verfertlichungen des Museums für Vor- und Frühgeschichte Potsdam* **21** 37-45.

WATERBOLK, H.T. (1968) Food production in prehistoric Europe. *Science* **162** 1093-102.

WATSON, D.M.S. (1931) Animal bones. In: *Skara Brae, a Pictish village in Orkney.* (Childe, V.G.) London: Kegan Paul, Trench, Trubner and Co. pp. 198-204.

WHITEHEAD, G.K. (1953) *Ancient white cattle of Britain and their descendants.* London: Faber and Faber.

WHITEHOUSE, R.D. (1968) Settlement and economy of southern Italy in the Neothermal period. *Proceedings of the Prehistoric Society* **34** 332-67.

WHITEHOUSE, R.D. (1986) Siticulosa Apulia revisited. Antiquity 60 36-44.

WIJNGAARDEN-BAKKER, L.H. VAN (1974) The animal remains from the Beaker settlement at Newgrange, Co. Meath: first report. *Proceedings of the Royal Irish Academy* **74** 313-83.

WIJNGAARDEN-BAKKER, L.H. VAN (1976) Animal bones from Gorsey Bigbury. In: Gorsey Bigbury: radiocarbon dating, human and animal bones, charcoals, archaeological reassessment. (ApSimon, A.M., Musgrave, J.H., Sheldon, J., Tratman, E.K. and Wijngaarden-Bakker, L.H. van) *Proceedings of the University of Bristol Speleological Society* **14** 155-83.

WOODS, H. (1839) *Description of the fossil skull of an ox discovered in May 1938 at Melksham, Wiltshire.* London: Edwards.

ZEUNER, F.E. (1963) *A history of domesticated animals.* London: Hutchinson.

10

THE EVIDENCE FOR CEREAL CULTIVATION AND ANIMAL HUSBANDRY IN THE SOUTHERN BRITISH NEOLITHIC AND BRONZE AGE

Roy Entwistle and Annie Grant

ABSTRACT

We suggest that the role of cereal cultivation during the Neolithic and earlier Bronze Age has assumed undue importance in archaeological reconstructions. The cultivation of cereals *per se* is not in dispute, but it should be seen as one component in a much more broadly based subsistence strategy. It is only in the later prehistoric period that cereal cultivation can be shown to have provided a major resource. The evidence that has been used to propose a specialised dairy emphasis in the cattle economy is also open to alternative interpretations. We argue that we should take a fresh look, both at the evidence and also at our attitudes towards economic reconstruction, which may be conditioned to too great an extent by our own preconceptions about agriculture and farming.

10.1 INTRODUCTION

Environmental science, using increasingly sophisticated techniques of analysis, has in recent years extended greatly the range and quality of information available to archaeologists. The wide range of specialisations in palaeoenvironmental science may be seen as a response to the increasing complexity of the subject matter within the individual disciplines. Unfortunately this has created communication difficulties both between specialists and to a greater extent between the specialists and those providing their data (cf. Edwards on the use of palynological data by archaeologists (1979: 255-70)). It is our contention that if we are to advance our understanding of early agricultural communities in Britain, a more critical evaluation of the data coupled with an approach to interpretation that places more emphasis on the social context of economic activity is required. As Tilley (1981: 145) has put it, 'economic activities can only be explained by the non-economic'.

10.2 CEREAL CULTIVATION

Agriculture has long been regarded as a definitive characteristic of the Neolithic. The evidence for cereal cultivation has at times been interpreted as demonstrating more or less sedentary farming by communities carefully attuned to the agricultural potential of their locality (Barker and Webley 1978). We also read of 'substantial areas' being 'broken into arable cultivation' (Mercer 1981: x), and of the rearing of dairy herds as a major focus of neolithic and Bronze Age animal management (Legge 1981a: 179-80).

The direct archaeological evidence for cereal cultivation occurs in the form of charred seeds, cereal grain impressions in pottery, and pollen grains. However, compared to many other forms of vegetable food, charred cereal grains have a relatively high survival potential and thus it is very difficult to make accurate quantitative assessments of their relative importance. Even if we were able to overcome the problems of differential survival, the small size of most samples recovered from early prehistoric contexts reduces their value. Estimates of the frequency with which different cereals were cultivated using tiny samples surviving as grains or impressions frequently ignore fundamental sampling principles. The equivocal nature of this type of evidence is well illustrated by Jones' (1980) analysis of Grooved Ware pottery. Previous studies had recorded an absence of cereal impressions in the pottery, encouraging an assumption that the economy of the people using Grooved Ware pottery was a pastoral one. Now that cereal impressions have been recognised, cereal growing has been accepted as a component of that economy.

The direct evidence for cereal cultivation in southern England during the Neolithic and the earlier Bronze Age, is not an adequate foundation for its ready acceptance as a central feature in the economy. The paucity of data in the early part of our period is shown in Hillman's (1981) review and confirmed by a more recent summary by Moffett *et al.* (this volume). In contrast, there is abundant evidence for cereal cultivation at later Bronze Age and Iron Age sites where charred grains are commonly preserved. On many of these sites, in addition to the cereal remains themselves there is also structural evidence for extensive storage, much of which seems likely to have been for grain, in large pits and four-post structures (Gent 1983). Storage facilities are essential if cereal production is to be advantageous. Therefore it is surely no coincidence that during the Iron Age, when vast tracts of the southern downland are crossed by extensive networks of field systems, the large bell-shaped storage pits become a common feature at many sites. Pits are also a common feature of neolithic sites in southern Britain (Field *et al.* 1984), where some are claimed to have been used for cereal storage, but unlike their Iron Age counterparts they are mainly too small or the wrong shape to have been used effectively (see Reynolds 1979: 71-82). One notable example of an earlier neolithic pit, resembling in dimensions the storage pits on Iron Age sites, was discovered just outside the Coneybury henge near Durrington Walls, but despite flotation there was no evidence that this had ever held cereals (J. Richards pers. comm.). In general there is a lack of carbonised cereal residues from these neolithic pits but hazel-nut remains are commonly recovered (Moffett *et al.* this volume). It may be argued that in the earlier period archaeological evidence for cereal storage is lacking because flimsier, above ground structures could have been used for storage. This argument may be offered as an explanation for the absence of traces of storage facilities, but it cannot be used to support extensive cereal cultivation.

A comparison with the cereal evidence from *Linearbandkeramik* sites in northern Europe further highlights the paucity of the British evidence. Although there are wide regional variations in the range of cultivated and wild resources being exploited, here cereal remains are abundant and ubiquitous in the settlements (Bakels 1978) and clearly attest to the importance of cereal cultivation. In the concentration of carbonised grain, *Linearbandkeramik* sites are more comparable to some of the English Iron Age sites than to those of the neolithic and early Bronze Age (Mark Robinson, pers. comm.).

The development of widespread cereal cultivation in the later prehistoric period is attested by the extensive tracts of fields which still survive on the downland of southern England. Although it is notoriously difficult to date individual field systems this formal dividing up of the landscape seems to have originated some time towards the middle of the Bronze Age. Although examples of neolithic fields are known in East Anglia and Ireland (Pryor 1978; Caulfield 1983), there is no evidence that they were used for cereal cultivation, and the excavators have suggested that they were animal enclosures.

The later period also produces abundant evidence of soil erosion in the form of widespread hill-wash deposits in dry valleys and aeolian silts in the upper ditch stratigraphies of earlier monuments. Pedological evidence for early cultivation is ambiguous: buried *sols lessivés* beneath several neolithic monuments were once thought to have formed under the influence of forest clearance and cultivation (Evans and Jones 1979: 208-13; Limbrey 1975: 184-87). More recently it has been suggested that the process of lessivage could also occur under Flandrian climax woodland wherever the superficial deposits over the chalk were of sufficient depth (Fisher 1982). This study weakens the case for early agriculture operating on a scale, or intensity, sufficient to bring about significant pedological change.

The occurrence of ard marks beneath a number of long barrows in the south of England is another category of evidence for early cultivation that has been questioned recently. Citing European examples, Rowley-Conwy (1987) has pointed out that in several cases the extent of the ard mark pattern is not congruent with the area of the mound. Instead the distribution of marks is confined to a limited area beneath the mound, sometimes defined by internal structural features. The implications are that the scouring of the ground prior to the barrow construction was part of the funerary ritual and not surviving traces of previous cultivation preserved by burial. However, at South Street long barrow the ard marks on the chalk sub-surface pre-date the construction of the mound by a period sufficiently long to allow a turf to develop above them (Ashbee *et al.* 1979).

In summary, we are not attempting to discount the cultivation of cereals during the Neolithic and early Bronze Age. We are however questioning the validity of much of the evidence which has been used to support an acceptance of their cultivation on a large scale. The evidence seems at least as consistent with cereal cultivation being a part, but perhaps only a minor part of a diverse subsistence economy including the exploitation of wild plant resources (see Moffett *et al.* this volume).

10.3 ANIMAL HUSBANDRY

Turning to animal husbandry, we see the apparent acceptance of a view by some writers that there was a dairy emphasis to the neolithic cattle economy (for example, Barker 1985: 200, 206; Harrison 1985: 87-8). The basis of this view is the apparent predominance of female cattle in the bone assemblages from some causewayed enclosures. It has been argued that 'the female bias at ceremonial sites can be taken to predict a dairy basis to cattle husbandry in the Neolithic of Britain' (Legge 1981a: 179). This argument has been further supported by the mortality pattern of the cattle from the middle Bronze Age settlement at Grimes Graves (Legge 1981b). We would argue here that neither a predominance of female cattle at these sites, nor the kind of cattle mortality found at Grimes Graves provides any acceptable foundation to justify the assumption of a 'dairy basis' for the neolithic, or indeed the Bronze Age economy.

Let us first examine the cattle mortality pattern at Grimes Graves. One of the main arguments used for postulating a dairy economy here was the high proportion of young cattle in the archaeological bone assemblage. It was suggested that this was 'unlikely to represent other than a deliberate pattern of killing decided by the husbandman' whereby young animals, mainly males, were killed once their birth had established lactation, in order to release the milk for human consumption (Legge 1981b: 86). The reference to a high proportion of animals of under about six months of age makes an implicit assumption that the cattle mandibles recovered from the site is indicative of, in a general way at least, the mortality of a herd of cattle, presumably one that was capable of reproducing itself. However, detailed examination of the age at death of the cattle whose bones were found at Grimes Graves makes it clear that this cannot be the case (see Appendix) - some animals, in particular some adult cows, must have died or been killed elsewhere (or alternatively, young animals must have been brought in from other cattle populations and killed at the site). We cannot then know what proportion of the number of young animals born in any herd or herds managed by the people who used the Grimes Graves site died or was killed in their first six months.

Nor does a policy of the deliberate killing of very young animals to release the milk for human consumption seem a very likely husbandry policy. A more common situation seems to be one where the milk produced by the cows is shared between the animal and human populations (Krige 1936: Chapter 9). After weaning, the male calves are then reared for several years to provide meat, and the females to provide further calves. In primitive cattle, the calves are necessary not only to initiate lactation, but also to maintain it, as the cows will not readily release their milk without the stimulus, by the presence or suckling of a calf, of the milk ejection reflex (Amoroso and Jewell 1963). An English medieval manuscript illustration showing a cow licking her calf while she is being milked, nicely demonstrates the point (see Grant 1988: Figure 8.3).

Furthermore, while a modern dairy cow will produce milk for most of the year, even as late as the medieval period, cattle were usually milked for less than six months of the year and it was not common to keep a specialised dairy herd. Medieval records demonstrate low yields; the productivity in a neolithic 'dairy herd' seems unlikely to have supplied enough of the animal protein requirements of a community to justify a specialised management for milk production.

Other issues, such as that of lactose tolerance in the human population must also be considered. The ability to digest lactose is only found in human populations that have had long exposure to cows' milk. If a dairy economy is being operated this presupposes a population capable of digesting milk or milk products. While this issue has been raised by others in the past (Sherratt 1981; Clutton-Brock 1981) the implications for the human population of neolithic Britain have still not been considered sufficiently.

The predominance of females amongst the mature cattle whose bones are found in the causewayed enclosures, far from requiring explanation in terms of a specialised economy, is not in fact in any way surprising. If one takes as an example an animal that is never exploited for milk, the pig, in a typical herd the majority of the mature animals will still be female, because the males will have been killed off for meat before maturity, while many of the females will have been maintained to and beyond maturity as breeders. In many animal husbandry systems, a majority of adult females is usual. Relatively high infant mortality and low fertility are not uncommon features of 'primitive' animal husbandry systems. It may have been necessary to keep the majority of, or even all the fertile cows until their productive days were over, merely to ensure that the herd was maintained. Unless fertility is very high, natural mortality is very low and food resources for the animals very plentiful, there is little flexibility possible in the management of the female portion of an animal herd. Much more flexibility is possible in the management of the males, the only restriction being that there are enough potent males to service the breeding females (Grant in preparation).

The main limitation affecting any attempt to reconstruct the animal management strategies in southern Britain in the neolithic period is that the majority of our evidence comes from sites that clearly have a specialised function. The data that we have suggests that cattle were connected with social status and the ritual life of the community, and this does not necessarily equate directly with the subsistence value of these animals. The selection of cattle bones to accompany, and sometimes to replace, the dead in long barrows (Bradley 1984: 24) emphasises their role in funerary ritual and provides a connection with the causewayed enclosures. At these latter, large quantities of cattle bones, the occurrence of highly structured forms of artefact deposition, and at some locations the frequent appearance of human remains, suggest that they were places where the public life of the community was centred and its ritual conducted.

We cannot even be sure that we have an accurate impression of the relative abundance of the three main domesticates. The relative proportions of animals kept is generally assumed to be roughly proportional to the occurrence of their bones in archaeological contexts (e.g. Grigson 1981: 196-99). However, it is perhaps unwise to attempt to reconstruct an economy, or even the relative subsistence value of the three main domesticates, from bone assemblages recovered from the specialised contexts of funerary monuments and causewayed camps. The danger is well illustrated by the results of investigations at Kerma, a second millennium B.C. town in the northern province of the Sudan. Here, sheep and a very small number of goats and dogs were the only animals found within the graves and tombs of the necropolis. However, the contemporary town deposits contained 50% cattle bones and only 45% from ovicaprids, of which at least a third were goats (Chaix and Grant 1987).

Just as problematical as those arising from site interpretation are questions of sample size. For the whole of the neolithic period in southern Britain there have been far fewer animal bones recovered than from some single sites of the Iron Age.

10.4 CONCLUSIONS

A view that a more or less settled intensive cereal growing and dairying economy was established in parts of Britain during the third millennium B.C. does not seem to be adequately supported by the available evidence. The case for cereal cultivation as a major component in the economy may only be appropriate in later prehistory, and there is no secure evidence for a dairy emphasis in the cattle husbandry of the period. The 'agricultural model' for the third millennium B.C. is thus unsatisfactory. Furthermore, it models economic processes in isolation with no specific regard for their social context. It is in this setting that ethnographic studies assume such an important role, not merely as sources of analogy but as the means by which we can explore the complex relationships between economic resources and the social relations that arise from their exploitation. Without the integration of social and economic studies our ability to understand economic change is seriously limited.

Reference to the ethnographic record illustrates the wide range of management strategies for plants and animals which have been adopted by different societies. It would be very difficult to distinguish between many of these practices using archaeological evidence of similar quality to that available for the British Neolithic. It is however interesting to note that in some societies in which cattle assume a symbolic importance, cultivation is often a transient, low status activity. None the less, in some pastoral societies, the diet is mainly vegetarian and meat is eaten only at festivals and sacrifices (Krige 1936).

For the southern British Neolithic, a model of transient hoe based horticulture might fit the evidence better than one positing settled agriculture. This scale of practice and level of technology are more in keeping with other aspects of the archaeological record of communities living in Britain and the Atlantic fringe of mainland Europe.

By the later Bronze Age in southern Britain permanent field blocks, often with integrated enclosures, and evidence for soil erosion suggest much more intensive cultivation.

Cattle in the Neolithic may have been exploited to provide a range of products (Grigson 1981: 197), but we should be cautious of separating their subsistence value from their social and symbolic importance.

APPENDIX: Interpreting mortality patterns: a preliminary note

Annie Grant

The evidence for age at death of animals represented in faunal assemblages from archaeological sites is perhaps the most important and useful information for the reconstruction of animal husbandry practices in the past. This information is, however, extremely difficult to interpret. There are many methodological problems involved in the assessment of age at death (e.g. Grant 1978; Watson 1978; Chaix and Grant 1987) and many more in its interpretation. The assumption is often made, implicitly if not explicitly, that the mortality patterns reconstructed from archaeological data are representative of the 'average' mortality of complete herds or flocks of animals. Where this is indeed the case, it becomes possible to talk of relative proportions of young, juvenile, mature and elderly animals among the live animals and not just in the archaeological assemblage, and thus to make some assessment of the animal management practices of past communities, and of the importance of the various animal products. Where this is not the case, as for example, in many urban contexts, where the remains may be those of animals bought at market, and thus only a part of the flocks and herds of the surrounding countryside, interpretation of husbandry practices becomes much more complex.

It is thus vital to attempt to assess whether it is possible that the mortality patterns we detect in the archaeological record could be those of the mortality of complete, viable, self-reproducing herds, or whether it is more likely that some parts of the age range of the original herds may be under- or over-represented. For example in a society with urban centres and a developed trade in meat, we may expect to find juvenile animals over-represented in urban contexts, and under-represented in the settlements of the farmers that raised the animals for sale.

Any attempt at such an assessment of a mortality pattern must inevitably involve making a series of assumptions that we cannot substantiate, about, for example, fertility, average age at first calving, average numbers of calves produced per female, and natural mortality. However, it is possible to make informed estimates of these using data derived from our knowledge of modern animals, from ethnographic evidence, and from documentary evidence.

This brief note cannot fully discuss the methods currently being developed to analyse mortality patterns. However, it attempts to demonstrate some of the ways in which such analysis may be approached, using the cattle mortality data from Grimes Graves. Full details will be published later (Grant in preparation).

Legge (1981b) aged the cattle mandibles recovered at Grimes Graves using a method proposed by Higham (1967). The results (taken from the histogram shown in Figure 51 of the Grimes Graves report) are given in Table 1. This also shows the approximate ages that Higham gave to his tooth development stages, and, using these ages, groups the mandibles into approximate year classes. The absolute ages assigned can only be approximations, but for the sake of this example they are accepted here. Since the results were initially presented in the form of percentages, the following calculations assume a total of 100 animals. The absolute numbers are irrelevant to this analysis - it is the relative proportions of the different age groups that are crucial.

Higham's stage	Approx. Age	Number	Year	Total
1		0	0-1	46
2		4		
3		23		
4		12		
5	6 months	6		
6		1		
7		2	1-2	18
8	15-16 months	6		
9		7		
10		0		
11		3		
12	24 months	4	2-3	9
13		1		
14		2		
15		0		
16		1		
17		1		
18	36 months	1	3-4	16
19		3		
20		3		
21	40-50 months	9		
22	50+ months	1	4+	11
23		10		

TABLE 1 The age of cattle mandibles from Grimes Graves (data from Legge 1981b).

The figures used for the natural variables, such as fertility rates and natural mortality, that affect herd dynamics have been guided by those discussed by Dahl and Hjort (1983).

The object of the analysis is to assess whether it is likely that the cattle mandibles found at Grimes Graves could be representative of the mortality of a complete, self-reproducing herd, and, if this is not found to be the case, whether we can gain any insight into the human actions that led to the accumulation of such an archaeological assemblage. The results are shown in Table 2. The notes that follow attempt to explain the analysis.

(1) Two sets of calculations have been made, one assuming first calving at two years, the other at three years. Two years is a rather optimistic estimate of the average age at first calving, and three years might be considered a more likely average age.

(2) The number of adults, that is sexually mature animals represented in the Grimes Graves assemblage. If the cattle were sexually mature at two years, there are 36 adults, 9 that were between two and three years at death, 16 three to four year olds and 11 older than four years. If sexual maturity did not occur until three years the two to three year group must be excluded, leaving 27.

(3) The number of mature females. The male:female ratio was calculated by Legge as 1:6, on the basis of metapodial measurements.

(4) The maximum number of calves that could have been produced if the fertility rate was 100%. The females that died when between two and three years could have produced one calf each, those that died between three and four years two calves each etc. The 4+ age group has been assumed to include two 4-5 year olds, two 5-6 year olds, two 6-7 year olds, two 7-8 year olds and one 8-9 year old.

(5) The maximum number of calves that could have been produced assuming an 80% fertility rate. This is by no means an unduly low estimate of likely average fertility.

(6) The number of female calves produced, assuming a 1:1 male to female ratio at birth.

(7) The number of animals dying in their first year. This will be conditioned both by natural mortality (here assumed to be 10%, a relatively low figure) and by the husbandry policy. A prudent herdsman would keep all surviving females at least until their fertility and general state of health had been demonstrated. If males were not required for other purposes, many of them could be killed off, although sufficient would need to be retained to maintain an appropriate male:female ratio, here taken as 1:6. (see 8)

(8) The number of one year old calves remaining, assuming the mortality calculated in 7. This gives a male:female ratio similar to that found at Grimes Graves. The number of calves remaining if 46 had died, as at Grimes Graves, is also calculated. With first calving at three years, this leaves a negative figure.

Note 1	Age at first calving	2 years	3 years
2	No. adults	36 (9+16+11)	27 (16+11)
3	No. mature females (male:female = 1:6)	31 (8+14+9)	23 (14+9)
4	No. calvings (100% fertility)	76 (8x1)+(14x2) +(2x3)+(2x4) +(2x5)+(2x6) +(1x7)	48 (14x1)+(2x2) +(2x3)+(2x4) +(2x5)+(1x6)
5	No. calves (80% fertility)	61	38
6	No. female calves (80% fertility)	31	19
7	No. dying in first year Females (10% natural mortality) Males (85% culled) Total dead (Grimes Graves = 46)	3 25 28	2 16 18
8	No. 1 year old calves remaining Female Male Total Male:female ratio (No. remaining if 46 had died)	28 5 33 1:5.6 15	17 3 20 1:5.6 (-28)
9 4	No. dying in second year (7% natural mortality) (Grimes Graves = 18)	2	1
10	No. 2 year calves remaining Females Males Total Male:female ratio No. remaining if 18 had died	27 4 31 1:6.7 (-3)	16 3 19 1:5.3 (-46)

TABLE 2 An analysis of the cattle mandibles from Grimes Graves (see text for explanation of notes).

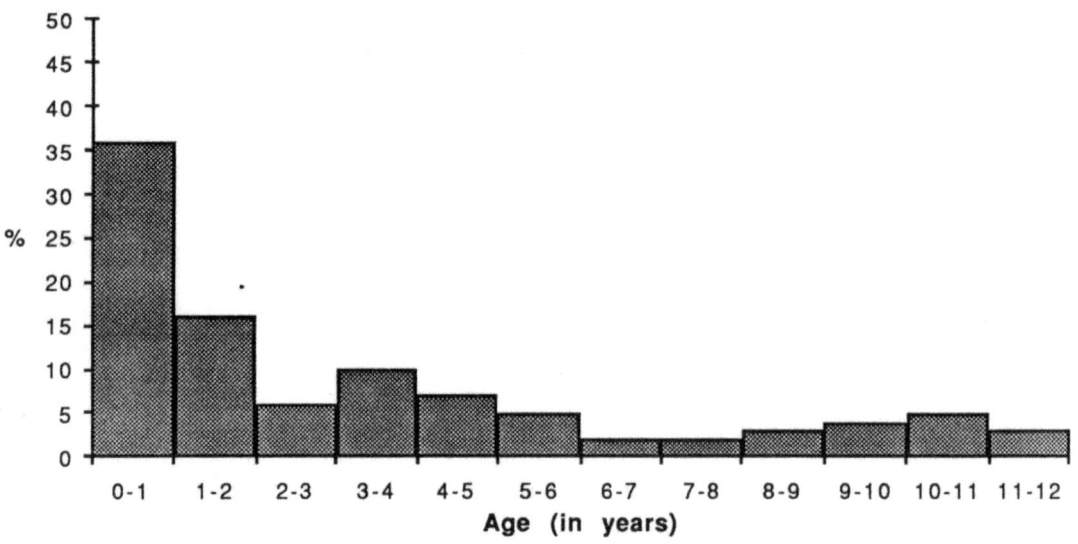

FIGURE 1 Simulated mortality where a high percentage of male calves are killed in year one and year two.

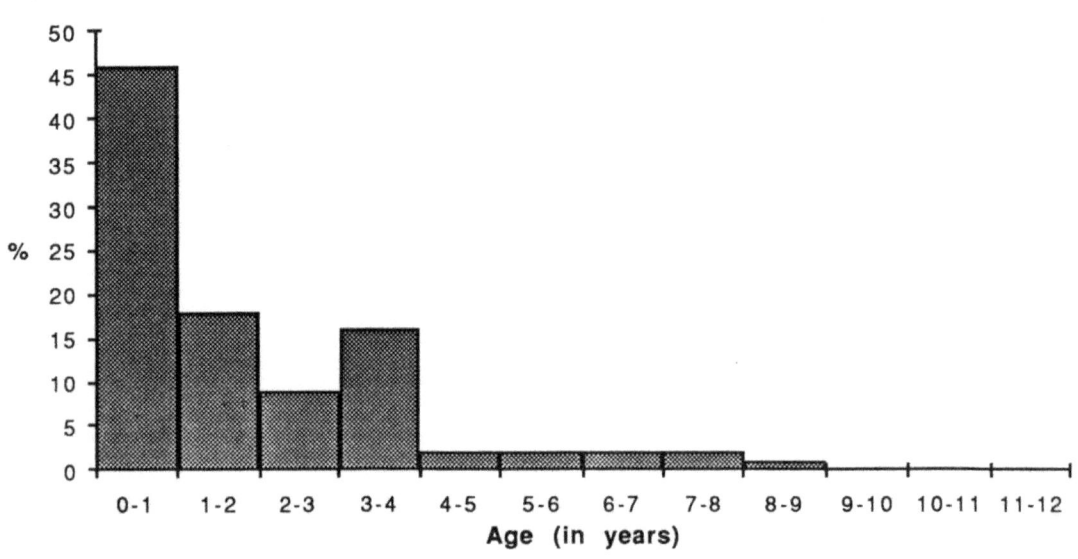

FIGURE 2 Suggested annual mortality at Grimes Graves.

(9) The number dying in their second year. This assumes only natural mortality, since, even with very low mortality the number of animals left to carry on into adulthood is already too low to be able to cull any more.

(10) The number of two year old calves remaining, assuming the mortality calculated in 9. The male:female ratio is still similar to that calculated at Grimes Graves, but there are only 16 or at the most 25 females to carry on. This would not be enough to replace the 31 cows who died at older ages. If 18 had died in their second year, as at Grimes Graves, even assuming the earlier age at first calving, there would be none left to carry on to maturity.

It is thus clear that there are not enough adult females represented in the Grimes Graves assemblage to have produced the number of immature animals recovered. Furthermore, even if we assume only natural mortality in the female population, it seems unlikely that the age structure of the mature cows found in the Grimes Graves assemblage could be that of a self-reproducing herd. Another way of looking at mortality is by simulating different herd dynamics, and then looking at the mortality patterns that would result. The analysis can be fairly simply undertaken with the aid of a computer. This approach has been attempted by Cribb (1985) and a rather different method has been developed by the present writer, and will be published with full details (Grant in preparation). Figure 1 shows a simulated mortality for a herd with a high male kill-off rate in the first and second years. The adult male:female ratio is just over 1:6, and in this simulation, an annual growth rate of a little under 1% could be maintained. Higher mortality rates in the first two years would produce negative growth rates and/or an extremely low male:female ratio. The pattern of mortality for the first four years is very similar to that seen at Grimes Graves, but the rates for these years are all somewhat lower. But the most significant and important difference lies in the adult group. At Grimes Graves, only 11% of the adults were over four years old. In the simulation, 31% were in this group.

Both approaches confirm the likelihood that the Grimes Graves mortality is not that of a complete herd. Adult animals are under-represented, or alternatively, young animals are over-represented. This is not the place to discuss the full significance of this finding, but it is clearly important to an understanding of the cattle management at this site, and has wider implications for our understanding of the social and economic functioning of the site.

BIBLIOGRAPHY

AMOROSO, E.C. AND JEWELL, P.A. (1963) The exploitation of the milk-ejection reflex by primitive peoples. In: *Man and Cattle*. (eds Mourant, A.E and Zeuner, F.E.) Royal Anthropological Institute Occasional Paper 18 pp. 126-38.

ASHBEE, P., SMITH, I.F. AND EVANS, J.G. (1979) Excavation of three long barrows near Avebury, Wiltshire. *Proceedings of the Prehistoric Society* 45 207-300.

BAKELS, C.C. (1978) *Four linearbandkeramik settlements and their environment: a paleoecological study of Sittard, Stein, Elsloo and Hienheim*. Leiden: University Press (also published as *Analecta Praehistorica Leidensia 11*).

BARKER, G. (1985) *Prehistoric farming in Europe*. Cambridge: University Press.

BARKER, G. AND WEBLEY, D. (1978) Causewayed camps and early neolithic economies in central southern England. *Proceedings of the Prehistoric Society* **44** 161-86.

BRADLEY, R. (1984) *The social foundations of prehistoric Britain.* London: Longmans.

CAULFIELD, S. (1983) The neolithic settlement of North Connaught. In: *Landscape archaeology in Ireland.* (eds Reeves-Smyth, T. and Hamond, F.) Oxford: British Archaeological Reports 116 pp. 195-215.

CHAIX, L. AND GRANT, A. (1987) A study of a prehistoric population of sheep (*Ovis aries* L.) from Kerma (Sudan). Archaeozoological and archaeological implications. *Archaeozoologica* **1** 77-92.

CLUTTON-BROCK, J. (1981) Discussion. In: *Farming practice in British prehistory.* (ed. Mercer, R.J.) Edinburgh: University Press pp. 218-20.

CRIBB, R. (1985) The analysis of ancient herding systems: an application of computer simulation in faunal studies. In: *Beyond domestication in prehistoric Europe.* (eds Barker, G. and Gamble, C.) London: Academic Press pp. 75-106.

DAHL, G. AND HJORT, A. (1983) *Having herds.* Stockholm Studies in Social Anthropology 2.

EDWARDS, K.J. (1979) Palynological and temporal inference in the context of prehistory, with special reference to the evidence from lake and peat deposits. *Journal of Archaeological Science* **6** 255-70.

EVANS, J.G. AND JONES H. (1979) The environment. In: *Mount Pleasant, Dorset: Excavations 1970-1971.* (Wainwright, G.J.) London: Society of Antiquaries pp. 190-213.

FIELD, N.H., MATTHEWS, C.L. AND SMITH, I.F. (1964) New neolithic sites in Dorset and Bedfordshire, with a note on the distribution of neolithic storage pits in Britain. *Proceedings of the Prehistoric Society* **30** 352-81.

FISHER, P.F. (1982) A review of lessivage and neolithic cultivation in southern England. *Journal of Archaeological Science* **9** 299-306.

GENT, H. (1983) Centralized storage in later prehistoric Britain. *Proceedings of the Prehistoric Society* **49** 243-67.

GRANT, A. (1978) Variation in dental attrition in mammals and its relevance to age estimation. In: *Research problems in zooarchaeology.* (eds Brothwell, D.R., Thomas, K.D. and Clutton-Brock, Juliet) London: Institute of Archaeology Occasional Publication No. 3 pp. 103-6.

GRANT, A. (1988) Animal resources. In: *The countryside of Medieval England.* (eds Astill, G. and Grant, A.) pp. 149-87.

GRANT, A. (in preparation) Livestock reconstruction from death assemblages.

GRIGSON, C. (1981) Fauna. In: The Neolithic. (Smith, A.G.) In: *The environment in British prehistory.* (eds Simmons, I. and Tooley, M.) London: Duckworth pp. 191-99.

HARRISON, R.J. (1985) The 'Policultivo Gandero', or the Secondary Products Revolution in Spanish agriculture, 5000-1000 bc. *Proceedings of the Prehistoric Society* **51** 75-102.

HIGHAM, C.F.W. (1967) Stock rearing as a cultural factor in prehistoric Europe. *Proceedings of the Prehistoric Society* **33** 84-106.

HILLMAN, G.C. (1981) Crop husbandry: evidence from macroscopic remains. In: *The environment in British prehistory.* (eds Simmons, I.G. and Tooley, M.J.) London: Duckworth pp. 183-91.

JONES, M. (1980) Carbonised cereals from Grooved Ware contexts. *Proceedings of the Prehistoric Society* **46** 61-63.

KRIGE, E.J. (1936) *The social system of the Zulus.* London: Longmans.

LEGGE, A.J. (1981a) Aspects of cattle husbandry. In: *Farming practice in British prehistory.* (ed. Mercer, R.J.) Edinburgh: University Press pp. 169-81.

LEGGE, A.J. (1981b) The agricultural economy. In: *Grimes Graves, Norfolk: excavations 1971-72, volume 1.* (Mercer, R.J.) Department of the Environment Research Report 11 London: Her Majesty's Stationery Office pp. 79-103.

LIMBREY, S. (1975) *Soil science and archaeology.* London: Academic Press.

MERCER, R.J. (1981) Introduction. In: *Farming practice in British prehistory.* (ed. Mercer, R.J.) Edinburgh: University Press pp. ix-xxvi.

PALMER, R. (1976) Interrupted ditch enclosures in Britain: the use of aerial photography for comparative studies. *Proceedings of the Prehistoric Society* **42** 161-86.

PRYOR, F. (1978) *Excavations at Fengate, Peterborough, England: the second report.* Toronto: Royal Ontario Museum Archaeology Monograph No. 5.

REYNOLDS, P. (1979) *Iron Age farm.* London: British Museum Publications Ltd.

ROWLEY-CONWY, P. (1987) The interpretation of ard marks. *Antiquity* **61** 263-66.

SHERRATT, A.G. (1981) Plough and pastoralism: aspects of the secondary products revolution. In: *Pattern of the past: studies in honour of David Clarke.* (eds Hodder, I., Isaac, G. and Hammond, N.) Cambridge: University Press pp. 261-305.

TILLEY, C. (1981) Economy and society: what relationship? In: *Economic archaeology: towards an integration of ecological and social approaches.* (eds Sheridan, A. and Bailey, G.) Oxford: British Archaeological Reports S96 pp. 131-48.

WATSON, J.P.N. (1978) The interpretation of epiphyseal fusion data. In: *Research problems in zooarchaeology.* (eds Brothwell, D.R., Thomas, K.D. and Clutton-Brock, Juliet) London: Institute of Archaeology Occasional Publication No. 3 pp. 97-101.

11

MILKING THE EVIDENCE: A REPLY TO ENTWISTLE AND GRANT

A.J. Legge

11.1 INTRODUCTION

I have been given the opportunity to comment upon the paper by Entwistle and Grant and upon the appendix to it by Grant. The central theme of the first paper is that the role of arable cultivation in the domestic economy of the British Neolithic has been overstated. They suggest that many archaeologists have too readily accepted the idea that cereal cultivation was important during the Neolithic in southern England, and suggest that cereal foods provided but a modest dietary input. The economy is seen in terms of '... a much more broadly based subsistence strategy' which, to an unstated degree, they see as directed towards the harvesting of wild plants (p. 203). The latter part of the paper questions the interpretation (Legge 1981a and b) that cattle were used as the providers of milk in the Bronze Age. I shall consider the points raised in their paper under a series of headings that bear upon the issues raised.

11.2 THE SCALE OF CULTIVATION AND THE IMPORTANCE OF CEREALS IN THE DIET

The main problem here is that the authors do not make wholly clear what they mean by 'large scale' (p. 205). If the intention is to imply that most of the land surface of southern England was not under grain crops at 3000 bc, I would agree. If, on the other hand, the implication is that the inhabitants of neolithic settlements - no matter how widely the settlements were spaced - had but a small part of their diet from arable crops, then one must pause for thought.

With regard to either the importance of cereals in the neolithic diet or to the scale of neolithic agriculture, it seems to me that most authors have been rather vague on the topic and they have usually skated speedily over the very thin data. The single reference cited by Entwistle and Grant in support of their contention is that of Mercer (1981), whose paper introduces the contents of the book in question. The 'substantial areas' which were 'broken to the plough' according to Mercer were actually inferred as restricted to certain classes of lighter soils; again, the question of the importance of cereals in the diet was not raised, but the paucity of evidence was remarked upon. The shortage of hard data has also been noted by many other writers, among whom Bradley (1978: 32) was even thrown into depression by contemplating the problem.

While it has been widely recognised that direct evidence in the form of neolithic charred plant remains is sparse, it must also be noted that routine efforts for its recovery are still, regrettably, far from being standard at excavations in Britain. At how many dry land neolithic and Bronze Age sites has large scale flotation been attempted? Even were the answer to that question satisfactory and excavation technique perfect, it is also important to consider the nature of the sites that have been dug.

Almost all of the neolithic sites that have been excavated in southern England appear to have a ritual, ceremonial or social function - barrows, causewayed camps and henges, along with the occasional cursus. The *Linearbandkeramic* sites to which they refer are domestic settlements. At a few neolithic sites in Britain some charred seeds have been recovered, while other data comes from impressions on pottery. In order to understand the meaning of these samples, consideration must first be given to the manner in which cereal crops would have been processed, and how seeds come to be charred. The important work of Hillman (1981, 1984a and b) in elucidating traditional cereal processing methods emphasises that the composition of charred seed samples allows the interpretation of function for the feature or structure from which the seeds were recovered. However, this is only straightforward where the samples are found in their primary position. When charred wastes were originally moved from their primary position and discarded during cleaning or maintenance, or moved in sedimentary processes, differing degrees of admixture will have resulted. While such samples may still be identifiable as crop processing waste, midden samples seldom reflect the composition of products or by-products which result from a single stage of crop processing, or the function of a single feature.

It follows from this that the most careful definition of context is required during excavation in order to ensure that sediments are not included from adjacent features or levels. Charred seeds must also be recovered from a wide range of context types. Those that matter for the understanding of the domestic economy were situated at or near where the processing work of the harvest was done and where the crop processing wastes would commonly be used as fuel. Samples in their primary position will be found in features such as house floors, hearths or in other well-defined structures which may also contain seeds accidentally charred due to the overheating of a desired product, such as in the familiar 'grain drying ovens'. Now, how likely is it that such features will be characteristic of ritual or ceremonial sites? Even when a fire or hearth was made in such a setting, would the weed seeds and cereal stems and chaff be available for use as fuel, or be accidentally spilt into the flames?

Yet even in causewayed camps there are charred seeds if you look for them. During Mercer's excavations of the massive Hambledon Hill-Stepleton causewayed camp complex in Dorset (Mercer 1980), I made an effort during several seasons of excavation to process as much sediment as possible by water flotation from the different features (Jones and Legge 1986, forthcoming). These were of two main sorts; ditch fills and the fills of small shallow pits (for example, Mercer 1980: Figure 14). Ditch fills had few or no charred seeds. On the other hand, of more than 80 samples from pit fills, almost all yielded charred grain (from a few seeds in some to many thousands in one case). Most is emmer wheat and barley, usually associated with fragments of hazel-nut shell which, as Entwistle and Grant note, are quite common at such neolithic sites. They also say that '... compared with many other forms of vegetable food charred cereals have a relatively high survival potential.' Of course, the survival potential of plant foods is virtually zero on dry land sites unless charred. Yet hazel-nut shell also has a very high probability of survival, exceeding that of grain. The shell is dense and woody and it can survive an intensity of charring that would destroy many other plant remains, including the seeds of cereals. It is also the case that when the nut had been eaten, the shell was waste and likely to have been discarded into the fire. On the other hand, people would always seek to limit the destruction of cereal seeds by accidental charring. Pieces of hazel-nut shell are also large enough to be found even with the trowel and by dry sieving, while cereal grains are not.

The nature of the flotation samples from Hambledon Hill-Stepleton was briefly described by Legge (1981b: 174-75) and Jones and Legge (1987), showing that there is a marked lack of weed seeds associated with the cereals. A more extensive analysis has shown that spikelet forks of emmer wheat are present in certain contexts. Emmer wheat, having a hulled grain, was traditionally processed at the farming settlement to the point where it could be stored in spikelet form, and then further processed as required on a day to day basis. I see no reason to infer other than that grain was taken to the causewayed camp in that semi-finished state or even fully de-hulled (Hillman 1981: Figures 5-7). The only process remaining that would require the use of fire was the cooking itself, as most of the processes likely to make charred residues had already taken place. The evidence of sample composition and the low counts of weed seeds is consistent with the interpretation that the primary processing site was elsewhere (Legge 1981b). As causewayed camps and henges do not appear to be residential sites, it seems improbable that their use was year-round. If used but periodically (in the case of causewayed camps), the hazel-nut shells could simply be indicative of late summer or autumn use. Were hazel-nuts ever more than a minor seasonal component of the diet, then some form of bulk storage again needs to be proposed for these. Yet it appears that hazel-nuts are more difficult to store in the long term than cereals and require to be fully ripe, dry and cool to last in an edible state beyond March (Howes 1948).

The indications that we have suggest that nuts are not a highly efficient food resource for large scale use. The point is made by Keene (1981) that nut harvests are irregular and unpredictable and this variability is further recorded for another potentially gathered food, the acorn, in Britain (Rackham 1986). Keene further suggests that nuts, in spite of their high calorific value, rank rather low in terms of efficiency due to the lengthy gathering and processing times involved in their exploitation.

The evidence for grain storage at neolithic sites in England is indeed slight and so too is that for nuts or any other plant food. This may well be due to both the nature of the sites that have been excavated and the limitations imposed by working only with material preserved by charring. The potential for macro-plant remains as an aid to understanding the food economy can only be understood by examples from elsewhere. G. Jones and Rowley-Conwy (1984) were able to examine plant remains from the Bronze Age lake village of Fiavé in Italy. At this site abundant material was preserved by waterlogging, having been dumped into the lake during the occupation of the village. Later the village was destroyed by fire and food stores were charred and then fell into the lake. The combination of excellent preservation by waterlogging and charring has shown that substantial quantities of nuts and fruits were stored as well as cereals and legumes. Concentrations were found on the lake bed of the cornelian cherry (*Cornus mas*), the apple and pear, acorns and hazel-nuts, suggesting that each food had its own place of storage within the houses. A range of other fruits was found, most of which were probably eaten, but not stored. Many of these species would usually be interpreted as being wild (such as strawberry, raspberry, blackberry etc.) but could be easily cultivated as part of the village economy. We must remember that the term 'domestic' when applied to plants is a statement about the morphology of specimens and not the actual manner of their exploitation. The presence of abundant wheat, barley, millet and pea led the authors to conclude that agriculture was the major part of the food economy. This site shows that hazel-nuts were indeed stored as food, at least in the Italian Bronze Age. However, consideration of the problem shows that under more usual circumstances, where food wastes are deliberately burned, hazel-nuts might well become over represented, as has been argued above.

Unfortunately such remarkable sites are uncommon in Britain, but when found certainly require exceptional resources to secure their proper study. Only in such cases can the range of plant foods be properly understood and the contribution of non-cereal plant foods begin to be evaluated. The limited charred samples that we have from British neolithic and Bronze Age sites do not show this degree of diversity and are normally restricted to the familiar cereals and rather uncommon peas and beans (M. Jones 1980; Legge 1981a; Murphy 1983, 1988), though Hillman (1982) has added the raspberry to the list of neolithic species from a Grooved Ware site in Wales. Legumes too are probably under represented among charred plant remains, as the processing of seeds that are held in pods does not require the use of fire, and the by-products of leaf and stem were probably used as hay rather than as fuel.

Putting a figure upon the importance of cereals and legumes in the diet, even when remains are abundant, is still problematical. We have all grappled for years with the notion that every bone is, in dietary terms, a success and something was eaten. On the other hand, every edible seed that was charred was a miniature disaster. Hazel-nut shell is an exception to this as it represents the waste of food consumption. So what is the relationship, in terms of food value, between our count of bones and charred seeds identified at any archaeological site? Some idea of the importance of arable farming can be gained from a consideration of the setting of the site, but beyond that all is supposition. Most judgements of the importance of agriculture to prehistoric communities therefore rest, to a significant degree, on data other than that of the charred seeds, such as the size and complexity of settlements or the scale of monumental construction.

11.3 THE SIGNIFICANCE OF NEOLITHIC PITS

The charring of seeds takes place both by accident and through deliberate intent, usually, in the latter case, when crop processing residues were used as fuel. Either way, the ashes were likely to be discarded into middens and, at many archaeological sites, such deposits often contain many charred seeds (Legge 1981a). Some such midden dumps might be placed in pits, whether specially dug or abandoned from some other purpose. The Bronze Age middens at Grimes Graves, in the pits left by slumped fills of the old neolithic mineshafts, are a case in point.

In the light of knowledge about charred seed samples and their archaeological context, it is quite remarkable to find that Entwistle and Grant equate the presence or absence of charred grain in pit fills with the function of the pit (p. 204). The absence of charred grains from the single example that they cite does not prove anything in relation to the reason why such pits were dug, or the purpose for which they were used. Neither does it provide any information about the amount of cereals grown or used in the Neolithic of England. In certain cases Iron Age pits have been found with the charred remnants of what was apparently a stored crop, but such residues will only be found where fire has been employed as a means of sterilising the pit (Reynolds 1974). As Reynolds cautions, most pits do not show such signs of burning and this treatment seems to have been exceptional. Most charred grain found in pits is thus in a secondary context and bears no relationship to the function of the pit.

This is illustrated by the well known find of charred grain at the late Bronze Age site of Itford Hill in Sussex (Helbaek 1957). The excavators reported the finding of 11.5lb (5.2kg) of charred barley in one pit found inside a house at the site. According to Helbaek (op. cit.), all was of hulled six-row barley and the illustrations (Plate XXII) show that some at least was charred in the ear. The find consisted of '... a cone-shaped pile in the centre of the floor of the pit...' and was found placed on the clean pit floor.

Finds like this led Helbaek (1952) to suggest that such pits were themselves associated with grain drying, though this is hardly likely in the light of subsequent work. Reynolds (1974) found that a skin of dead, sprouted grain remained adhering to the walls of pits used experimentally for grain storage and remaining fragments of such material might be charred if pits were burned to sterilise them. However, this could not explain the Itford Hill finding. The cone of charred grain was obviously poured into the pit after it had been accidentally charred elsewhere, and the pit was used for refuse disposal. Reynolds (op. cit.) also concluded that pit storage was ideally suited for seed corn, as the anaerobic conditions retained a high percentage of germination. Such pits could not be opened and closed for the periodic use of the contents as food due to the loss of the carbon dioxide which prevented germination. He suggested that cereals for daily use would be stored above ground in clay containers; certainly above ground storage is a common practice in the ethnographic record.

Entwistle and Grant argue that the possibility of above ground storage cannot be used to support the idea of extensive cereal cultivation because the four post structures known from Iron Age sites are generally absent from neolithic sites. Yet the absence of either putative storage pits or evidence for above ground storage structures from neolithic sites is hardly surprising, bearing in mind their nature. The scarcity of known neolithic settlements itself shows that agriculture (and human population) at that time cannot be directly compared with that of the Iron Age. The problem of food storage makes it worthwhile to look briefly at the amount of stored food that would be needed by a neolithic family.

One ton of wheat fills a volume of about $1.3m^3$ and one ton of barley slightly more at $1.4m^3$. As emmer was probably stored in spikelet form, the volume for this species might be nearer to $2.0m^3$ per ton. The calorific value of whole wheat flour is in the region of 330kcal/100g at 12% moisture. This means that a cubic metre of clean wheat grain represents about 2.5 million kcal. The 'standard nutritional unit' (SNU) by which human food supplies are calculated (and not only by archaeologists) is 1,000,000kcal, which represents a little under 3000kcal per adult per day for one year. This daily need could alone be provided by 0.9kg of grain. In other words, a cubic metre of stored grain would feed two people for one year if they ate nothing else, with enough left over at Reynolds' (1981) yields to supply the seed corn. The area cultivated to yield an adequate amount for a family would not need to be more than a few acres (Reynolds 1981; Rowley-Conwy 1981). Hazel-nuts, in common with most nuts, have a higher energy value, at about 671kcal/100g of the edible part. However, though the calorific value is high, storage in the shell would contribute to a considerable bulk.

Given that other foods, from animals and no doubt some plant foods collected from the wild, were also eaten, the total volume of storage would be hardly visible even had we dug neolithic domestic sites. To compare the massive *Linearbandkeramic* settlements of northern Europe (where the charred plant remains are also scant in most cases; P. Halstead, pers. comm.) with either the scant traces of neolithic settlements or the massive ritual monuments in England is indeed setting chalk against cheese.

It seems increasingly probable that even neolithic pits were used for a ritual purpose. M. Jones (1980) reported the presence of charred seeds in flotation samples (not from impressions of seeds on pot sherds as suggested by Entwistle and Grant, p. 204) from three neolithic pit groups in which Grooved Ware was found. Cereal grains were present in all cases, though in small numbers. Weeds seeds were few though hazel-nut shell, when counted simply as fragments, was very common in the Mount Farm group in Oxfordshire. The Down Farm pit group had few cereals and traces only of hazel-nut shell. Jones describes the pit fills as containing '... what appeared to be domestic refuse.'

I have examined the animal bones from the same neolithic pits that were excavated at Down Farm in Dorset (Barrett *et al.* 1981; Legge, forthcoming) and it is evident that these are not simply dumps of domestic refuse; almost all of the bone is in large, identifiable lumps (even though some is dog gnawed) with a very small proportion of unidentifiable fragments, and the assemblages also show an evident selection of skulls and jaws. The other pit contents also have the appearance of purposeful arrangement (Brown, forthcoming). Whatever such pit groups were intended to mean in symbolic terms, it would stretch credibility to see their contents as a direct reflection of the prevailing domestic economy. If this is so for the animal bones, then it is so for the charred seeds. It might be thought that the charred seeds were derived from the firing of grain storage pits, which were then later used for the placement of deliberate deposits of bone, antler and pottery. However, the range of plant species found in the Down Farm examples and other Grooved Ware pits (including hazel-nut shell and apple pips; Jones op. cit.: Table 1) makes the charring of grain storage detritus less likely than that the charred seeds were simply adventitious, scattered in the sediments that fill the pits and present simply as the result of burning household food wastes. In that case the seeds relate neither to the pit function (were such pits ever dug during the Neolithic for storage) nor to their ritual use. Indeed, Hillman (1982) pointed out that these finds did not even demonstrate local cultivation of cereals, as the range of by-products associated with crop processing was absent from the pits.

Yet the findings of hazel-nut shell and crab apple at two of the sites, combined with the 'high occurrence of pig and deer among contemporary bone samples' led Jones to remark upon the '... exploitation of woodland resources by Grooved Ware cultures', the statement that appears to have stimulated the paper by Entwistle and Grant. Putting aside the fact that hazel-nuts and crab apples were as likely to be cultivated as wild in the Neolithic, even the contemporary bone remains can be seen in other ways. Firstly red deer, considered as a forest resource, are not especially common in the bone samples. In the Down Farm pits they make up 9.5% of the identified limb bones; if jaws are considered as well they fall to about 5%. Such low proportions of wild mammals are not high when compared with those from neolithic domestic sites in other countries in Europe (Higham 1968; Legge in press). Secondly, the relative abundance of pigs (and possibly deer too) in Grooved Ware sites is much more likely to be the function of social rather than environmental forces. As Rackham (1986: 122) notes '... many archaeologists unthinkingly equate pigs with woodland.' That observation refers to domestic pigs in the light of the fact that the acorn and beechmast harvest in the British climate is irregular and appears only about one year in three. Pigs in the Down Farm pits are on average less common than cattle, but are rather better represented by skulls and jaws. The evidence for the exploitation of forest resources is in fact rather slight at both Grooved Ware and other British neolithic sites; yet again, the nature of such sites is such that neither the bones nor charred seeds are likely to represent all aspects of the agricultural economy.

11.4 THE IMPORTANCE OF WILD PLANT SPECIES IN THE DIET

Most peoples of the world consume carbohydrate as the essential provider of energy, whether they are hunter-gatherers or farmers. A few highly specialised groups such as the Inuit traditionally used fats, such as seal blubber, instead of carbohydrates. Even nomadic herders supplement the animal fats in their diet by cultivation, or by exchanging animal products for grain with settled farmers. The weight of ethnographic evidence shows that plants provide the staple energy among peoples in temperate latitudes. If Entwistle and Grant are to suggest that the neolithic economy was more 'broadly based' than on arable cultivation (and they wish to dismiss the use of fats obtained from milk), then the source of the dietary carbohydrate can only have been plant gathering. The evidence for this is slight.

It is well known that plant remains will become charred only when exposed to fire, and the likelihood of this happening will vary from species to species. Even in pre-agricultural sites where plant use is commonly inferred, charred seeds or other plant tissues are often rather sparse (Legge 1986). I suspect that at many sites this is due to poor preservation rather than the failure to encounter the processing area, or that few plant remains were ever charred. On the other hand, at certain late palaeolithic and mesolithic sites in other regions the charred plant remains can be of staggering quantity and variety (Hillman 1975, 1989; Hillman *et al.* 1989).

At waterlogged sites, the situation is different. But the problem there is that you get everything. How many of the potential food plants at Star Carr were actually eaten? Ethnography shows that some of the species found in the peat and mud at that site were indeed used as food in the historic past (Clark 1952, 1954), though in some species only after detoxification (Stahl 1989). Were rhizomes of *Phragmites* there because they were collected as food, or did they just grow? This is always the main problem in the reconstruction of diet where there are no particular indications of the storage, processing or consumption of potential wild plant foods. Any speculation on their use as food also brings to mind investigations on the diet of the Tollund and Grauballe men (Helbaek 1950, 1958); here at least was direct evidence of consumption. In consequence, archaeological thinking in the 1950s and 1960s had the Iron Age citizens of Denmark creeping over the ground harvesting the varied and sometimes toxic vegetation and seeds found in the guts of the two unfortunates. A much more economical explanation for the gut contents was again provided by Hillman (1981, 1986), who sees the food as the wastes of grain cleaning, and the diversity of species as that of the weed infestation of the fields. Such crop residues were traditionally stored for reprocessing in an emergency and it seems to be this waste that was fed to the bog men; the diet, perhaps, of dishonour.

We all know that studies of the energy capture among hunter-gatherers (in terms of the energy return of calories gained for calories expended in work) tend to show an advantageous return in the order of 1:10, and in consequence the idea of an indolent affluence among such peoples is widespread, though such calculations do not always take account of food processing time (Cane 1989). At first sight it would seem that non-mechanised farmers do not do vastly better with a simple technology and manage from about 1:14 to 1:60 (Leach 1976). The point is that they manage with a very much smaller area, so that there tends to be many more people about. Food collecting is extensive and commonly involves considerably more than 10km^2 of foraging area per individual (Clark and Haswell 1970; Lee and DeVore 1968). Land use estimates of about 3km^2 per individual are given even for sedentary hunter-gatherers in very diverse and productive environments (Watanabe 1972).

On the other hand cultivation is intensive and can be seen in terms of hectares per family (Clark and Haswell 1970). After all, the point of cultivation is simply to produce a monoculture of the largest amount of the desired crop in the smallest area and to thus maximise energy capture and minimise the energy cost of its production. Gathering, by its nature, is the exploitation of more dispersed resources. Although these can sometimes equal the energy return of non-mechanised agriculture, especially where nuts are exploited (Russell 1988), the diversity of vegetation ensures that such opportunities are local and restricted, and in consequence foraging imposes great demands of time and land area. The result of large scale plant food gathering in the Neolithic would be a dispersed population of low density, with a considerable degree of mobility. Yet the archaeological evidence argues otherwise. The lack of adequate samples relating to economic practice needs to be seen in the light of the abundant evidence for the complexity of social patterns that are manifest in the standing monuments of the period.

The conclusion is obvious. While it is easy to postulate that wild plant foods were important, evidence for the use of such foods in the archaeological record of Britain and northern Europe is very limited. Any consideration of the importance of wild plant foods can only be sustained by actual data, or by ethnographic example. The modest achievements of mesolithic peoples in Britain are a reflection of an economy based upon hunting and gathering. The major achievements of the neolithic peoples of Britain were not.

11.5 THE MANNER OF CULTIVATION

The authors also question the evidence for ard marks as a sign of agricultural technology, though noting the examples found beneath the South Street long barrow as at least pre-dating the mound by some interval. It is worth remembering that Reynolds (1981) describes the manner of using an ard, and found that furrows were shallow and the subsoil was seldom marked. Even if the preserved ard-marked surfaces at the base of shallow soil profiles beneath barrows are not the product of agricultural practice, it is reasonable to presume that the ard itself was the product of agricultural practice; the use of an ard to prepare barrow sites is at variance with the suggestion that:

> '... a model of transient hoe based horticulture might fit the evidence better than one positing settled agriculture. This scale of practice and level of technology are more in keeping with other aspects of the archaeological record of communities living in Britain and the Atlantic fringe of mainland Europe.' (p. 208)

What would the builders of Stonehenge, Avebury and Carnac think of that? What are the aspects of the archaeological record with which that statement is in keeping? The European parallels for shifting cultivation are few, and the soils appear to be able to sustain an adequate yield of cereals even in the absence of fertiliser or of crop rotation (Reynolds 1981; Rowley-Conwy 1981). Indeed, Rowley-Conwy (op. cit.) argued that the historical examples of slash-and-burn cultivation in Europe were of very local extent and restricted to the occasional use of outfields by permanently settled farmers in situations where cultivation was, to say the least, highly marginal.

11.6 OTHER ENVIRONMENTAL CONSIDERATIONS

One other aspect of environmental archaeology - that of the pollen evidence - is also left out of consideration by Entwistle and Grant. While the limitations of pollen analysis with regard to the actual representation of vegetation in the landscape is perhaps now more widely recognised, the marked changes seen in the pollen record from *circa* 3500 bc do need to be explained. The probability of mesolithic interference with the natural landscape is well established in our thinking (Simmons and Innes 1987; Bush 1988) and, in the light of the impact of hunter-gatherers on the landscape in other parts of the world, some evidence for pre-neolithic forest disturbance is not surprising.

On the other hand, we must remember that England is a wet country, with mainly deciduous tree species and these do not burn until felled and dried (Rackham 1986, 1987). Forest clearance, whether for the propagation of desired wild species or for the creation of agricultural soils, is hard work. It is likely that sustained browsing by domestic livestock had a marked effect upon the forest cover but however clearance was achieved, the degree of change speaks for a human population (and livestock) on a scale capable of causing significant impact. Is it possible to see the large scale forest clearance of the pollen record as the product of an economy in which plant collecting produced the major dietary input? The relative paucity of cereal pollen in pollen profiles can also be explained by the fact that the pollen is trapped within the glumes of self-fertile species of cereals, to be released only where the crop was threshed (Robinson and Hubbard 1977). It is also improbable that the shifting, small plot horticulture suggested by Entwistle and Grant would be registered in the pollen record on the scale at which the clearance is manifest (Rowley-Conwy 1981).

11.7 ECONOMY AND SOCIETY

Early in Entwistle and Grant's paper there is a bow towards the social aspects of archaeology, with a quote from the work of Tilley, which could simply be reversed to read 'non-economic activities can only be explained by the economic', which is but another, if older, point of view. Yet in spite of the fact that the authors begin and end their paper with cautionary observations about the influence of social values on economic practice, it remains the case that their interpretation of the data is almost wholly economic; the fact that virtually all neolithic sites from which evidence can be drawn apparently have some form of social function and none has direct evidence of domestic activities, is ignored. Perhaps it would be best to recognise that either sphere of human activity can best be explained by some understanding of both.

The notable characteristic about the British Neolithic was its concern with great works. The classes of site listed above (barrows, causewayed camps, cursuses and henges) all speak of a population with time and labour to spare, whether given by communal agreement or due to central direction. I find it very hard to believe that all this is a product of a sort of nuts-and-berries neolithic. It is perfectly evident that even among hunter-gatherers the elaboration of culture is associated with food storage and the concentration of population that this practice allows. However the neolithic farmers organised their economy and, in response to whatever social forces, their public works do imply that their economy could generate adequate stored food, and that the neolithic population existed at a much greater density than is found among hunter-gatherers. Even the limited evidence that we have inclines me to believe that this population could only be supported by cereal based agriculture and not by a more generalised collecting and gathering economy.

11.8 NEOLITHIC-BRONZE AGE CATTLE AS MILKING ANIMALS?

When it comes to the question of the exploitation of cattle for milk (as the major, but by no means the only output; Legge 1981a: 98-100), both Entwistle and Grant, and Grant present extensive criticisms of the work of Legge (1981a and b), in which the hypothesis was advanced that cattle were indeed exploited in this manner at the Bronze Age settlement of Grimes Graves and possibly at other prehistoric sites too. Their objections rest upon two main counts. Firstly, they regard this as impossible in the light of the behaviour of primitive cattle. Secondly, they argue, the herd structure shows that the herd could not have been self-sustaining.

The issue of cattle behaviour and the tolerance that prehistoric cattle might have for milking rehearses old arguments. These questions were raised in part by Clutton-Brock (1981), who also advanced the idea that cattle will not let down their milk in the absence of the calf. This suggestion is based upon the paper by Amoroso and Jewell (1963) in which methods among African cattle herders for the initiation of milk let-down were discussed. It is necessary to look in some detail at what was actually said in that paper. The authors make clear that the response of the cow to the 'let-down' of milk is a conditioned reflex; the presence of the calf is commonly the stimulus that initiates the reflex. They make clear that in modern cattle, a lesser stimulus will normally suffice; for example, the sounds associated with milking, or manipulation of the udder. Yet they go on to say:

> 'It is the ability to condition this response [i.e. the let-down of milk] which, since the most ancient times, farmers have exploited, and they have conditioned their animals to a variety of visual and auditory stimuli other than those that arise directly from the presence of the calf.' (Amoroso and Jewell op. cit.: 129, note in brackets inserted)

They describe these artificial stimuli as:

(1) feeding the cow while milking;
(2) the use of calf models or dummies during milking;
(3) the use of a stuffed dead calf skin during milking;
(4) use of a substitute living calf, smeared with the milk cow's urine;
(5) to allow the calf all of the milk for a period of some days or weeks before it is removed or killed;
(6) to use a small boy dressed in the skin of the dead calf; in this case, the cow can become habituated to the presence of the boy at milking;
(7) stimulation of the vulva by inflation or other means.

None of these methods would be beyond the wit or technology of a neolithic or Bronze Age community. Entwistle and Grant cite a medieval illustration of a cow being milked in the presence of a calf, and several such illustrations accompany the paper by Amoroso and Jewell, including one example of a cow milking in the absence of the calf but while being fed as the alternative stimulus. Of course, there is also no way of knowing whether or not the calf depicted in those illustrations was the progeny of the cow being milked. Other instances are cited in Legge (1981a, b and c) of early English husbandry manuals in which the killing or removal of the calf is recommended on the grounds that the milk is the more profitable output. These add up to show that cows can be, and were, milked in the absence of their calves. This can be accomplished by simple trickery, or by habituation.

In her Appendix, Grant emphasises the importance of ethnographic evidence in the understanding of prehistoric husbandry. Yet this idea is perhaps less novel than it may seem. The reader will find examples of this approach in Legge (1981a), but with a reminder of the caution required in transferring ethnographic examples between continents. I have long felt that the application of data derived from African cattle husbandry to northern European circumstances may be unwise. This is again discussed elsewhere (Legge 1981c: 221-22), especially in the light of herd structure; as many sources make evident, calves are carried on in many African herding societies as an insurance against the uncertainties of drought and the inevitable effect of that upon milk yield (see Dahl and Hjort 1976 for a review of this). Equally, extracting mortality expectations from the review by Dahl and Hjort of African circumstances may not be the best foundation for extrapolation into prehistoric Britain. Some farmers here still maintain dairy herds in a traditional way, using no bought-in prepared feeds, raising their own female calves, and keeping cattle outdoors all year. Information that I have been given suggests that such herds are characterised by considerably higher fertility rates and lower mortality rates than is found among the African examples cited by Grant. African cattle husbandry is characteristic of areas too arid for cultivation, while pasture in Britain is more usually found in regions too wet for effective cultivation, or on soils with impeded drainage or of a very intractable kind. The unsatisfactory nature of any direct parallel between these very different areas is obvious.

The question of the 'primitive' nature of prehistoric cattle can also usefully be considered. By the mid-late Bronze Age, cattle had been domesticated in Britain for about 2500 radiocarbon years, which is as long as the time that has lapsed since. Bronze Age cattle skulls from Down Farm (Legge forthcoming a) are small and show a considerable variability, with one having very reduced horns and a markedly 'peaked' intercornual ridge. A similar specimen is described by Jackson (1957) from the late Bronze Age site of Itford Hill in Sussex. Others from Down Farm and from Bronze Age Grimes Graves are rather different. The intercornual ridge shows a bow shape and the cattle there are quite uniform in that respect. The small size of Bronze Age cattle and the beginnings of evidence as to their diversity of form implies that not only had selection been practised upon the animals over a very long period, but also that their form was very much changed from the wild prototype. What is 'primitive' in this respect? It is only quite recently that cattle have been purposefully selected according to herd books (and often for points of bodily conformation that seem unrelated to performance), lived in automated houses and been fed artificial diets. At what point did they become advanced? As with so many domestic plants and animals, the variation that is shown, serving many different human needs and modes of exploitation, was achieved before recorded history began and agricultural advancement is not only the product of 20th century experimentation. In this sense, the term 'primitive' is more of a value judgement than an observed or known fact.

The question of lactose intolerance is again well rehearsed and the literature is too extensive to enter into here. This issue was also raised by Clutton-Brock (1981), who pointed out that little lactose remains in processed milk products. It is the general practice among herders in Africa, Europe and the Near East to consume milk soured, fermented, or processed into a wide variety of products that can be stored and, in consequence, traded (Martin 1980; Russell 1988; Ryder 1983), so that lactose intolerance is a non-issue. Gallagher *et al.* (1974) found that even those who are deficient in the enzyme lactase exhibited no symptoms of intolerance to lactose when eating fermented dairy products. It is also of particular interest to see that a find of butter from the Bronze Age has been reported from the underwater excavations by Nicholas Dixon at the site of Loch Tay in Scotland (*The Independent* 17/11/88) though, as with charred grain, sparse finds speak little of the frequency of use. In a recent review of the literature, Russell (1988: 29-32) thought that the problem of lactose intolerance was '... overly stressed and misunderstood.'

Grant also claims that the population structure model she proposes shows that the Bronze Age cattle herd at Grimes Graves could not be self-sustaining. Earlier excavations by Manning (1866), Greenwell (1870) and in this century by Armstrong (Legge forthcoming b), also found Bronze Age middens that had high proportions of juvenile cattle; Greenwell even inferred that milk was the desired output from the cattle at Grimes Graves. The pattern found in the bones from the Mercer excavation (Legge 1981a) has been duplicated from those found in a second midden (Legge forthcoming b) excavated by the British Museum. In my first study, only about one quarter of the cattle showed full dental maturity at death, judged on the basis of the third permanent molar coming into wear on the third cusp. The proportion from the British Museum excavation (based on a further 154 mandibles) is virtually identical. The slaughter pattern in both samples is also characterised by a very high incidence of infantile deaths. The earlier sample also indicated a female:male ratio among the adults of about 6:1. From the British Museum excavations there are ratios of 8:0 on the metatarsal and 5:1 on the metacarpal (Figure 1). Combining all the samples that I have from Bronze Age Grimes Graves suggests that about 6:1 does seem to be about the correct representation of the sexes among the more adult cattle.

The striking characteristic of the slaughter pattern is the high proportion of very young animals that were killed or died within the first month or two. Again, the two populations are very similar in this part of the slaughter pattern. While some proportion of the cattle were dead quite soon after birth, the larger proportion died at a few weeks of age (see point 5 in the discussion of Amoroso and Jewell's paper above). After this, and to about six months, the mortality is low and this falls still lower after six months. In both instances the sheep do not show high juvenile mortality, which provides some support for the suggestion that high mortality is not a reflection of incompetent husbandry. In the original publication (Legge 1981a) I used the method devised by Higham (1967) for recording the tooth eruption and wear of cattle mandibles. Without in any way diminishing Higham's pioneering studies of animal husbandry, the recording system suffers the limitation of being unable to estimate age beyond the appearance of wear on all cusps of M_3. Work in hand is directed towards producing a system that will allow a closer estimate of age throughout the animal's life (Beasley, Brown and Legge forthcoming). From Grimes Graves the pattern that emerges from the analysis of two large and well recovered samples agree more closely than might reasonably be expected of archaeological data.

In her paper, Grant allows assumed mortalities of 10% among the female calves and a calving efficiency of 80%. As she notes, these can at best be estimates and are largely based upon African examples. Taking a standing herd of 100 animals and with the first calving taking place at two years of age, the actual deficit of calves is 2. If the mortality was lower and the calving efficiency higher (both well within the realm of possibility) even the deficit of female calves at a proposed calving age of three years is relatively small. Yet there is one other aspect of the husbandry pattern that needs to be considered. Due to the limitations of the Higham system of tooth eruption recording, the ages of the older cows cannot be estimated from Legge (1981a) and these ages have again been assumed by Grant. Such ageing has been done in the second Grimes Graves study (Legge forthcoming b). In both samples of cattle mandibles from Grimes Graves, most of the females were somewhat older at death than Grant estimates (p. 210 point 4). It seems likely that the cattle were managed by retaining females to quite advanced ages

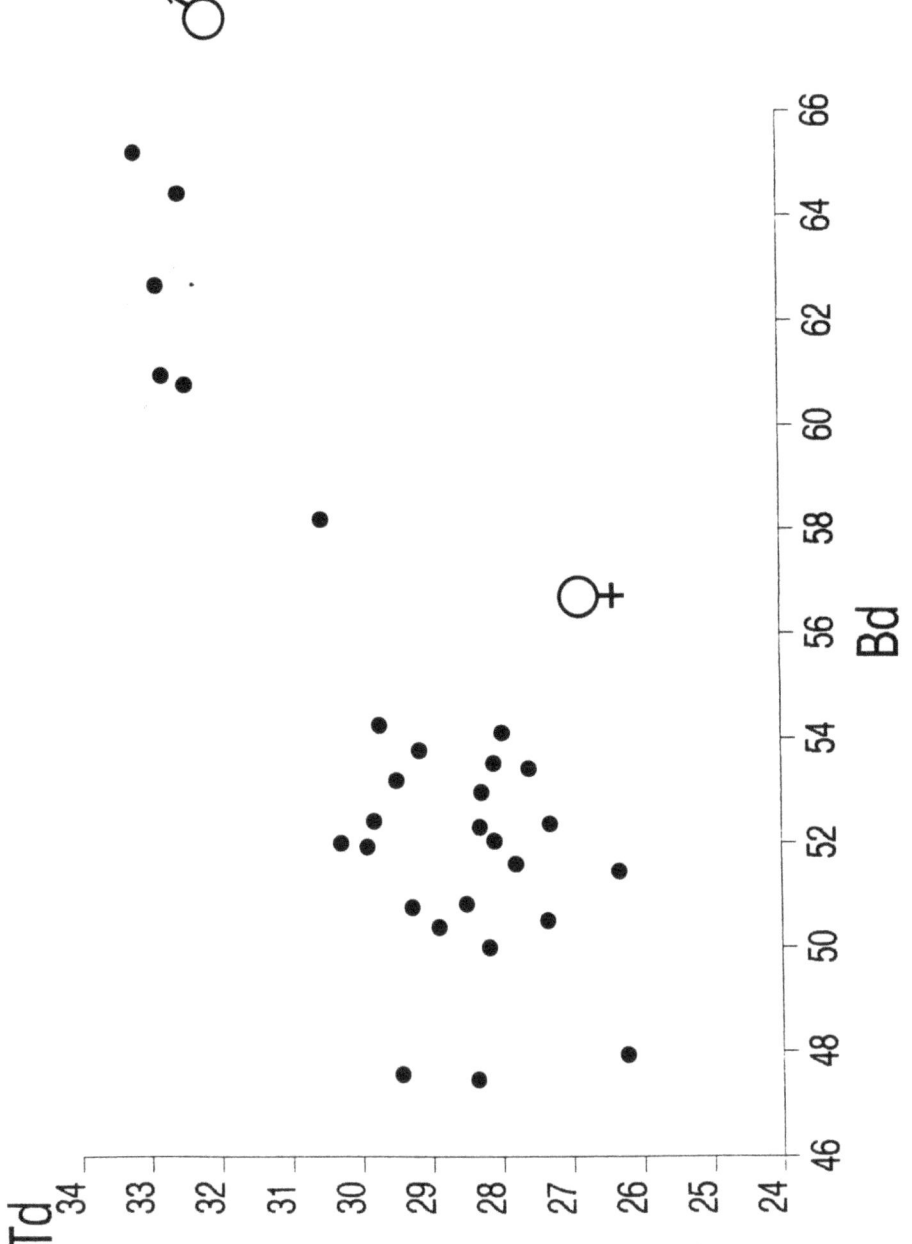

FIGURE 1 Distal metacarpal dimensions of cattle from three Bronze Age middens at Grimes Graves. The suggested division into females and males is the author's interpretation.

and with a relatively low rate of replacement. In this manner, the small proportion of adult females would each have contributed more calves to the midden before their own death.

Entwistle and Grant also argue that a predominance of female cattle at archaeological sites '... is not in any way surprising.' This is indeed so, as long as it is accepted that the herder was content with a small meat output from the cull of very young animals and that other products were of importance. While most sites in Britain appear to show a majority of adult females, sites are found elsewhere in which many males were carried on to greater ages. This can be determined from measurements of the fused distal metacarpal, as in cattle this bone seems to reflect sexual dimorphism to the greatest degree. While we now recognise that certain bones grow after fusion has taken place, (Legge and Rowley-Conwy 1988: Figure 20), evidence for the degree of such growth is still limited. However, it is evident that the earlier a bone fuses the greater will be its post-fusion growth. The metacarpal fuses at 2-2.5 years, which is relatively late in the growth period of young cattle. Yet by that time males are significantly larger than females. This is confirmed in studies of modern cattle metacarpals where the width of the distal articular surface (measurement Bd) quite satisfactorily separates females and steers (Jewell 1963; Higham 1968; Legge 1981b, Figure 4g and f). It is also relevant to note that these metacarpal samples were the products of the meat trade of recent times. In all but the most intensive systems of meat production steers are killed at ages of 30 months and upwards (Taylor 1974), a time by which most bones, including the metacarpal, would be fused (Habermehl 1975; Silver 1969).

The point of my argument was that in some archaeological sites there is evidence that males were retained to effective meat weight. Where such males were retained for several years the incidence of juvenile slaughter is low. In sites with few older males, the incidence of juvenile slaughter was high (Legge 1981a: 89 Table XIV). I have argued that meat production was given greater emphasis at the site of Moncin in Spain (Legge 1987) and in the neolithic site of Selevac in Serbia, Yugoslavia (Legge in press), where only 35% of the cattle were killed before 24-30 months (determined from tooth eruption) and in which 30% of the cattle having fused metacarpals were male (determined from bone measurement). This is an almost identical pattern to those found by Higham (1961, 1969) at Bronze Age sites in Switzerland. The evident difference between the two forms of economy was extensively discussed in Legge (1981a Figure 52 and Table XIV) and the conclusion was presented that two different methods of exploitation were reflected by these patterns. In one group of sites milk was a more important yield and in the other meat production was more important. Both groups would have had both milk and meat, but gave a different emphasis to the production of each. For Entwistle and Grant (p. 208) to cite Grigson (1981) as suggesting that 'cattle in the neolithic may have been exploited to provide a range of products' seems to imply that I had missed that point; Legge (1981a) offered the interpretation that there was a difference in the degree to which the potential outputs were emphasised, not that such outputs were exclusive. Even the argument that the retained males were steers kept as work oxen doesn't work as an explanation for the proportion of retained males in the above examples; from the slaughter patterns there were simply too few animals killed at more advanced ages for these to be both old cows and worn out work oxen (Legge 1981a: 87).

The published metacarpal and metatarsal measurements from Danebury show a ratio of about 3 females to each male (Grant 1984 Fiche 17; this paper Figures 2 and 3). It was suggested that few animals were killed at '... the optimum age for meat yield...', though that age is not defined. The higher proportion of adult males was explained as due to the retention of oxen for traction. It is at least as likely that most of these were steers retained for meat production. The constraints imposed by the densely settled and farmed Iron Age landscape would limit meat production and, of course, encourage the use of milk.

Entwistle and Grant further argue the predominance of adult females among domestic animals by reference to the pig. This species, they say, shows a majority of adult females among the bone remains because '... the males would have been killed off for meat before maturity...' Yet it is doubtful if the two species can be so directly compared. The large litters produced by pigs, and their rapid growth potential means that most young animals of both sexes would be eaten early, usually well under one year of age and before most bones were fused. In cattle the most efficient slaughter age for meat results in at least a partly fused skeleton, while in pigs it does not. Putting aside the difficulties of determining the sex from pig bones by measurements, the whole problem comes down to the bone that is considered and the extent to which the species under consideration shows sexual dimorphism. The example cited by Entwistle and Grant is simply not comparable and is thus irrelevant to the discussion.

It has been shown that in other parts of Europe examples can be found where meat production is seen as the main output from the herd. The fact that most fused limb bones of cattle from sites in Britain prove to be mainly of females raises an important question as to why this should be so. If meat yield or traction needs from steers were the paramount considerations in husbandry, why are there so few steers at British prehistoric sites? This contrasts sharply with practice in later urban communities; here the demands of an organised meat trade does indeed seem to have required the retention of more males beyond the age of metacarpal fusion and this is reflected in the published measurements (e.g. Maltby 1979: Figure 5).

The question of how the 'major output' from a cattle herd is to be defined could be fruitfully discussed at length. Put simply, this might be seen in terms of relative energy values of the food outputs. Alternatively, the willingness of consumers to pay higher prices for a given unit of food energy as meat or fat than for starch is widespread and this was probably so in prehistoric times as well. This is evident in the economies of many peoples who exchange milk products for grain. The Gavli nomads of the Western Ghats in India exchange milk products for grain from settled farmers at the rate of 1kg of butter for between 12kg and 20kg of grain (Malhotra and Gadgil 1988), which represents an advantageous gain in terms of kilocalories of between 6:1 and 10:1.

It is interesting to note that in Britain now this relationship is blurred due to artificial manipulation of costs and by the massive subsidies of fossil energy in the food producing system. A given amount of food energy derived from bread or butter costs about the same, at about £0.30-0.32/1000kcal. On the other hand, hard cheeses of the cheddar type are much more expensive at about £1.05/1000kcal. Meat is more variable in price and energy value depending upon which species and body part is considered, but typically 1000kcal of energy would cost about £1.5-3.5 (energy values from Documenta Geigy 1959, costs at average shop prices). Butter is obviously under priced in Britain and doubtless to the cost of the nation's health.

In earlier times consumers were perhaps less aware of the health risks, but even now there is none the less a strong desire for fats and oils in the human diet. We perhaps need to remind ourselves that the year round variety in our own food supplies is provided by energy intensive methods of preservation and the use of extensive air freight. Traditional diets, both in Britain and elsewhere, consisted of a much more limited range of foods according to the season of the year. This is most marked in regions where summer and winter, or wet season and dry season, exhibit pronounced differences. The diet in such circumstances can indeed be found to be bland and monotonous for much of the year (Casimir 1988). There is then a very important role for animal fats in achieving both dietary balance and enhanced palatability of a limited range of farinaceous staples. It is therefore very easy to surmise every incentive for prehistoric populations to exploit their animals for fat at least as much as the meat that we more usually anticipate. Milk is, of course, a ready supply of that fat with the added advantage that it can be easily converted for long term storage with a simple technology, and does not require the slaughtering of the adult animal.

In quite recent times it has been thought desirable that up to 35% of energy intake should be in the form of fats (Geigy 1959), most of which is contributed by dairy products and meat. Even on high protein diets people develop a craving for fat where this is deficient in their diet (Speth and Spielmann 1983; Speth 1987; Noli and Avery 1988). The calorific values of the 'lean' meat of modern domestic mammals (about 220kcal/100g for beef and mutton, 290kcal/100g for pork) contrast markedly with the 130kcal/100g that is typical of wild meat such as hare or venison and show that a combination of husbandry and selection have combined to enhance fat deposition within the muscle. It is commonly suggested that the unimproved domestic animals of the prehistoric period would have been leaner than their modern equivalents and it follows, if this was indeed so, that fat supplies from slaughtered animals would have been very limited. It is this desire for fat that has resulted in the very fragmented states in which we find archaeological animal bone collections, where they have been systematically broken for its removal (Binford 1978: Table 1.6).

Entwistle and Grant (p. 206) suggest that '... the productivity in a neolithic 'dairy herd' seems unlikely to have supplied enough of the animal protein requirements of a community to justify a specialised management for milk production.' This statement raises two important points. Firstly, even a scant reference to the ethnographic data and a modest familiarity with modern practice shows that it is not protein but fat that matters in dairy outputs. The familiar methods of milk processing into butter and cheese produce popular foods because they extract and concentrate the milk fat. Secondly, a diverse diet of plant foods contains a good deal of protein; emmer wheat is particularly rich in protein, containing about 19-20% (Peter Reynolds pers. comm.). The human daily protein need has been suggested as 50-100g (Geigy 1959), much of which can readily be obtained from plant foods. Only a small proportion of this needs to be obtained from animal protein, mainly to supply the essential amino acids in which plant proteins are lacking.

It is obvious that in herds and flocks which are kept for milk production there are very low margins in excess livestock over needs and that is, of course, the whole point of such husbandry systems. They are designed to minimise the number of livestock and to maximise output; Legge (1981a: Table XV) showed that exploiting cattle for milk was 3-5 times as efficient as for meat, whether considered simply as kilocalories of food energy or as protein. In the Near East and such as we know of European traditional herding, most males were killed when very young. When herders are confronted with increasing markets for meat in urban centres, which in turn have a wider range of exotic products available for exchange, pressure will exist to retain more young and the herders would have to decide what proportion of milk could be given up in relation to the value of the retained offspring that will be raised on the milk (Barth 1961; K.J. Legge 1989).

It could be argued that the degree of husbandry specialisation is a reflection of the degree of self-sufficiency. When the self-contained subsistence needs of the group are paramount the greater efficiency in food will be the goal (as milk), but with rising market forces this priority will change towards obtaining the greatest income. In consequence we find among herders now every form of economy between the raising of very few young where milk is emphasised through to the raising of all young where meat production is the major priority. Many economies are somewhere in the middle of this possible range of choice and the bones from such a pattern would be less easy to interpret. It is my argument that Grimes Graves and other sites represent the most specialised 'dairy' end of a broad pattern of variation; the interpretation of such extreme examples is where we must begin.

At the Iron Age site of Danebury, Grant encountered a cattle bone sample in which there was a high mortality among juveniles (Grant 1984). Of the 64 cattle mandibles in the late (b) phase, some 19% were dead with an unworn or lightly worn third milk tooth and a further 17% by the time that the first permanent molar was in early eruption (calculated from Grant 1984: Figure 9.22a). Although this is a slightly lower rate of death or cull than I have found at Grimes Graves, the two slaughter patterns are quite similar. In Figure 9.22b (Grant 1984) the proportion of young is rather lower when the broken mandibles were attributed to probable wear stages. However, it is important to consider the probability of different degrees of attrition during the taphonomic processes at the two sites; while mandibles were scattered in many features at Danebury, those at Grimes Graves were abandoned into dense middens that underwent rapid sedimentation. In consequence even the most juvenile mandibles were well preserved at Grimes Graves and most can be fitted to precise eruption and wear states. At Danebury the proportion of broken mandibles from older cattle (Grant 1984: Figure 9.22b) suggests that juvenile jaws may therefore be under represented at that site, especially as Grant (1984: 496-97) notes a considerable loss of loose teeth as the spoil was not sieved.

Grant's explanation (1984) of the rather high mortality among infant cattle at Danebury in the late (b) phase was that calves were born at or near the hillfort. This implies that only natural mortality was the cause of infantile and juvenile deaths. The relatively low proportions of animals killed at the 'optimum age for meat yield' was further argued as showing that meat production was not an important consideration. The slaughter pattern is interpreted as the product of a herd in which the adult cattle were mainly cows and work oxen, retained to provide traction. The need for work animals is seen as a major influence upon herd structure. The female:male ratio was about 3:1 (Figures 2 and 3), which is a somewhat higher proportion of adult males than characterises either ceremonial neolithic sites (Legge 1981b), or the Bronze Age settlement at Grimes Graves. It has been noted above that cows too can be used for traction and an example from the paintings of the tomb of Sennedjem at Thebes in Egypt (Baines and Malek 1980: 190) shows just such a scene.

The use of cows for traction has also been tested experimentally (Reynolds 1981: Plate 3) with the use of an ard. In Serbia during the early 1970s pairs of dairy cows yoked to carts were being used for the transportation of the crops from the fields to the village. The farmers told me that such work did indeed reduce milk yield and milking was the primary reason for keeping the cows. However, their view was that the amount of milk lost was a lesser cost to their economy than that of keeping oxen.

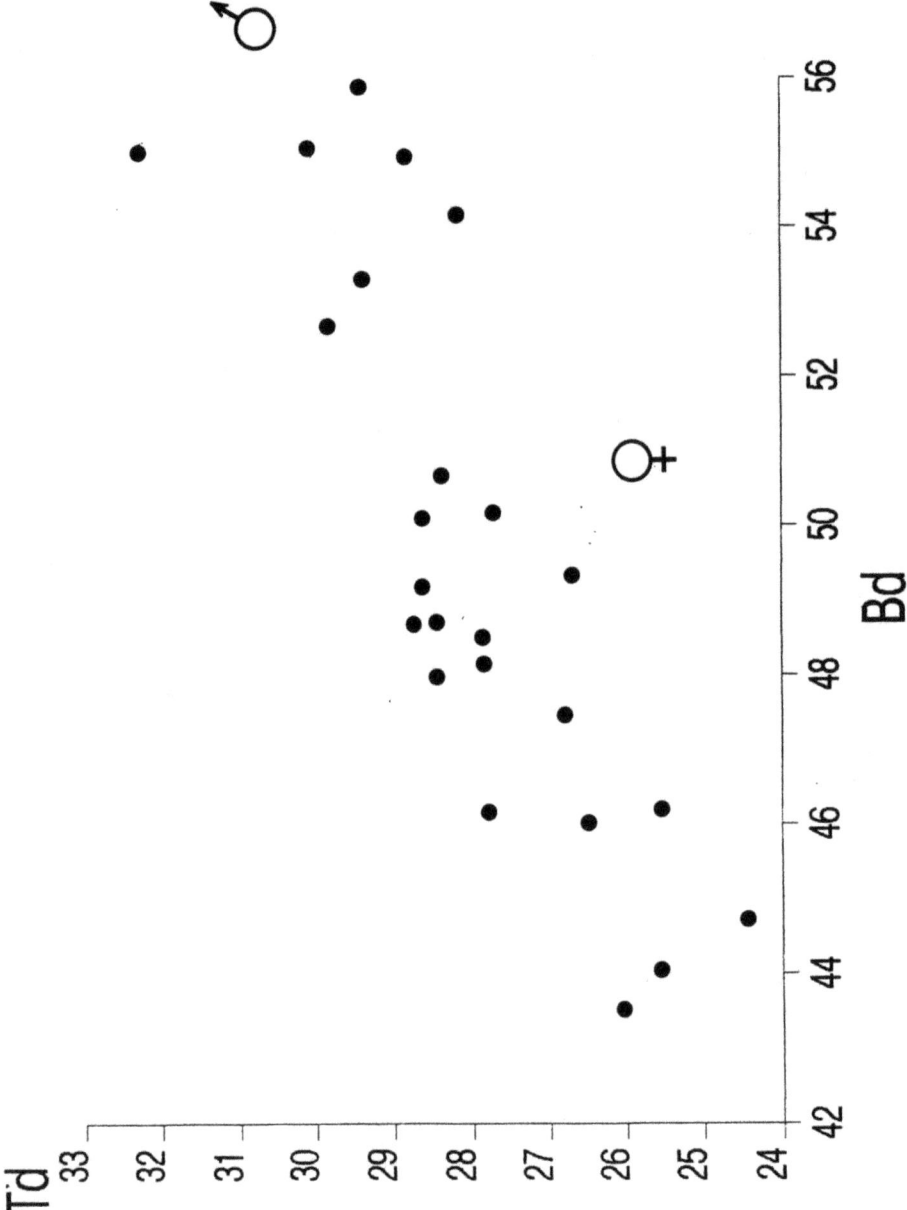

FIGURE 2 Distal metatarsal dimensions of Iron Age cattle from the later (b) phase at Danebury (Grant 1984). The division into the sexes is the author's interpretation.

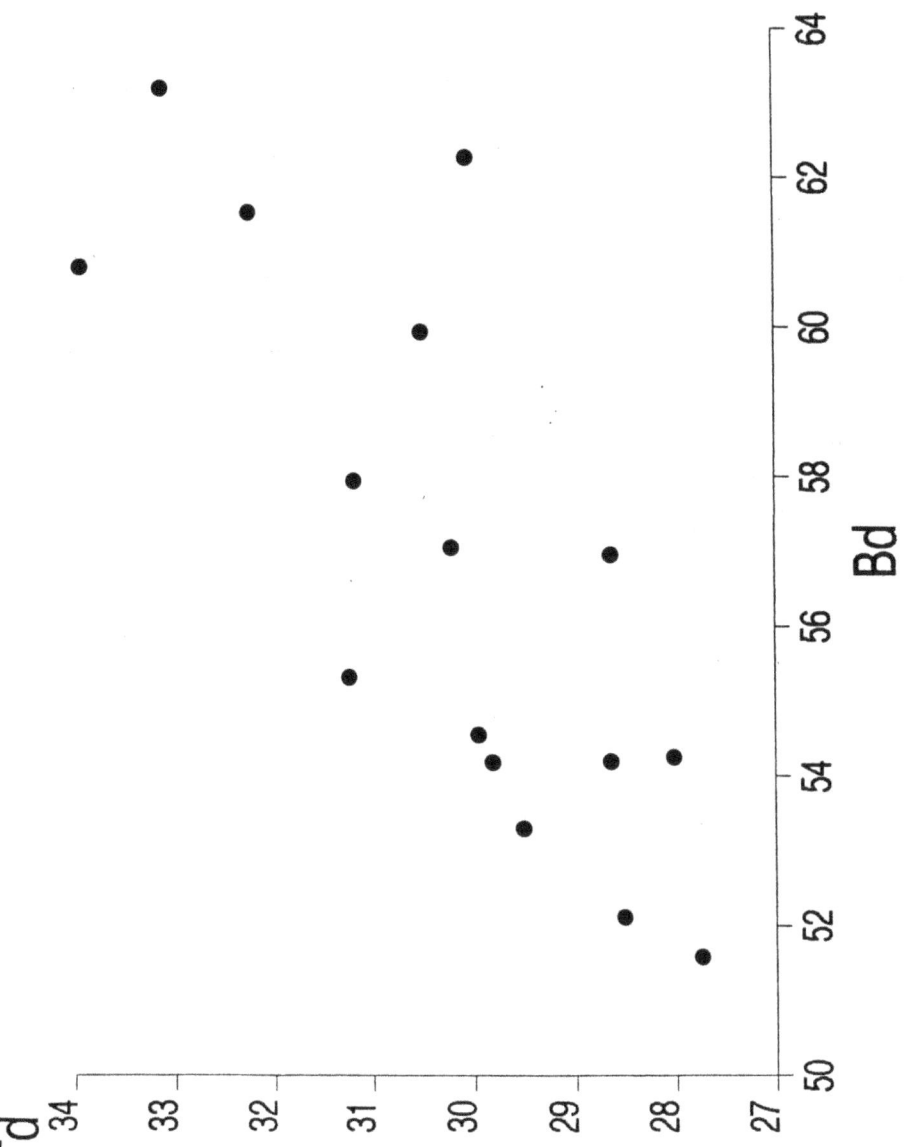

FIGURE 3 Distal metacarpal dimensions of Iron Age cattle from the later (b) phase at Danebury. The female:male division is less clear in this case, but the scatter of dimensions does show that most animals fall at the female end of the range, with perhaps 3 or 4 males represented.

So the Danebury cattle mortality pattern shows a substantial juvenile mortality, with cows predominating among the mature adults and with a moderate number of males retained into early adult life (Figures 2 and 3). This herd structure is found quite widely in prehistoric Britain. If the outputs of such a herd are considered (a calculation that is just as easy and with no more assumptions as is herd dynamics), the rather low output of meat that Grant (1984) inferred requires that the question be asked as to what outputs the herd provided. A significant use of traction animals needs to be demonstrated and the influence of traction needs on herd dynamics remains to be studied.

11.9 SUMMARY

The general scarcity of the seeds from domesticated cereals and legumes at neolithic sites in southern England has led Entwistle and Grant to postulate that agriculture was comparatively unimportant, and this implies that gathering was an important aspect of the economy. Yet few attempts at extensive flotation have been made at neolithic sites. The limited samples do not show a prevalence of wild plant foods and the range of wild species known is very narrow. Even the apparent local abundance of the hazel-nut can be understood in relation to the probability of shell fragments being exposed to charring, to surviving that process and to being found even without the use of flotation. The common use of a nut, extensively cultivated in Europe even now, hardly indicates a significant diversification of the neolithic economy in southern England. Much of the evidence that is further advanced in support of their hypothesis - such as the absence of assumed granaries and storage pits at neolithic sites - is to be expected when the excavated sites evidently had ceremonial or ritual functions.

That the neolithic agricultural complex may have been more broadly based than we imagine is suggested by the find of a morphologically domestic grape pip which has been accelerator dated as neolithic from Hambledon Hill-Stepleton (Jones and Legge 1987). In terms of simple efficiency, cereal agriculture would be required to support the population density shown by the scale of neolithic works. Only agriculture could provide the essential fuel for that degree of social elaboration.

With regard to animal husbandry, Grant's calculation of herd dynamics on the published data fits at one level of assumption. This hardly shows that the hypothesis of specialised milking is impossible. The meat outputs from such a herd would be low. The need for animals to provide traction at that time can readily be over-estimated and this work was often done by cows. Of the possible interpretations of herds showing high juvenile mortality, I see no reason to revise my view that husbandry primarily for milk production best fits the available data, nor is any alternative interpretation offered by Grant beyond the suggestion that cattle may have moved into and out of the system at Grimes Graves. Grant's suggestion (p. 213) that this '... has wider implications for our understanding of the social and economic functioning of the site' is best appreciated by a reading of the published work concerned with plant and animal husbandry of Grimes Graves.

Old debates about tolerance of cattle to milking, and people to lactose, are simply rebutted by reference to ethnographic and historical examples. The calculation of herd dynamics is based upon assumptions largely taken from recent African parallels and is of questionable relevance. While I agree that the interpretations of herd structure are not always easy, the data from Grimes Graves, from other sites in prehistoric Britain and in Europe lead me to agree that: '... cows that lost their calves could have provided milk for human consumption' (Grant 1984: 514).

BIBLIOGRAPHY

AMOROSO, E.C. AND JEWELL, P.A. (1963) The exploitation of the milk-ejection reflex by primitive peoples. In: *Man and cattle.* (eds Mourant, A.E. and Zeuner, F.E.) Royal Anthropological Institute Occasional Paper 18 pp. 126-38.

BARRETT, J., BRADLEY, R., GREEN, M. AND LEWIS, B. (1981) The earlier prehistoric settlement of Cranborne Chase - the first results of current fieldwork. *Antiquaries Journal* **61** 203-37.

BARTH, F. (1961) *Nomads of South Persia.* Oslo: University Press.

BAINES, J. AND MALEK, J. (1980) *Atlas of ancient Egypt.* Oxford: Phaidon.

BINFORD, L.R. (1978) *Nunamiut ethnoarchaeology.* New York: Academic Press.

BRADLEY, R. (1978) *The prehistoric settlement of Britain.* London: Routledge and Kegan Paul.

BROWN, A. (forthcoming) Structured deposition and technological change in the excavated Neolithic and Bronze Age lithic material from Cranborne Chase. In: *The prehistory of Cranborne Chase.* (eds Barrett, J. and Bradley, R.) Cambridge: University Press.

BULL, G. AND PAYNE, S. (1982) Tooth eruption and epiphyseal fusion in pigs and wild boar. In: *Ageing and sexing animal bones from archaeological sites.* (eds Wilson, B., Grigson, C. and Payne, S.) Oxford: British Archaeological Reports 109 pp. 55-72.

BURSTOW, G.P. AND HOLLEYMAN, G.A. (1957) Late Bronze Age settlement on Itford Hill, Sussex. *Proceedings of the Prehistoric Society* **23** 167-212.

BUSH, M.B. (1988) Early mesolithic disturbance; a force in the landscape. *Journal of Archaeological Science* **15** 453-62.

CANE, S. (1989) Australian aboriginal seed grinding and its archaeological record: a case study from the western desert. In: *Foraging and farming: the evolution of plant exploitation.* (eds Harris, D.R. and Hillman, G.C.) London: Unwin and Hyman pp. 99-119.

CASIMIR, M.J. (1988) Nutrition and socio-economic strategies in mobile pastoral societies of the Near East with special reference to the west Afghan Pashtuns. In: *Coping with uncertainty in food supply.* (eds De Garine, I. and Harrison, G.A.) Oxford: Clarendon Press pp. 337-59.

CLARK, J.G.D. (1952) *Prehistoric Europe; the economic basis.* London: Methuen.

CLARK, J.G.D. (1954) *Star Carr.* Cambridge: University Press.

CLARK, C. AND HASWELL, M. (1970) *The economics of subsistence agriculture.* (4th edition) London: Macmillan.

CLUTTON-BROCK, J. (1981) Discussion. In: *Farming practice in British prehistory.* (ed. Mercer, R.J.) Edinburgh: University Press pp. 218-20.

DAHL, G. AND HJORT, A. (1976) *Having herds: pastoral herd growth and household economy.* Stockholm Studies in Anthropology 2, University of Stockholm.

Documenta Geigy (1959) *Scientific Tables.* J.R. Geigy S.A., Basle.

GALLAGHER, C.R., MOLLESON, A.L. AND CALDWELL, J.H. (1974) Lactose intolerance and fermented dairy products. *Journal of the American Dietary Association* **65** 418-19.

GRANT, A. (1984) Animal husbandry. In: *Danebury; an Iron Age hillfort in Hampshire.* (ed. Cunliffe, B.W.) Council for British Archaeology Research Reports 52 pp. 526-49.

GREENWELL, W. (1870) On the opening of Grime's Graves in Norfolk. *Journal of the Ethnological Society* (New Series) **2** 419-39.

HABERMEHL, K.-H. (1975) *Die Altersbestimmung bei Haus- und Labortieren.* 2 Auflage. Berlin: Paul Parey.

HELBAEK, H. (1950) Tollund mandens sidste maaltid. *Aarboger for Nordisk Oldkyndighed og Historie* 311-41.

HELBAEK, H. (1952) Early crops in southern England. *Proceedings of the Prehistoric Society* **18** 194-233.

HELBAEK, H. (1957) Plant remains. In: Late Bronze Age settlement on Itford Hill, Sussex. (Burstow, G.P. and Holleyman, G.A.) *Proceedings of the Prehistoric Society* **23** 167-212.

HELBAEK, H. (1958) Grauballemandens sidste Maaltid. *Kuml* 83-116.

HIGHAM, C.F.W. (1967) Stock rearing as a cultural factor in prehistoric Europe. *Proceedings of the Prehistoric Society* **33** 84-106.

HIGHAM, C.F.W. (1968b) Patterns of prehistoric economic exploitation on the Alpine Foreland. *Vierteljahrsschrift der Naturforschenden Gesellschaft in Zürich* **113** (1) 41-92.

HIGHAM, C.F.W. AND MESSAGE, M. (1969) An assessment of a prehistoric technique of bovine husbandry. In: *Science in archaeology.* (eds Brothwell, D. and Higgs, E.) (2nd edition) London: Thames and Hudson pp. 315-30.

HILLMAN, G.C. (1974) The plant remains. In: The excavation of Tell Abu Hureyra in Syria; a preliminary report. (Moore, A.M.T., Hillman, G.C. and Legge, A.J.) *Proceedings of the Prehistoric Society* **41** 50-69.

HILLMAN, G.C. (1981) Reconstructing crop husbandry practices from charred remains of crops. In: *Farming practice in British prehistory.* (ed. Mercer, R.J.) Edinburgh: University Press pp. 123-62.

HILLMAN, G.C. (1982) Appendix 6: charred remains of plants. In: The excavation of two round barrows at Trelystan, Powys. (Britnell, W.) *Proceedings of the Prehistoric Society* **48** 133-201.

HILLMAN, G.C. (1984a) Interpretation of archaeological plant remains; the application of ethnographic models from Turkey. In: *Plants and ancient Man.* (eds Zeist, W. van and Casparie, W.A.) Rotterdam: A.A. Balkema pp. 1-42.

HILLMAN, G.C. (1984b) Traditional husbandry and processing of archaic cereals in recent times; the operations, products, equipment which might feature in Sumerian texts. Part I; the glume wheats. *Bulletin of Sumerian Agriculture* **1** 114-52.

HILLMAN, G.C. (1986) Plant food in the ancient diet; the archaeological role of palaeofaeces in general, and Lindow Man's gut contents in particular. In: *Lindow Man: the body in the bog.* (eds Stead, I., Bourse, J. and Brothwell, D.) London: British Museum Publications pp. 99-115, 198-202.

HILLMAN, G.C. (1989) Late palaeolithic plant foods from Wadi Kubbaniya; dietary diversity, infant weaning and seasonality in a riverine environment. In: *Foraging and farming: the evolution of plant exploitation.* (eds Harris, D.R. and Hillman, G.C.) London: Unwin and Hyman pp. 207-33.

HILLMAN, G.C., COLLEDGE, S.M. AND HARRIS, D.R. (1989) Plant food economy during the epipalaeolithic period at Abu Hureyra, Syria: dietary diversity, seasonality and modes of exploitation. In: *Foraging and farming: the evolution of plant exploitation.* (eds Harris, D.R. and Hillman, G.C.) London: Unwin and Hyman pp. 240-66.

HOWES, F.N. (1948) *Nuts; their production and everyday use.* London: Faber.

JACKSON, J.W. (1957) Report on the skeleton of an ox. In: Late Bronze Age settlement on Itford Hill, Sussex. (Burstow, G.P. and Holleyman, G.A.) *Proceedings of the Prehistoric Society* **23** 167-212.

JEWELL, P.A. (1963) Cattle from British archaeological sites. In: *Man and cattle.* (eds Mourant, A.E. and Zeuner, F.E.) Royal Anthropological Institute Occasional Paper 18 pp. 80-91.

JONES, G. AND LEGGE, A.J. (1987) The grape (*Vitis vinifera* L.) in the Neolithic of Britain. *Antiquity* **61** 233, 452-55.

JONES, G. AND ROWLEY-CONWY, P.A. (1984) Plant remains from the north Italian lake dwellings of Fiavé (1400-1200 bc). In: *Scavi Archeologici nella zona Palafitticola di Fiavé - Carera.* (ed. Perini, R.) Trento: Servizio Beni Culturali della Provincia Autonoma di Trento pp. 323-55.

JONES, M. (1980) Carbonised cereals from Grooved Ware contexts. *Proceedings of the Prehistoric Society* **46** 61-63.

KEENE, A.S. (1982) Optimal foraging in a non-marginal environment: a model of prehistoric subsistence activities in Michigan. In: *Hunter-gatherer foraging strategies: ethnographic and archaeological analysis.* (eds Winterhalder, B. and Smith, E.A.) University of Chicago Press pp. 171-93.

LEACH, G. (1976) *Energy and food production.* Guildford: IPC Science and Technology Press.

LEE, R.B. AND DEVORE, I. (1968) Problems in the study of hunter-gatherers. In: *Man the hunter.* (eds Lee, R.B. and DeVore, I.) Chicago: Aldine pp. 3-12.

LEGGE, A.J. (1981a) The agricultural economy. In: *Grimes Graves, Norfolk: excavations 1971-72.* (Mercer, R.J.) Department of the Environment Research reports 11 London: Her Majesty's Stationery Office pp. 79-103.

LEGGE, A.J. (1981b) Aspects of cattle husbandry. In: *Farming practice in British prehistory.* (ed. Mercer, R.J.) Edinburgh: University Press pp. 169-81.

LEGGE, A.J. (1981c) Discussion. In: *Farming practice in British prehistory.* (ed. Mercer, R.J.) Edinburgh: University Press pp. 220-22.

LEGGE, A.J. (1986) Seeds of discontent: accelerator dates on some charred plant remains from the Kebaran and Natufian cultures. In: *Archaeological results from accelerator dating.* (eds Gowlett, J.A.J. and Hedges, R.E.M.) Oxford University Committee for Archaeology Monograph 11.

LEGGE, A.J. (1987) La fauna en la economia prehistorica de Moncin. In: Moncin: Poblado Prehistorico de la Edad del Bronce (I). (Harrison, R.J., Moreno-Lopez, G. and Legge, A.J) *Noticiario Arqueologico Hispanico* 29. Madrid: Ministerio de Cultura.

LEGGE, A.J. (in press) Animals, economy and environment at Selevac. In: *Selevac: socioeconomic transformations in the Neolithic of southeast Europe.* (Tringham, R.) Berkeley Institute of Archaeology Press, UCLA.

LEGGE A.J. (forthcoming a) Animal remains from Neolithic and Bronze Age sites. In: *The prehistory of Cranborne Chase.* (eds Barrett, J. and Bradley, R.) Cambridge: University Press.

LEGGE A.J. (forthcoming b) *Animal remains from Bronze Age middens at Grimes Graves.* British Museum Publications.

LEGGE, A.J. AND ROWLEY-CONWY, P.A. (1988) *Star Carr revisited: a re-analysis of the large mammals.* University of London, Centre for Extra-Mural Studies, Birkbeck College.

LEGGE, K. J. (1989) Changing responses to drought among the Wodaabe of Niger. In: *Bad year economics.* (eds Halstead, P. and O'Shea, J.) Cambridge: University Press.

MARTIN, M. (1980) Pastoral production: milk and firewood in the ecology of Turan. *Expedition.* (Journal of the University Museum of Archaeology/Anthropology, Pennsylvania) **22** 24-29.

MALHOTRA, K.C. AND GADGIL, M. (1988) Coping with uncertainty in food supply; case studies among pastoral and non pastoral nomads in western India. In: *Coping with uncertainty in food supply.* (eds De Garine, I. and Harrison, G.A.) Oxford: Clarendon Press pp. 379-404.

MANNING, C.R. (1872) Grime's Graves, Weeting. *Norfolk Archaeology* 7.

MALTBY, M. (1979) *The animal bones from Exeter.* Department of Prehistory and Archaeology, University of Sheffield.

MERCER, R.J. (1980) *Hambledon Hill: a Neolithic landscape.* Edinburgh: University Press.

MERCER, R.J. (1981) Introduction. In: *Farming practice in British prehistory.* (ed. Mercer, R.J.) Edinburgh: University Press pp. ix-xxvi.

MURPHY, P. (1983) Carbonised beans from feature 11, Frog Hall Farm, Fingrinhoe, Essex. *Ancient Monuments Laboratory Report No. 2033.*

MURPHY, P. (1988) Plant macrofossils. In: A late Bronze Age enclosure at Lofts Farm in Essex. (Brown, N.) *Proceedings of the Prehistoric Society* 54 281-94.

NOLI, D. AND AVERY, G. (1988) Protein poisoning and coastal subsistence. *Journal of Archaeological Science* 15 395-401.

RACKHAM, O. (1986) *The history of the countryside.* London: Dent.

RACKHAM, O. (1988) Wildwood. In: *The archaeology of the flora of the British Isles.* (ed. Jones, M.) Oxford Committee for Archaeology Monograph 14 pp. 3-6.

REYNOLDS, P. (1974) Experimental Iron Age storage pits: an interim report. *Proceedings of the Prehistoric Society* 40 118-31.

REYNOLDS, P. (1981) Deadstock and livestock. In: *Farming practice in British prehistory.* (ed. Mercer, R.J.) Edinburgh: University Press pp. 97-122.

ROBINSON, M. AND HUBBARD, R.N.L.B. (1977) The transport of pollen in the bracts of hulled cereals. *Journal of Archaeological Science* 4 197-99.

ROWLEY-CONWY, P. (1981) Slash-and-burn in the temperate European neolithic. In: *Farming practice in British prehistory.* (ed. Mercer, R.J.) Edinburgh: University Press pp. 85-96.

RUSSELL, K. (1988) *After Eden: behavioural ecology of early food production.* Oxford: British Archaeological Reports S391.

RYDER, M.L. (1983) Milk products. In: *Integrating the subsistence economy.* (ed. Jones, M.) Oxford: British Archaeological Reports S181 pp. 239-50.

SILVER, I.A. (1969) The ageing of domestic mammals. In: *Science in archaeology.* (eds Brothwell, D. and Higgs, E.) (2nd edition) London: Thames and Hudson pp. 283-302.

SIMMONS, I.G. AND INNES, J.B. (1987) Mid-Holocene adaptation and later mesolithic forest disturbance in northern England. *Journal of Archaeological Science* **14** 385-403.

SPETH, J.D. (1987) Early hominid subsistence strategies in seasonal habitats. *Journal of Archaeological Science* **14** 13-29.

SPETH, J.D. AND SPIELMANN, K.D. (1983) Energy source, protein metabolism and hunter-gatherer subsistence strategies. *Journal of Anthropological Archaeology* **2** 1-31.

STAHL, A.B. (1989) Plant food processing: implications for dietary diversity. In: *Foraging and farming: the evolution of plant exploitation.* (eds Harris, D.R. and Hillman, G.C.) London: Unwin and Hyman pp. 171-94.

TAYLOR, R.E. (1974) Management of beef cattle. In: *Animal agriculture.* (eds Cole, H.H. and Ronning, M.) San Francisco: Freeman pp. 549-62.

WATANABE, H. (1972) The Ainu. In: *Hunters and gatherers today.* (ed. Biccieri, M.) New York: Holt, Reinhart and Winston pp. 451-84.

12

CEREALS, FRUIT AND NUTS: CHARRED PLANT REMAINS FROM NEOLITHIC SITES IN ENGLAND AND WALES AND THE NEOLITHIC ECONOMY

L. Moffett, M.A. Robinson and V. Straker

ABSTRACT

The evidence from the recovery of charred plant remains from twenty-six neolithic sites in England and Wales is reviewed in this paper. The general paucity of cereal evidence relative to evidence for collected food resources suggests that the conventional view of the neolithic economy as relying primarily on the farming of cereals may be incorrect. Present evidence suggests that there was considerable reliance on collected plant resources in addition to cultivated cereals and that a fully settled way of life based on arable husbandry may not have been widely adopted until well into the Bronze Age.

12.1 INTRODUCTION

The conventional view of the neolithic economy throughout most of England and Wales is that it was based on the rearing of domestic animals and the cultivation of cereals (Smith 1974: 103). The low proportion of wild animal bones from most neolithic bone assemblages certainly suggests that hunting was of secondary importance to herding domestic stock (Grigson 1981a: 196, 1981b: 223-24; Smith 1974: 104). The work of Helbaek (1952) on impressions in pottery provided reliable evidence for the cultivation of emmer wheat, bread-type wheat, six-row naked barley and six-row hulled barley, as well as flax. Apple pip impressions were noted in pottery from Windmill Hill, but he did not find any other examples of food gathering so assumed that agricultural output in the Neolithic was sufficient for feeding the population (Helbaek 1952: 200).

Pottery impressions, however, show a bias against the larger items which tend to make up collected edible wild plants. Since Helbaek's work, discoveries of charred remains of wild food plants including hazel-nuts and crab-apples were often reported (e.g. Houlder 1963: 27; Field et al 1964: 366) but cereals were only occasionally noticed (e.g. Reaney 1968: 71). Bradley (1978: 88) comments that until the middle Bronze Age, so-called storage pits are more likely to yield nut shells in their backfill than grain and Hillman (1981: 189) suggested that most of the early agricultural communities of Britain were substantially dependent on a wide range of wild food resources. However, the systematic recovery of carbonised plant remains by flotation and wet sieving on neolithic sites, which enables the biases towards larger items and rich deposits inherent in hand picking items noticed during excavation to be overcome, only became standard practice within the past ten years. Hillman (1981: 186) could only cite a single published report on carbonised seeds which had been recovered by sieving.

Although flotation is now widespread, neolithic charred plant remains have the

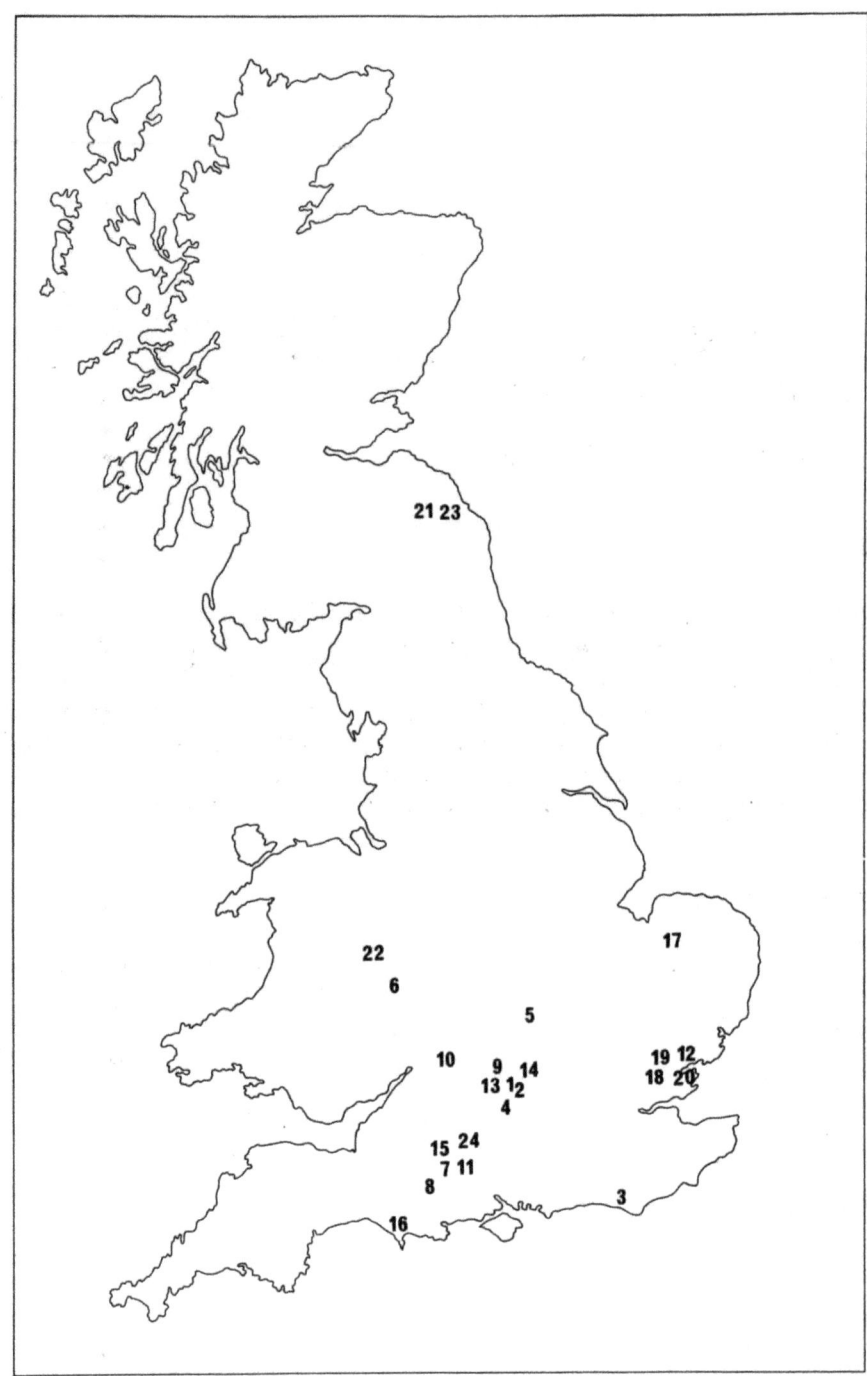

FIGURE 1 Location of sites in Table 1 and Appendix.

1.	Barrow Hills	2.	Barton Court Farm	3.	Bishopstone
4.	Blewbury	5.	Briar Hill	6.	Bromfield
7.	Coneybury	8.	Down Farm	9.	Gravelly Guy
10.	Hazleton	11.	King Barrow Ridge	12.	Lofts Farm
13.	M.G. Abingdon	14.	Mount Farm	15.	Robin Hood's Ball
16.	Rowden	17.	Spong Hill	18.	Springfield Barnes
19.	Springfield Lyons	20.	The Stumble	21.	Thirlings
22.	Trelystan	23.	Whitton Hill	24.	Wilsford Down

reputation of being sparse. For example, processing 104 samples with a total volume of about 3000l from the later neolithic features at Maxey produced the paltry total of 14 cereal grains, 3 weed seeds and 5 hazel-nut shell fragments (Green 1985). Some of these grains were probably contaminants from more recent deposits. The archaeological features on many neolithic sites, however, are solely related to ceremonial, symbolic or funerary purposes, as was the case at Maxey. When deposits contain occupation debris, flotation is often more successful though the concentration of carbonised plant remains rarely reaches the level that is usual for later prehistoric sites.

The first large-scale study to be published was by Jones (1980) and concerned Grooved Ware sites in Oxfordshire and Dorset. The main thrust of the paper was to show that cereal remains indeed occur in Grooved Ware contexts but he also drew attention to the strong representation of woodland food plants and the evidence for the exploitation of woodland resources. Similar results were obtained by van der Veen (1985: 206-9) from Grooved Ware sites in Northumberland, and she drew similar conclusions.

Full results are now available from twenty-four neolithic sites with some evidence for occupation and there are preliminary results from several more sites. They enable an assessment to be made of the importance of collected food plants in the earlier Neolithic as well as on sites of later neolithic date.

12.2 THE RESULTS

A summary of the information obtained from the sites located in Figure 1 is given in Table 1. Context information and relevant radiocarbon dates are presented in the Appendix.

In all instances the remains were recovered by water flotation onto a mesh of 0.5mm or less. On some sites sediment conditions were such that the mineral residue required drying and refloating or sorting.

Different workers took their identifications to different levels, so for brevity, some of the firm and tentative identifications have been combined to make single entries in Table 1. Wheat spikelet forks have been recorded as two glume bases. The term *Triticum aestivum s.l.* has been used to describe those free threshing apparently hexaploid wheat grains that some workers refer to *Triticum aestivocompactum* Schiem. The shape of the grains, which is short and broad, is characteristic of the possession of a recessive S locus in the genotype for modern hexaploid wheats (Ellerton 1939). The crab-apple remains could all have come from *Malus sylvestris* Mill. ssp. *sylvestris*.

Some workers quantified the number of hazel-nut shell fragments in their samples, others just recorded presence or absence. At both Bromfield and Hazleton, the nut shell fragments were very abundant.

12.3 DISCUSSION OF THE RESULTS

On all but one of the sites listed in Table 1, the remains seem to represent waste from food preparation or the accidental burning of crops during processing activities which involve heat, for example the parching of hulled cereals prior to dehusking. There were probably many minor incidents of charring which resulted in the accumulation of the larger of these assemblages. The later neolithic assemblage from Coneybury, however, probably resulted from the accidental (or deliberate) burning of large quantities of grain in a single event. The concentration of grain was about ten times greater than on any of the other sites, it was from a single context and only comprised barley. Other caches of neolithic charred grain have been noticed at Aston-on-Trent, Derbyshire (Reaney 1968: 71); Hembury, Devon (Liddell 1932: 171); and Woodham Walter, Essex (Boyd and Cartwright 1987). Hazel-nut shell fragments were also recorded from these three sites.

Cereals were identified from all but one of the sites listed in Table 1 although their concentration was low on most of them. They comprised the same four varieties recorded by Helbaek from impressions in neolithic pottery (see Introduction). Chaff remains were very sparse and other crop species were absent. Further work on the Stumble, however, has discovered significant quantities of neolithic chaff (P. Murphy pers. comm.). Weed seeds were not very abundant in most of the assemblages: they included both species which can grow as arable weeds and a few of grassland or scrub. In no instance could a plough arable flora be identified with certainty, as all the arable weed species found in these assemblages will also grow readily under conditions of hoe or spade horticulture.

Wild food plant remains were also identified from all but three of the sites. Hazel-nut shell fragments were recovered in particularly large quantities from some of the sites and were present on all but five. Remains of crab-apple were abundant on a couple of the sites, but the occurrence of raspberry, blackberry, sloe and hawthorn stones was sporadic and slight. All these collected food resources are from trees and shrubs. The hazel and the apple are the only ones likely to be very productive under fully wooded conditions, but all are favoured by some opening of the tree canopy, or by growing at the edge of woodland or in scrub. Plum (*Prunus domestica* including ssp. *insititia*) and wild cherry (*Prunus avium*) for which Godwin (1975: 197) gives British neolithic records, were absent. However, all the pre-Roman Flandrian records for these species cited by Godwin are suspect, as they were based on charcoal, stones of dubious identity, sites of uncertain stratigraphy, or were clerical errors. Plum is regarded as an introduction but wild cherry appears to be a native member of the British woodland flora.

P. Murphy has kindly drawn our attention to his discovery of many neolithic carbonised roots, rhizomes and tubers, some of which are likely to have been collected for food, at the Stumble, Essex.

It is not possible to give an accurate assessment of the relative importance of cereals and wild plants in the neolithic economy. A hazel-nut kernel weighs about twenty times as much as a cereal grain but the hazel-nuts were enumerated as shell fragments whereas the cereals were recorded as whole grains.

The various plant remains would have had different susceptibilities to charring. Cereals come into contact with fire if they are parched, hazel-nut shells might be discarded onto fires after the kernels had been extracted. (Indeed it was noticeable that all the hazel remains were nut shell fragments rather than intact nuts.) Apples might have been dried over fires. There does not seem to be any particular reason for the other fruits to come into contact with fire except for cooking.

The apparent importance of collected fruits and nuts shown by most of the sites in Table 1 is not just an artefact of preservation. Numerous waterlogged fragments of hazel-nut shell were noticed by one of the authors in a sample from a pit within the causewayed enclosure at Etton, Cambridgeshire.

Five sites in Table 1 had significant quantities of cereal remains that were not greatly outnumbered by fragments of woodland food plants: Bishopstone, both periods of Coneybury, Down Farm and Rowden. All are on chalk and all but Bishopstone are in Dorset or Wiltshire. Bishopstone was a coastal site where the collection of marine molluscs played an important part in the neolithic economy (Bell 1977: 42). The only site where cereal remains were not discovered was Trelystan, Powys which, at an altitude of about 370m was probably the least suitable site for cereal cultivation.

There are no sites from the fourth millennium bc in England and Wales for which full results from flotation are available. The earliest neolithic site from which macroscopic charred plant remains other than charcoal have been recovered is Hembury causewayed camp, Devon, although flotation was not used. Radiocarbon dates of 3150±150 bc, 3240±150 bc and 3330±150 bc were obtained for the site (Fox 1963). A brief re-examination of the plant remains from the site (now in the Albert Memorial Museum, Exeter), by Straker has so far revealed a large number of hazel-nut shell fragments as well as much grain. Spelt wheat was not among the grain or chaff purportedly shown to Helbaek (1952: 207-8) or any other samples seen yet. An occupation deposit containing carbonised food plant remains beneath a Cotswold-Severn tomb at Gwernvale, Powys, gave a radiocarbon date of 3100±75 bc (Britnell and Savory 1984: 50). A preliminary investigation of the remains recovered by flotation of large quantities of soil revealed many hazel-nut shell fragments, scant cereal remains and a few weed seeds (Hillman and Moffett in prep.), somewhat similar results to the slightly younger occupation deposit from the Hazleton Cotswold-Severn tomb (Table 1).

There does not seem to be any evidence for a decrease in the importance of collected food plants throughout the third millennium bc, indeed Table 1 shows some of the highest ratios for the number of fragments of the gathered plants to cereals for the later neolithic Grooved Ware sites.

It is difficult to relate site status to the general composition of their carbonised plant assemblages. Carbonised grain only occurred very sparsely in the features of the Etton causewayed enclosure whereas hazel-nut shells were frequently noticed during the flotation programme (C. French and F. Pryor pers. comm.). The one discovery that is likely to be related to status is a single grape pip dated to 2710±80 bc from the causewayed enclosure at Stepleton (Jones and Legge 1987).

The following conclusions can be drawn from the evidence of carbonised food plant remains discovered by flotation of soil samples from neolithic sites.

> (1) The use of cereals was part of the neolithic economy over much if not all of England and Wales.
>
> (2) The collection of wild food plants from woodland or scrub, particularly hazel-nuts but also various fruits, was also a usual aspect of the neolithic economy.
>
> (3) The collection of wild food plants in addition to the cultivation of cereals was not just a feature of the earlier Neolithic but continued alongside cereal cultivation throughout the later Neolithic.
>
> (4) There does seem to be some regional variation, with arable agriculture making a more important contribution to the economy on the Wessex chalklands than elsewhere.

		Hazleton	Rowden	Bromfield
		2970bc	2910bc	2730bc
Triticum cf. *monococcum* (einkorn)	glume base	1	0	0
T. dicoccum and *T.* cf. *dicoccum* (emmer)	grain	49	2	0
T. dicoccum and *T.* cf. *dicoccum* (emmer)	glume base	1	0	0
T. dicoccum or *spelta* (emmer/spelt)	grain	0	6	0
T. aestivum s.l. and *T.* cf. *aestivum* s.l. (bread-type wheat)	grain	33	0	0
Triticum sp. (wheat)	grain	186	0	1
Triticum sp. (wheat)	glume base	5	0	0
Hordeum sativum (hulled six-row barley)	grain	0	0	0
H. sativum var. *nudum* (naked six-row barley)	grain	0	0	0
H. sativum and *H.* cf. *sativum* (barley)	grain	3	2	3
Indeterminate cereals	grain	11	31	0
Total cereal grain		283	41	4
Rubus spp. (raspberry and blackberry)	stone	0	0	0
Prunus spinosa L. (sloe)	stone	0	0	0
Crataegus sp. (hawthorn)	stone	0	1	0
Malus sylvestris Mill. (crab apple)	seed	0	0	0
M. sylvestris Mill. (crab apple)	endocarp frags.	0	0	0
Malus or *Pyrus* sp. (crab apple or pear)	seed frags.	0	0	1
Malus or *Pyrus* sp. (crab apple or pear)	calyx	0	0	1
Malus or *Pyrus* sp. (crab apple or pear)	epidermal frags.	0	0	0
Corylus avellana L. (hazel)	nut shell frags.	1000+	0	*
Weed seeds		42	0	3
Unidentified seeds (mostly weeds)		10	0	0
Number of samples		61	23	2
Total volume of samples (litres)		550	66	7

*: present

TABLE 1 Carbonised cereal, fruit and nut remains from floated neolithic samples.

CEREALS, FRUIT AND NUTS

	Earlier Neolithic								Later Neolithic	
									Grooved Ware	
	Briar Hill 2635bc	Bishop-stone 2510bc	The Stumble	Coneybury	Wilsford Down	Robin Hood's Ball	Lofts Farm	Spong Hill	Trelystan 2180bc	Barton Court 2020bc
	0	0	0	0	0	0	0	0	0	0
	0	6	30	0	0	0	0	0	0	2
	0	*	9	1	0	0	0	0	0	1
	0	0	0	10	0	0	0	0	0	0
	0	0	0	0	0	0	0	0	0	3
	0	9	0	0	0	0	0	2	0	9
	0	0	0	0	0	0	0	0	0	0
	0	0	0	0	0	0	0	0	0	0
	1	0	0	0	0	0	0	0	0	0
	0	10	0	0	0	0	0	0	0	3
	2	0	25	40	0	1	2	2	0	4
	3	25	55	50		1	2	4		22
	0	0	0	0	0	0	0	0	1	0
	1	0	0	0	0	0	cf1	0	0	0
	0	0	0	0	0	0	0	0	0	0
	0	0	0	0	0	0	0	0	cf.1	23
	0	0	0	0	0	0	0	0	0	0
	0	0	0	0	0	0	0	0	0	0
	0	0	0	0	0	0	0	0	0	0
	0	0	0	0	0	0	0	0	0	0
	5	0	*	0	2	26	*	2	*	84
	13	14	1	0	0	0	1	0	1	8
	7	0	2	0	0	0	0	0	2	3
	3	1	6	6	1	8	17	4	5	4
	6			61	10	102	292	15	114	115

		Grooved Ware		
		Down Farm 2160bc	Barrow Hills	Blewbury
Triticum cf. *monococcum* (einkorn)	glume base	0	0	0
T. dicoccum and *T.* cf. *dicoccum* (emmer)	grain	0	0	0
T. dicoccum and *T.* cf. *dicoccum* (emmer)	glume base	0	0	0
T. dicoccum or *spelta* (emmer/spelt)	grain	0	0	0
T. aestivum s.l. and *T.* cf. *aestivum* s.l. (bread-type wheat)	grain	0	0	6
Triticum sp. (wheat)	grain	1	2	3
Triticum sp. (wheat)	glume base	0	0	0
Hordeum sativum (hulled six-row barley)	grain	0	0	0
H. sativum var. *nudum* (naked six-row barley)	grain	0	0	0
H. sativum and *H.* cf. *sativum* (barley)	grain	11	0	2
Indeterminate cereals	grain	13	8	8
Total cereal grain		25	10	19
Rubus spp. (raspberry and blackberry)	stone	0	0	0
Prunus spinosa L. (sloe)	stone	0	0	0
Crataegus sp. (hawthorn)	stone	0	0	0
Malus sylvestris Mill. (crab apple)	seed	1	1	0
M. sylvestris Mill. (crab apple)	endocarp frags.	0	106	0
Malus or *Pyrus* sp. (crab apple or pear)	seed frags.	0	39	0
Malus or *Pyrus* sp. (crab apple or pear)	calyx	0	6	0
Malus or *Pyrus* sp. (crab apple or pear)	epidermal frags.	0	171	0
Corylus avellana L. (hazel)	nut shell frags.	2	340	76
Weed seeds		1	6	15
Unidentified seeds (mostly weeds)		5	7	1
Number of samples		14	22	2
Total volume of samples (litres)			260	25

*: present

TABLE 1 (CONTINUED) Carbonised cereal, fruit and nut remains from floated neolithic samples.

CEREALS, FRUIT AND NUTS

	Later Neolithic								Beaker	
M.G.	Thirlings	Whitton Hill	Coney-bury	Mount Farm	King Barrow Ridge	Spring-field Barnes	Spring-field Lyons	Gravelly Guy	Spong Hill
0	0	0	0	0	0	0	0	0	0
0	0	1	0	1	0	2	0	0	0
0	0	0	0	0	0	0	0	0	0
0	0	0	0	0	1	0	0	0	0
0	0	0	0	1	0	4	0	0	0
0	0	0	0	1	0	1	3	0	0
0	0	0	0	0	0	0	0	0	0
0	0	0	102	0	0	0	0	0	0
0	1	1	159	0	0	0	0	0	0
0	0	2	43	8	0	0	0	2	0
1	1	1	27	0	0	6	2	6	1
1	2	5	331	11	1	13	5	8	1
0	1	0	0	0	0	0	0	0	0
0	0	0	0	0	1	1	0	0	0
0	1	0	1	0	0	2	0	0	0
0	0	0	0	0	0	0	0	0	0
0	0	0	0	0	0	0	0	0	0
0	0	0	0	0	0	0	0	0	0
0	0	0	0	0	0	0	0	0	0
0	0	0	0	0	0	0	0	0	0
66	1587	9	0	425	160	*	0	156	1
1	18	5	5	4	1	8	2	14	0
0	4	1	0	7	0	1	0	4	0
3	4	3	3	2	14	18	4	16	1
28	168	21	26		125	440	32	170	

(5) It is not possible to make an accurate estimate of the relative importance of cultivated and wild plants in the neolithic diet but it is possible that on some of the sites hazel-nuts were at least as important a source of food as grain.

12.4 NEOLITHIC LANDSCAPE AND AGRICULTURE

The results from the carbonised plant remains do not conflict with the palynological and molluscan evidence for the environment of the British Isles during the Neolithic. A very slight presence of cereal pollen in some pre-elm decline deposits provides the earliest possible evidence for 'neolithic' type activities and such sites are distributed throughout the British Isles (Edwards and Hirons 1984). Unfortunately carbonised food plant remains have not yet been recovered from any very early sites. From the elm decline at *circa* 3100 bc onwards, pollen diagrams from both lowland and to a lesser extent highland zones of England and Wales often show a slight presence of cereals along with ruderal species of cultivated ground.

Even well into the Bronze Age, however, much woodland remained from which fruit and nuts could be collected. Tinsley (1981: 231) wrote that 'At the start of the Bronze Age the British Isles were still largely forested, although the agricultural activities of successive Neolithic cultures had taken their toll and in some places the composition of the Atlantic forest had been permanently altered. ... In contrast to the national picture, the regional view shows that in some parts of the country the forest had been very much reduced by the start of the Bronze Age'.

In those areas which remained largely wooded until after the end of the Neolithic, pollen sequences often show evidence for a series of small scale clearances during the first half of the third millennium bc, each episode sometimes lasting only a few decades and being followed by regeneration. Larger scale clearances do occur but their ultimate fate is usually abandonment and woodland regeneration. In the second half of the third millennium bc there is sometimes an intensification of clearance activity and larger scale clearances are more frequent. On many sites, however, periods of regeneration continue to occur, and even a gradual rise in herb pollen may be the result of an increase in the proportion of the landscape experiencing small scale clearance followed by abandonment, rather than an inexorable rise of permanently open areas. Not all neolithic clearance was for arable: cereal pollen is often absent and the range of herb taxa more suggestive of pasture than disturbed ground. A typical example of a pollen sequence from an area which was wooded at the start of the Bronze Age is provided by Scaife (Tomalin and Scaife 1979; Scaife 1980), who sampled a peat mire with a small pollen catchment area adjacent to a neolithic occupation site at Gatcombe Withy Bed, Isle of Wight. Cereal pollen first occurred at the elm decline, along with other herbaceous taxa which suggest that tree cover was no longer complete. There followed a series of peaks and troughs of cereal and ruderal pollen suggestive of ephemeral cultivation against a background of open woodland. In the middle of the Neolithic there was a brief phase of local forest clearance possibly related to the occupation site, with an associated rise in the pollen of cereals and ruderals. During the later Neolithic, however, regeneration occurred of secondary woodland which survived into the Bronze Age.

Parts of Cumbria and the chalk of Wessex were more or less permanently cleared during the Neolithic (Tinsley 1981: 231-32). However, it has been established that the

Breckland of East Anglia, a region formerly thought to have been permanently cleared in the Neolithic, did not experience widespread clearance until the start of the Iron Age (Bennett 1983: 482). Some upland areas also lost their trees during the Mesolithic and Neolithic. It is possible that there were also large permanent clearances on the terraces of the Thames and some of the major lowland river valleys of the south Midlands.

When considering the economy, it is important to distinguish between areas which remained open following clearance because they were ecologically fragile and areas which were kept open because they were agriculturally productive. The initiation of blanket peat formation on the uplands of mid Wales, Exmoor and elsewhere in England has been in part related to neolithic clearance activities (Moore 1973, 1975).

The analysis of land snail assemblages particularly from the Avebury area of Wiltshire, provides evidence for large, permanent clearances on parts of the Wessex chalk from the beginning of the third millennium bc onwards (Evans 1971, 1972, 1975: 116-19). The chalk of Sussex, however, largely retained its woodland cover throughout the Neolithic (Thomas 1982). The soil and molluscan sequence from beneath the South Street long barrow, Avebury, suggested clearance, cross ploughing, a possible phase of grassland, further disturbance (possibly caused by spade or hoe cultivation), and finally short-turfed grassland before the construction of the barrow (Ashbee *et al.* 1979: 290-95). A radiocarbon date of 2810±130 bc was obtained on charcoal from the surface of the buried soil.

Evidence from the soil buried beneath the banks of some henge monuments dating to around 2000 bc (for example, Marden and Durrington Walls, Evans 1972: 230, 364), suggests that they had been built on grassland that had been in existence for about five hundred years after clearance. Although many of the neolithic sites in Wiltshire produced evidence for episodes of cultivation, these were generally of short duration and arable land does not seem to have been at a premium (Evans 1972: 364). The open landscape seems predominantly to have been grassland. Grazing even seems to have been relaxed for periods on some sites, as for example at Avebury (Evans *et al.* 1985).

The higher proportion of carbonised cereals compared with wild plant remains from some of the sites on the Wessex chalk is consistent with the evidence for a more open agricultural landscape in that part of the country. However, hazel-nuts are by no means absent from sites in Wessex and much woodland must have remained for exploitation. Assemblages from the upper Thames Valley sites were all dominated by gathered food plant remains, but even if there were large open areas in the later Neolithic around the religious centres on the gravel terraces, the valley sides probably remained wooded. Analysis of insect remains from later neolithic riverine sediments at Runneymede, Berkshire, on the middle Thames suggests a landscape that had only been partly cleared (Robinson, in prep.). Carbonised remains from later neolithic occupation deposits on the site contain many hazel-nut shell fragments (S. Needham, pers. comm.).

It is easy to relate the prevalence of gathered woodland plant resources on neolithic sites to the extensively wooded landscape. It is much harder to explain why gathered food plants should still have played such an important part in the economy at least one and a half millennia after the introduction of cereal-based agriculture.

One might reasonably expect an economy based primarily on woodland resources during the fourth millennium bc, when pollen evidence suggests agriculture was becoming established in the British Isles. Given the available organisation and technology, progressive clearance and intensification of food production should have been possible leading to an increased sedentary population and the creation of a permanent mixed agricultural landscape. By the beginning of the third millennium bc, neolithic societies in England and Wales were sufficiently well organised to construct long barrows and causewayed enclosures. The discoveries at South Street show that both cross-ploughing and spade or hoe cultivation methods were known at this date. Given the available organisation and technology, a rapid rise in population based on progressive clearance might have been expected, creating a permanent mixed agricultural landscape in many areas.

This pattern of agricultural expansion, however, does not seem to have begun over much of the British Isles until the late Bronze Age or Iron Age, even in regions that at a later date were agriculturally productive (Robinson 1984: 9-10).

Jones (1980) was struck by the contrast between the plentiful occurrence of hazel-nut fragments in the Grooved Ware deposits he was investigating and their almost total absence from Iron Age and Roman samples from which he had recovered several tens of thousands of cereal grains. This contrast has still been found to hold true between neolithic assemblages and assemblages from the later Bronze Age onwards.

Little evidence is available for the first half of the Bronze Age. The earliest Bronze Age assemblages seem to have a similar character to neolithic assemblages. For example, a later neolithic and early Bronze Age site at Hunstanton, Norfolk, produced the familiar range of charred cereals and hazel-nut shells (P. Murphy pers. comm.) as did the Beaker settlement at Gravelly Guy. Hazel-nut shells were almost as frequent as cereal remains at the earlier Bronze Age settlement at West Row, Mildenhall, Suffolk (P. Murphy pers. comm.). A middle Bronze Age midden at Grimes Graves, Norfolk, however, contained many charred cereal grains but no hazel-nuts or apples (Legge 1981: 90-94), and waterlogged seeds from the middle Bronze Age shaft at Wilsford, Wiltshire, included a significant component from a ploughed field weed flora (Robinson unpublished).

There was no need for slash-and-burn shifting cultivation in neolithic Britain and the temporary clearances for which there is evidence were of a much longer duration than the period when the crops would benefit from the practice (Rowley-Conwy 1981). Even if declining soil fertility or increasing problems with perennial weeds on old cultivation plots made cultivation of newly cleared areas more economical than continuing to use the old plots, they could still have been used for pasture. Smith (1984) presents a spectre of bracken invading the agricultural plots of the Avebury region, forcing their abandonment and the clearance of new areas, but his hypothesis is based on a misunderstanding of the ecology of bracken. The buried soils of the sites from which he draws his evidence were sufficiently calcareous for molluscan shell to survive in them. Adult bracken plants are capable of tolerating a high pH but sporeling colonisation is impossible under such conditions (Conway and Stephens 1957).

12.5 CONCLUSIONS

The impression gained from the evidence is that throughout the Neolithic, the landscape was not being exploited to its full agricultural potential over much of England and Wales even if spades or hoes were the only tools normally available for cultivation.

This is in strong contrast with parts of Europe. On the loessic soils of central Europe the *Linearbandkeramik* sites (second half of the fifth millennium bc), which predate most of the British Neolithic, already suggest a heavy commitment to arable agriculture with permanent land clearance. These sites produce substantial quantities of cereal remains with reasonably convincing arable weed floras, while collected food plants appear to play a relatively small role (e.g. Bakels and Rousselle 1985: Table 4; Knörzer 1977). A rather extreme example is Hienheim (4300-3900 bc), which produced several thousand peas, cereal grains and chaff fragments as well as lentils and flax but only a single fragment of hazel-nut shell (Bakels 1978: Tables 15-16).

This pattern is continued on later sites such as the middle neolithic site at Hochdorf (Küster 1985) and late neolithic Ehrenstein (Hopf 1968), where large quantities of charred cereals dominate the plant assemblages on both sites, although hazel and collected fruits are still in evidence. Gathered plant remains are only abundant on sites in mountainous regions. The permanently open areas of the later Neolithic in England and Wales however, seem as much to be organised around ceremonial or religious sites as for agriculture. Remains of settlement sites are usually insubstantial.

There is no evidence of conditions which would have made neolithic British agriculture inherently less successful than contemporary agriculture in central Europe. However an economy totally dependent on agriculture requires reliable and fairly large-scale methods of crop storage to be successful, and even then is more vulnerable to drastic famine stress in the event of successive crop failures than the hunting-gathering economy. Hunter-gatherers, with their broad subsistence base, are unlikely to experience total resource failure although they may be forced to utilise less favoured foods.

Pastoralism can be incorporated more readily into the way of life of woodland hunter-gatherers than settled agriculture. Domestic stock can be herded in partly cleared woodland whereas cereals require levels of light only achieved by complete clearance of the cultivated area and any shade-casting trees beyond. If current theories that agriculture is at least partly an adaptation to population pressure are correct, then what we may be seeing in the British Neolithic is the reluctant adoption, perhaps by way of pastoralism and horticulture, of a less desirable, more labour intensive economy under pressure from an increasing population.

Isolation from continental Europe and a low population level, perhaps deliberately constrained, was possibly the reason that there was no need for a rapid transition to a fully agricultural, settled society. Indeed it was not until the late Bronze Age that the wave of arable intensification began which resulted in large areas of the British Isles being exploited to their full potential to support high population levels during the Iron Age, although the importance of gathered food plants had declined much earlier.

ACKNOWLEDGEMENTS

We are extremely grateful to Wendy Carruthers and Peter Murphy for providing us with their unpublished results for nine of the sites in Table 1. The analyses of carbonised plant remains from most of the sites were funded by the Historic Buildings and Monuments Commission (England).

APPENDIX: The sites

BARROW HILLS, Oxfordshire (SU513981). A group of Grooved Ware pits on the second gravel terrace of the Thames excavated by C. Halpin. Identifications by L. Moffett.

BARTON COURT FARM, Oxfordshire (SU510978). A group of Grooved Ware pits on the second gravel terrace of the Thames excavated by D. Miles (1986). Radiocarbon dates of 2080±70 bc (HAR 2387) and 1960±70 bc (HAR 2388). Identifications by M. Jones (1980; in Miles 1986: microfiche 9: A2-B3, F2-F4).

BISHOPSTONE, Sussex (TQ467007). A neolithic pit on the chalk downs above the coast excavated by M. Bell (1977). Radiocarbon date of 2510±70 bc (HAR 1622). Identifications by J. Arthur (in Bell 1977: 273-75).

BLEWBURY, Oxfordshire (SU533862). A pair of Grooved Ware pits on the chalk of the Berkshire Downs excavated by C. Halpin. Identifications by L. Moffett and M. Robinson.

BRIAR HILL, Northamptonshire (SP736592). Phase VII of the causewayed enclosure on sandy ironstone excavated by H. Bamford (1985). Radiocarbon dates for this phase of 2650±90 bc (HAR 3208), 2660±90 bc (HAR 4071), 2710±70 bc (HAR 4075) and 2470±90 bc (HAR 5217). Identifications by A. Perry and M. Robinson (in Bamford 1985: 126, microfiche 3 249-51).

BROMFIELD, Shropshire (SO485775). A pair of pits on the terrace of the Onny excavated by S. Stanford (1982). Radiocarbon date of 2730±80 bc (HAR 3968). Identifications by S. Colledge (in Stanford 1982: 286-87).

CONEYBURY, Wiltshire (SU134416). Early neolithic pit and later neolithic henge ditch on chalk excavated by J. Richards. Identifications by W. Carruthers.

DOWN FARM, Woodcutts, Dorset (SU000147). Grooved Ware pits on chalk near the Dorset cursus excavated by J. Barrett and R. Bradley. Identifications by M. Jones (1980). Radiocarbon dates of 2190±60 bc (BM2406) and 2130±50 bc (BM2407).

GRAVELLY GUY, Oxfordshire (SP402053). Pits containing Beaker pottery on the second gravel terrace of the Thames near the Devil's Quoits henge excavated by G. Lambrick. Identifications by L. Moffett.

HAZLETON, Gloucestershire (SP073189). Occupation debris beneath a Cotswold-Severn chambered tomb on the limestone of the Cotswolds excavated by A. Saville (forthcoming; Saville *et al.* 1987). Radiocarbon dates of bone from the buried soil of 2925±80 bc (OXA646), 3020±80 bc (OXA738) and 2965±80 bc (OXA739). Identifications by V. Straker (in Saville forthcoming).

KING BARROW RIDGE, Wiltshire (SU135425). Later neolithic pits on chalk excavated by J. Richards. Identifications by W. Carruthers.

LOFTS FARM, Heybridge, Essex (TL867093). Early neolithic features on terrace gravels excavated by N. Brown. Identifications by P. Murphy.

M.G., ABINGDON, Oxfordshire (SU483971). Grooved Ware pits on the second gravel terrace of the Thames excavated by C. Halpin. Identifications by L. Moffett.

MOUNT FARM, Oxfordshire (SU582967). Two pits containing late neolithic flints on the third gravel terrace of the Thames near the Dorchester late neolithic ceremonial complex. Excavated by G. Lambrick. Identifications by M. Jones (1980).

ROBIN HOOD'S BALL, Wiltshire (SU103461). Early neolithic pits on chalk excavated by J. Richards. Identifications by W. Carruthers.

ROWDEN, Dorset (SY616891). Pit on chalk excavated by P. Woodward (forthcoming). Radiocarbon date of 2910±70 bc (HAR 5248). Identifications by W. Carruthers.

SPONG HILL, Norfolk (TF981195). Four earlier neolithic and one Beaker pit on glacial outwash gravels excavated by C. Hills. Identifications by P. Murphy.

SPRINGFIELD BARNES, Essex (TL732069). Late neolithic cursus on terrace gravels with silty loessic deposits excavated by J. Hedges and D. Buckley. Identifications by P. Murphy.

SPRINGFIELD LYONS, Essex (TL735084). Isolated late neolithic pit on terrace gravels with some silty loessic deposits excavated by D. Buckley and J. Hedges. Identifications by P. Murphy.

THE STUMBLE, Essex (TL901072). Neolithic occupation site (28A) on redeposited London Clay of the early third millennium bc, on what is now the Blackwater Estuary, excavated by T.J. Wilkinson and P. Murphy (1987: 19-39). Identifications by P. Murphy (in Wilkinson and Murphy 1987: 71-73).

THIRLINGS, Northumberland (NT956324). Four pits containing Grooved Ware excavated by C. O'Brien (1982). Identifications by M. van der Veen (1985: 207-8).

TRELYSTAN, Powys (SJ277070). Five Grooved Ware contexts within a late neolithic structure on the siltstone of Long Mountain, excavated by W. Britnell (1982). Radiocarbon dates of 2310±70 bc (CAR 272), 2185±65 bc (CAR 273) and 2035±70 bc (CAR 274). Identifications by G. Hillman (in Britnell 1982: 198-200).

WHITTON HILL, Northumberland (NT933347). Site 1 ring ditch and central cremation containing Grooved Ware and Peterborough Ware pottery, on sand and gravel in the Millfield Basin excavated by R. Miket (1985). Identifications by M. van der Veen (1985: 206-7; in Miket 1985: 143).

WILSFORD DOWN, Wiltshire (SU108408). Early neolithic pit on chalk excavated by J. Richards. Identifications by W. Carruthers.

BIBLIOGRAPHY

ASHBEE, P., SMITH, I.F. AND EVANS, J.G. (1979) Excavation of three long barrows near Avebury, Wiltshire. *Proceedings of the Prehistoric Society* **45** 207-300.

BAKELS, C.C. (1978) *Four linearbandkeramik settlements and their environment: a paleoecological study of Sittard, Stein, Elsloo and Hienheim.* Leiden: University Press (also published as *Analecta Praehistorica Leidensia 11*).

BAKELS, C.C. AND ROUSSELLE, R. (1985) Restes botaniques et agriculture du néolithique ancien en Belgique et aux Pays-Bas. *Helenium* **25** 37-57.

BAMFORD, H.M. (1985) *Briar Hill excavation 1974-1987.* Northampton: Development Corporation.

BELL, M. (1977) Excavations at Bishopstone, Sussex. *Sussex Archaeological Collections* **115** 1-291.

BENNETT, K.D. (1983) Devensian late glacial and Flandrian vegetational history at Hockham Mere, Norfolk, England. I: pollen percentages and concentrations. *New Phytologist* **95** 457-87.

BOYD, P. AND CARTWRIGHT, C. (1987) Carbonised seeds and charcoal. In: Excavation of a cropmark enclosure complex at Woodham Walter, Essex, 1976. (eds Buckley, D.G. and Hedges, J.D.) *East Anglian Archaeology* **33** 41.

BRADLEY, R. (1978) *The prehistoric settlement of Britain.* London: Routledge and Kegan Paul.

BRITNELL, W.J. (1982) The excavation of two round barrows at Trelystan, Powys. *Proceedings of the Prehistoric Society* **48** 133-201.

BRITNELL, W.J. AND SAVORY, H.N. (1984) *Gwernvale and Penywyrlod: two neolithic long cairns in the Black Mountains of Brecknock.* Cardiff: Cambrian Archaeological Association Monograph 2.

CONWAY, E. AND STEPHENS, R. (1957) Sporeling establishment in *Pteridium aquilinium*: effects of mineral nutrients. *Journal of Ecology* **45** 389-99.

EDWARDS, K.J. AND HIRONS, K.R. (1984) Cereal pollen grains in pre-elm decline deposits: implications for the earliest agriculture in Britain and Ireland. *Journal of Archaeological Science* **11** 71-80.

ELLERTON, S. (1939) The origin and geographical distribution of *Triticum sphaerococcum* Perc. and its cytogenetical behaviour in crosses with *T. vulgare* Vill. *Journal of Genetics* **38** 307-24.

EVANS, J.G. (1971) Habitat change on the calcareous soils of Britain: the impact of neolithic man. In: *Economy and settlement in neolithic and early Bronze Age Britain and Europe.* (ed. Simpson, D.D.A.) Leicester: University Press pp. 27-73.

EVANS, J.G. (1972) *Land snails in archaeology.* London: Seminar Press.

EVANS, J.G. (1975) *The environment of early man in the British Isles.* London: Elek.

EVANS, J.G., PITTS, M.W. AND WILLIAMS, DIANE (1985) An excavation at Avebury, Wiltshire, 1982. *Proceedings of the Prehistoric Society* **51** 305-10.

FIELD, N.H., MATTHEWS, C.L. AND SMITH, I.F. (1964) New neolithic sites in Dorset and Bedfordshire, with a note on the distribution of neolithic storage pits in Britain. *Proceedings of the Prehistoric Society* **30** 352-81.

FOX, A. (1963) Neolithic charcoal from Hembury. *Antiquity* **37** 228-29.

GODWIN, H. (1975) *History of the British flora*. (2nd edition) Cambridge: University Press.

GREEN, F.J. (1985) Evidence for domestic cereal use at Maxey. In: *Archaeology and environment in the Lower Welland Valley. Volume 1*. (Pryor, F., French, C., Crowther, D., Gurney, D., Simpson, G. and Taylor, M.) Norwich: East Anglian Archaeology Report 27 pp. 224-32.

GRIGSON, C. (1981a) Fauna. In: The Neolithic. (Smith, A.G.) In: *The environment in British prehistory*. (eds Simmons, I. and Tooley, M.) London: Duckworth pp. 191-99.

GRIGSON, C. (1981b) Fauna. In: The Bronze Age (Tinsley, Heather M.) In: *The environment in British prehistory*. (eds Simmons, I. and Tooley, M.) London: Duckworth pp. 217-30.

HELBAEK, H. (1952) Early crops in southern England. *Proceedings of the Prehistoric Society* **18** 194-233.

HILLMAN, G.C. (1981a) Crop husbandry: evidence from macroscopic remains. In: The Neolithic. (Smith, A.G.) In: *The environment in British prehistory*. (eds Simmons, I. and Tooley, M.) London: Duckworth pp. 183-91.

HOPF, M. (1968) Fruchte und Samen. In: Das jungsteinzeitliche Dorf Ehrenstein (Kreis Ulm) Ausgrabung 1960 (Zurn, H.) *Veroffentlichungen des Staatlichen Amtes für Denkmalpflege Stuttgart, Reihe A.* **10** (2) 7-77, Plates 1-50.

HOULDER, C.H. (1963) A neolithic settlement on Hazard Hill, Totnes. *Transactions of the Devon Archaeological Exploration Society* **21** 2-28.

JONES, G. AND LEGGE, A.J. (1987) The grape (*Vitis vinifera* L.) in the Neolithic of Britain. *Antiquity* **61** 452-55.

JONES, M. (1980) Carbonised cereals from Grooved Ware contexts. *Proceedings of the Prehistoric Society* **46** 61-63.

KNÖRZER, K.-H. (1977) Pflanzliche Grossreste. In: Der bandkeramische Siedlungsplatz Langweiler 9, Geminde Aldenhoven, Kreis Duren. *Beitrage zur neolithischen Besiedlung der Aldenhovener Platte 2 (Rheinische Ausgrabungen 18)* 279-303.

KÜSTER, H.J. (1985) Neolithische Pflanzenreste aus Hochdorf, Gemeinde Eberdingen (Kreis ludwigsburg). In: *Hochdorf I*. (Küster, H.J. and Körber-Grohne, U.) Stuttgart: Konrad Theiss.

LEGGE, A.J. (1981) The agricultural economy. In: *Grimes Graves, Norfolk: excavations 1971-72, volume 1*. (Mercer, R.J.) Department of the Environment Research Report 11 London: Her Majesty's Stationery Office pp. 79-103.

LIDDELL, D.M. (1932) Report on the excavations at Hembury Fort. Third Season 1932. *Proceedings of the Devon Archaeological Exploration Society* **1** 162-90.

MIKET, R. (1985) Ritual enclosures at Whitton Hill, Northumberland. *Proceedings of the Prehistoric Society* **51** 137-48.

MILES, D. (ed.) (1986) *Archaeology at Barton Court Farm, Abingdon, Oxon.* London: Council for British Archaeology Research Report 50.

MOORE, P.D. (1973) The influence of prehistoric cultures upon the initiation and spread of blanket bog in upland Wales. *Nature* **241** 350-53.

MOORE, P.D. (1975) Origin of blanket mires. *Nature* **256** 267-69.

O'BRIEN, C. (1982) Excavations at Thirlings. *Archaeological Reports for 1981*. Universities of Newcastle and Durham pp. 44-55.

REANEY, D. (1968) Beaker burials in South Derbyshire. *Derbyshire Archaeological Journal* **88** 68-81.

ROBINSON, M.A. (1984) Landscape and environment of central southern England during the Iron Age. In: *Aspects of the Iron Age in central southern Britain*. (eds Cunliffe, B.W. and Miles, D.) Oxford: University Committee for Archaeology Monograph 2 pp. 1-11.

ROWLEY-CONWY, P. (1981) Slash-and-burn in the temperate European Neolithic. In: *Farming practice in British prehistory*. (ed. Mercer, R.J.) Edinburgh: University Press pp. 85-96.

SAVILLE, A. (forthcoming) *Hazleton North 1979-1982: the excavation of a Cotswold-Severn tomb in Gloucestershire.*

SAVILLE, A., GOWLETT, A.J. AND HEDGES, R.E.M. (1987) Radiocarbon dates from the chambered tomb at Hazleton (Glos.): a chronology for a neolithic collective burial. *Antiquity* **61** 108-19.

SCAIFE, R.G. (1980) *Late Devensian and Flandrian palaeoecological studies in the Isle of Wight*. Ph.D. thesis, University of London.

SIMMONS, I.G. AND TOOLEY, M.G. (eds) (1981) *The environment in British prehistory*. London: Duckworth.

SMITH, I.F. (1974) The Neolithic. In: *British prehistory, a new outline*. (ed. Renfrew, C.) London: Duckworth pp. 100-36.

SMITH, R.W. (1984) The ecology of neolithic farming systems as exemplified by the Avebury region of Wiltshire. *Proceedings of the Prehistoric Society* **50** 99-120.

STANFORD, S.C. (1982) Bromfield, Shropshire - neolithic, Beaker and Bronze Age sites 1966-79. *Proceedings of the Prehistoric Society* **48** 279-320.

THOMAS, K.D. (1982) Neolithic enclosures and woodland habitats on the South Downs in Sussex, England. In: *Archaeological aspects of woodland ecology.* (eds Bell, M. and Limbrey, S.) Oxford: British Archaeological Reports S146 pp. 147-70.

TINSLEY, H.M. (1981) The Bronze Age. In: *The environment in British prehistory.* (eds Simmons, I. and Tooley, M.) London: Duckworth pp. 210-49.

TOMALIN, D.J. AND SCAIFE, R.G. (1979) A neolithic flint assemblage and associated palynological sequence at Gatcombe, Isle of Wight. *Proceedings of the Hampshire Field Club and Archaeological Society* **36** 25-33.

VEEN, M. VAN DER (1985) Evidence for crop plants from north east England: an interim overview with discussion of new results. In: *Palaeobiological investigations.* (eds Fieller, N.R.J., Gilbertson, D.D. and Ralph, N.G.A.) Oxford: British Archaeological Reports S266 pp. 197-219.

WILKINSON, T.J. AND MURPHY, P. (1987) *The Hullsbridge Basin Survey 1986 interim report no. 7.* Chelmsford: Essex County Council.

WOODWARD, P.J. (Forthcoming) The South Dorset Ridgeway survey and excavation 1977-1983: the pre-Iron Age landscapes. *Proceedings of the Dorset Natural History and Archaeological Society.*

V

CONCLUDING REMARKS

CONCLUDING REMARKS

W. Groenman-van Waateringe

Writing up my concluding remarks on the Cardiff conference posed a serious problem, because several of the papers presented there are missing from the papers submitted for publication. This means that, to me, the highly satisfactory and well-balanced coverage of the topic in the conference is no longer there. However, I decided to write my concluding remarks more or less as they were delivered at the conference. Perforce, for those who were not in Cardiff, part of what I say will fall into a vacuum and will be unverifiable.

Harris' fear of another conference on the beginnings of agriculture is in my view unjustified. Perhaps a *contradictio in terminis*, even archaeology is moving fast. In ten to fifteen years much has changed, not only in pure knowledge of the field, but in our view of the problems, both methodological and theoretical. Moreover, the problems around the topic were far from solved.

Ken Thomas' offering of a substitute for the equilibrium model has to be seen in the wider context of the new ideas on equilibrium in ecology, or indeed, its non-existence. The big changes in theoretical concepts in cultural archaeology have not passed by environmental archaeologists, as was apparent especially in the lectures by Paul Halstead and Royston Clark. Although the long term development both in environmental and social processes in Clark's paper can already be found in ideas on domestication voiced by Higgs and Jarman (1969), it is now placed in the broader context of social mechanisms.

As more and more data on early agriculture is becoming available, studies such as those of Caroline Grigson and Kevin Edwards can lead us to new conclusions, while they show at the same time the importance of sharpening our criteria, the recognition of false conclusions, based on a shortage of data, and of their correction.

The paper by Angela Kreuz provided ideas on the deliberate exploitation of wood and the forest fringe by the *Linearbandkeramik* people, as did the paper by Gordon Hillman for the Near East. Forest management rather than exploitation in prehistoric times is becoming a very popular subject. David Robinson discussed the old ideas of Troels-Smith concerning forest exploitation in the Swiss Neolithic based on new data.

Two papers were based on British evidence, one by Barbara Noddle based on what the material *sec* can tell us, and one by Roy Entwistle and Annie Grant with a footing in ethnographic data on types of primitive agriculture. Both tried to specify the known data for the neolithic and Bronze Age periods, while the latter paper clearly took a highly alternative viewpoint, on which the last word has not been said. If meant to loosen the tongues and sharpen the ideas on the topic it will serve its function.

One can find a parallel between Harris' ideas concerning the existence of long held knowledge, which only surfaces on a large scale when truly necessary, and the use of iron. In our areas also, iron objects have already been dated rather early in the Bronze Age, but the use of iron remains more or less dormant until much later. Is this perhaps also the case with the cultivation of rye, for which much earlier evidence seems to be available now, long before it became a staple crop?

On the beginnings of agriculture there are three questions: why, when and how? All three have been dealt with during this conference and for widely different geographical areas. I expected many new ideas on the 'how', and I have not been disappointed. I mention here the lectures by Sebastian Payne, Paul Halstead and Royston Clark. 'When' was answered in papers by Tony Legge, Simon Davis and Hansjörg Küster.

Concerning, however, the 'why' I expected much less. All right, demographic pressure, sedentism, and over-exploitation were all mentioned by Harris in his splendid historical outline. But then came the lecture by Gordon Hillman and I will draw your attention specifically to one sentence in it: 'The triggering factor for the beginning of agriculture is a dramatic increase in food.' Thus not food shortage, but just the opposite: abundance. In a recently published article I came to the same conclusion (Groenman-van Waateringe 1988). The increase disrupts the social sub-system with its population control by means of abortion, prolonged lactation, infanticide etc. One can say that the changes in the environmental sub-system lull to sleep certain mechanisms of the social sub-system. Even in our society with a highly developed information system, the response to the economic crisis of the early 1970s came *circa* 10 years too late.

Much of the data presented at the conference has been the result of a tremendous input of labour (identifying thousands of pieces of charcoal, pollen etc.), but everyone has tried to translate the data into more general terms. The standard of the papers given was high and the general impression of the conference very satisfying. The chain of lectures dealing with data from Tell Abu Hureyrah (Harris, Legge and Hillman) gave me the feeling that this tell is becoming the Jarmo of the late 1980s. All the more pity that none of these lectures were submitted for publication.

While I did not dare at the end of the conference to formulate the essence of it, I will try to do so now. However, I will not pretend that this is the idea of its value, it simply is mine.

In contrast to earlier conferences on this topic, the emphasis was no longer on the Near East as the area of the origin of agriculture (and this is even less so in the publication). Therefore the title of the conference should perhaps have been 'Early agriculture and its social and environmental implications' rather than simply 'The beginnings of agriculture'. What happened in Europe, south, central and west, comes more to the fore and rightly so. The inhabitants of these areas are not any longer seen as only passively changing over to agriculture, either by invasion, diffusion or adaptation, but they are studied now in their own right. Their (pre)history and social organisation and their interaction with their environment are subjects of study and the question of whether the introduction of domestication happened through invasion, diffusion or adaptation seems not to the point. In fact it can be any of these three or a combination and was not necessarily the same over the whole of Europe.

Where are we going from here? Three topics seem to me to be most promising for future work. First there are the ideas of Ken Thomas on hierarchical ecosystems. I would like to see an evaluation of his ethnographical data within such systems and what it can teach us about past ecosystems and through these, societies. Secondly, Susan Limbrey's paper was, as far as I know, the first time that the pedological aspect of early agriculture had been emphasised, and although she decided not to publish her paper, I do hope that she will continue her line of thought and will come back to it later. The third topic which seems to be promising, is the application to the European situation of Gordon Hillman's ideas on an abundance of food as the main cause of a change-over to agriculture. The change in the population curve some time in the early Mesolithic (Constandse-Westermann and Newell 1984: Figures 1 and 2, Table 7) may well point to just such a situation of population growth as supposed by myself, and Hillman for the Near East, caused by an increase in primary producers and consumers. And is it just a coincidence that along the coasts of north western Europe the mesolithic way of life made place for the neolithic one at the time when the change in sea level came to a halt and the extension of rich coastal biotopes ended?

We may hope that within another ten to fifteen years it will be necessary to ask Harris to give a sequel to his lecture on paradigms and processes in the transition to agriculture and that again the feeling of *déja vu* can be dismissed.

BIBLIOGRAPHY

CONSTANDSE-WESTERMANN, T.S. AND NEWELL, R.R. (1984) Human biological background of population dynamics in the Western European Mesolithic. *Proceedings of the Koninklijke Nederlandse Akademie van Wetenschappen* B **87** (2) 139-223.

GROENMAN-VAN WAATERINGE, W. (1988) Interaction between environment and social subsystems. In: *Conceptual issues in environmental archaeology.* (eds Bintliff, J.L., Davidson, D.A. and Grant, E.C.) Edinburgh: University Press pp. 278-82.

HIGGS, E.S. AND JARMAN, M.R. (1969) The origins of agriculture: a reconsideration. *Antiquity* **43** 31-41.

www.ingramcontent.com/pod-product-compliance
Ingram Content Group UK Ltd.
Pitfield, Milton Keynes, MK11 3LW, UK
UKHW060200240426
12048UKWH00029B/1676